Artificial Organs

Artificial Organs

Gerald E. Miller

ISBN: 978-3-031-00483-4   paper
ISBN: 978-3-031-01611-0   e-book

DOI: 10.1007/978-3-031-01611-0

A Publication in the Springer series
SYNTHESIS LECTURES ON BIOMEDICAL ENGINEERING
Lecture #4

First Edition
10 9 8 7 6 5 4 3 2 1

Artificial Organs

Gerald E. Miller
Virginia Commonwealth University

SYNTHESIS LECTURES ON BIOMEDICAL ENGINEERING #4

ABSTRACT

The replacement or augmentation of failing human organs with artificial devices and systems has been an important element in health care for several decades. Such devices as kidney dialysis to augment failing kidneys, artificial heart valves to replace failing human valves, cardiac pacemakers to reestablish normal cardiac rhythm, and heart assist devices to augment a weakened human heart have assisted millions of patients in the previous 50 years and offers lifesaving technology for tens of thousands of patients each year. Significant advances in these biomedical technologies have continually occurred during this period, saving numerous lives with cutting edge technologies. Each of these artificial organ systems will be described in detail in separate sections of this lecture.

KEYWORDS

Artificial heart and ventricular assist, Artificial heart valve, Cardiac pacemaker, Dialysis, Human heart and heart surgery, Human kidney

Contents

CHAPTER 1

Artificial Heart Valves

The replacement or augmentation of failing human organs with artificial devices and systems has been an important element in health care for several decades. Such devices as kidney dialysis to augment failing kidneys, artificial heart valves to replace failing human valves, cardiac pacemakers to reestablish normal cardiac rhythm, and heart assist devices to augment a weakened human heart have assisted millions of patients in the previous 50 years and offers lifesaving technology for tens of thousands of patients each year. Significant advances in these biomedical technologies have continually occurred during this period, saving numerous lives with cutting edge technologies. Each of these artificial organ systems will be described in detail in separate sections below.

1.1 CARDIAC ANATOMY AND PATHOPHYSIOLOGY

The human heart consists of two pumping chambers (ventricles) and two filling chambers (atria). The heart is a pulsatile pump, operating via muscular contraction of both the ventricles and atria, and is designed to produce positive displacement of blood through two circulatory systems. The right ventricle pumps blood into the pulmonary circulation, where blood becomes oxygenated and the left ventricle pumps into the systemic circulation that allows oxygenated blood to reach tissues throughout the body, where oxygen is transported to these tissues.

In order for blood to flow in the proper direction from each of the ventricles as well as from each atrium to the associated ventricle, the heart contains heart valves to prevent backflow of blood. The atrioventricular valves (between the atria and the ventricles) prevent backflow into the atria when the ventricles contract during systole. The outflow valves (from each ventricle) are designed to prevent backflow during ventricular diastole. The human heart valves are shown in Figure 1, with the bileaflet mitral valve located between the left atrium and the left ventricle and the trileaflet aortic valve located between the left ventricle and the aorta (outflow tract). The trileaflet tricuspid valve is located between the right atrium and the right ventricle, while the trileaflet pulmonary valve is located between the right ventricle and the pulmonary artery. Each of these valves consists of an annulus from which valve leaflets project into the orifice opening. There are no valves located between the atria and the respective veins feeding blood into the atria. Thus, there is no valve from the vena cava to the right atrium, nor a valve from

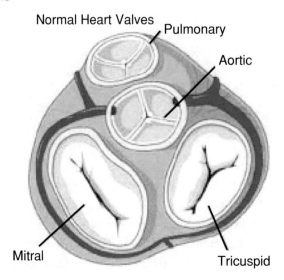

FIGURE 1: Human heart valves.

the pulmonary vein to the left atrium. All heart valves operate by means of a pressure gradient, moving in the direction of decreasing pressure at all times, whether toward the open or closed position.

Human heart valves are designed to close in an overlapping fashion with the valve leaflets abutting over each other in a fashion noted in Figure 2. The leaflets overlap to ensure adequate

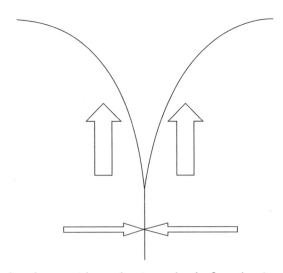

FIGURE 2: Human valve closure with overlapping valve leaflets showing pressure gradient acting upward against closed valve and lateral frictional force created by overlap of leaflets.

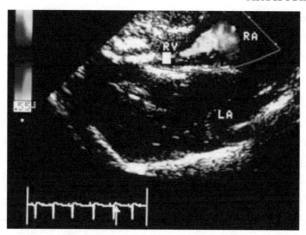

FIGURE 3: Doppler ultrasound image of a heart indicating jetting from the right ventricle toward the right atrium, resulting from reversing flow across the tricuspid valve.

valve closure with the vertical forces created by a pressure gradient across the valve being transferred (in part) to lateral (friction) forces. Thus, each valve leaflet is actually longer than one half of the radius of the valve orifice, which provides sufficient length for the leaflet overlap.

There are two potential sources of heart valve failure, each of which requires years to develop. One such mechanism is valve *stenosis* or narrowing of the valve orifice (best seen with the valve in an open position). Valve stenosis is often related to atherosclerosis and creates a narrowed outflow area. This not only reduces the area for blood flow, but also creates an added pressure drop due to the existence of a contraction in the flow pathway. This situation may be diagnosed with an echocardiogram that uses 2-D ultrasound and a Doppler flow probe. However, symptoms of such a narrowing would not normally appear until the orifice is significantly reduced (up to a 50% reduction in area). With the reduced area, it is possible that the flow across the valve orifice may become turbulent, which can also be visualized with Doppler ultrasound or even with the aid of a stethoscope. A color Doppler profile of a heart is shown in Figure 3.

Another source of valve failure is via weakening of the valve leaflets. In such a case, the leaflets begin to bulge upwards with a portion of the overlapping leaflets protruding above the valve orifice. Such an occurrence is termed valve *prolapse*. In the extreme case, the leaflets bulge so far that the overlap of the valve leaflets during closure is in jeopardy. If a gap appears between the leaflets in the (purportedly) closed position, there is a regurgitation of the blood backwards from its normal intended path (see Figure 3). Since the gap between the leaflets is initially small, the backflow is typically turbulent, which can be easily heard via a stethoscope or imaged via Doppler ultrasound. The sounds heard are called heart murmurs. In advanced heart valve failure from either mechanism, the use of a stethoscope followed by ultrasound (as needed) may indicate a need to replace the failing valve. The mitral valve and the aortic valve (both in the left heart) are the most prevalent valves to be replaced. This is not unusual as the left heart experiences greater pressures than the right heart.

1.2   PROSTHETIC HEART VALVES

As with any implantable device, issues related to biocompatibility, longevity, and function of replacement heart valves are important. Artificial heart valves have been clinically available for over 50 years with early designs consisting of either a caged ball or a tilting disk (within a caged or strut assembly). Early successful designs of each type include the Starr Edwards ball valve and the Bjork Shiley tilting disk valve. The ball valve, as seen in Figure 4, consists of a silastic or silicone rubber ball encased within a stainless steel cage and annular ring. The metallic ring (used for strength) was covered with a Dacron sewing ring. All materials are biocompatible with ample animal and clinical trials in support of these choices. The sewing ring allows attachment of the valve to the surrounding tissue of the original valve annulus.

The ball valve had an important advantage, at the time, in that it is structurally strong and durable. However, it closes with considerable force that can cause hemolysis and thrombosis, as seen in Figure 5. In addition, the ball acts as a central occluder with the valve in the open position, thus producing a greater pressure drop across the open valve that reduces the downstream

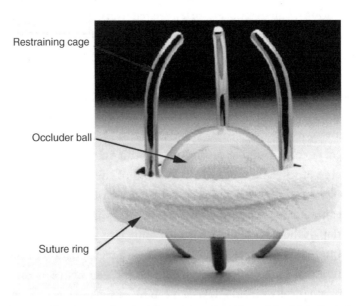

FIGURE 4: Prosthetic ball valve in the closed position showing with the ball sitting in the dacron-covered annular ring and the upper cage protecting the ball in the open position.

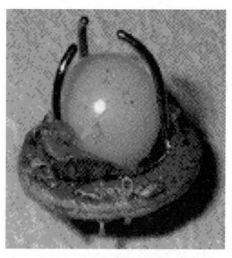

FIGURE 5: Thrombus formation at the seat of a ball valve as a result of the closing force of the ball against its seat.

pressure available to propel blood through a circulation. This type of valve also has a large vertical profile that makes implantation, particularly in the mitral position, somewhat difficult for many patients whose anatomical space is limited.

The tilting disk valve, as seen in Figure 6, opens to 60° as constrained by a stainless steel strut. The disk is constructed of pyrolytic carbon, given the trade name pyrolite.

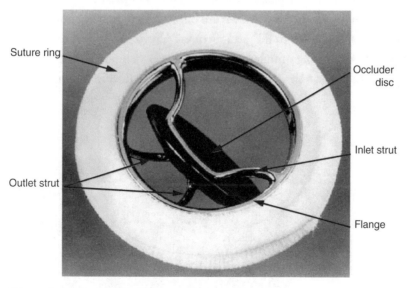

FIGURE 6: Tilting disk valve in the open position.

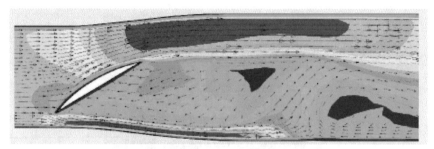

FIGURE 7: Flow patterns around a tilting disk valve generated by computational methods. Note the reversal of the velocity vector behind the tilted disk and the higher velocities at the periphery.

The tilting disk valve has an open profile that offers blood flow around the tilted disk as well as to the outer sides, which are between the disk and the blood chamber or artery. As such, it is far less of a central occluder to blood flow than the ball valve and has a smaller pressure drop across the open valve, which is also advantageous. This valve only opens to 60° due to the stainless steel strut assembly. The valve is not actually affixed to the struts. A flow pattern across the open valve is shown in Figure 7, which indicates a partial central occluder (the tilted disk) as well as peripheral flow around the disk. There is a bit of recirculation behind the disk and the peripheral flow is at relatively high velocities as compared to the central region. As is the case with most flow images, the red (or dark gray) colors in the image refer to higher velocities, the yellow (or lighter gray) refer to moderate velocities, the green (or lighter gray) refer to small velocities, and the blue (or very light gray) refer to extremely small (near zero) velocities. Arrows within the flow field indicate individual vector lines of fluid particles within the flow field.

Many modern prosthetic heart valves are bileaflet in nature as shown in Figure 8.

The bileaflet valves open to 80° and pivot via pins that connect to the valve leaflets and protrude into slots within the annular ring. The ring is composed of either stainless steel or titanium, often with a pyrolytic carbon coating. There is a dacron sewing ring attached to the

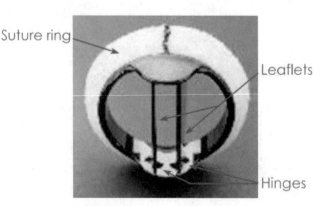

FIGURE 8: Bileaflet valve showing flanges where leaflet pins pivot within annular ring.

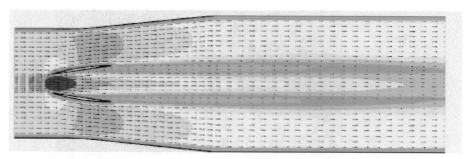

FIGURE 9: Flow through an open bileaflet valve with a large central flow region as well as a large peripheral flow region. Trailing edge vortices are seen emanating from the tips of the leaflet edges.

metal substrate for ease in attachment to the valve orifice. The leaflets are often composed of pyrolytic carbon for strength and durability. With two leaflets, the closing force is less than that of a single tilting disk valve, and certainly far less than that of a ball valve. As the pin assembly can be a site for thrombus formation, such valves normally have a washout zone near the pins to flush that area and avoid stagnant blood leading to thrombus formation. As such, this valve has mild regurgitation as a design element. Figure 9 depicts flow across an open bileaflet valve, indicating far more central flow than the tilting (monoleaflet) valve. There is no central occluder in the bileaflet valve as there was (partially) with the tilting (monoleaflet valve). Thus, there is a greater central flow region in this type of valve.

Ideally, prosthetic heart valves require low levels of hemolysis (red cell destruction), low levels of thrombosis, and a long life span (20–30 years). Most patients must receive antico-agulants indefinitely to deter thrombosis. Pyrolytic carbon and silicon rubber as valve or ball materials have been shown to be relatively antithrombogenic and their surfaces are often cov-ered with proteins over time. Dacron polyester is sometimes used for valve leaflets. Such a material is more flexible than the pyrolite leaflets and thus closes with far less force and less hemolysis. However, such flexible leaflets also take longer to open and close, as they flex during the leaflet movement. An undue delay in valve closure produces a slight backflow, not unlike that of a natural human valve, which produces the incisura (dip) in the aortic pressure wave-form. However, too much delay can adversely affect net forward flow, thus reducing net cardiac output.

Bioprosthetic valves, often a preferred alternative, are composed of natural valves from a pig, which has a cardiovascular system most similar to humans. These valves must be "treated" with glutaraldehyde to resist antigenicity or rejection of a foreign substance. As such, these biological valves are stiffer than natural heart valves due to the treatment process. In some cases, the bioprosthetic valve has a dacron sewing ring attached along its base and may even have a stiffening ring of stainless steel attached along its base as well. In some cases, dacron is attached along the outer face of the leaflets to provide greater strength to the leaflets. This does not

Natural heart
tissue

Polymer
scaffold

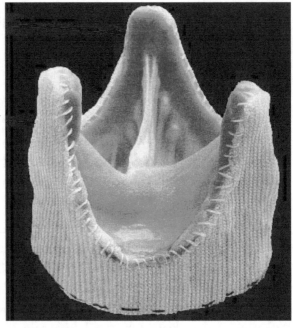

FIGURE 10: Bioprosthetic (pig) valve with dacron along the base and outer edges of the leaflets.

significantly add to the stiffness of the leaflets as compared to the glutaraldehyde treatment. Figure 10 depicts a bioprosthetic valve with a dacron covering.

Surgical implantation of prosthetic heart valves has evolved into an efficient and rapid procedure. The orifice diameter of the patient is evaluated with the aid of pulse echo ultrasound,

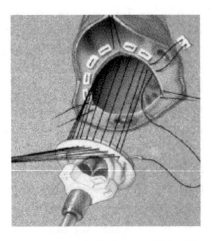
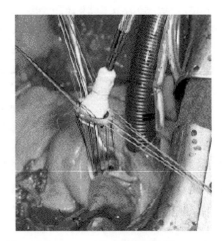

FIGURE 11: Prosthetic heart valve within valve holder with sutures preplaced within sewing ring and valve orifice.

the 2-D version often called echocardiography. The appropriately sized prosthetic valve is then placed into a valve holder and sutures are threaded through the sewing ring of the valve. The patient is placed on cardiopulmonary bypass (heart–lung machine) and the original valve is removed. The prosthetic valve is then placed into the orifice by one surgeon and another continually sews the sutures around the valve orifice using the curved needle attached to the sutures. The valve holder is then lowered into place, and the holder eventually depressed to release the valve—and the empty holder is then removed. The orifice is checked for potential leaks and additional sutures are placed on site as needed. The patient is then removed from bypass and the incision site closed as is the chest incision. The entire procedure often is completed within 1 h and the patient is normally on cardiopulmonary bypass for as little as 15 min. Figure 11 depicts a prosthetic heart valve within a valve holder with sutures in place.

1.3   EVALUATION OF PROSTHETIC VALVES

The performance and efficacy of prosthetic heart valves are normally evaluated by means of pressure gradients across the open valve, long-term durability, and prevalence of adverse physiological conditions such as calcification, stenosis, thrombosis, or hemolysis. Pressure gradients can be determined via direct pressure sensors or via Doppler ultrasound (indirect determination by means of flow/pressure relationship). Thrombosis, hemolysis, and calcification are determined via direct biochemical measures or via imaging modalities such as ultrasound. Such determinations are on mechanical test beds or blood flow loops in mock circulatory systems, or with animal models. Computer modeling techniques are also employed to evaluate valve performance using computational fluid dynamics. An excellent review of the fluid mechanics of prosthetic heart valves was produced by Yoganathan *et al.* (2004). Other groups involved with the evaluation of prosthetic and bioprosthetic heart valves include Labrosse *et al.* (2005), Jamieson *et al.* (2005), Medart *et al.* (2005), Pierrakos *et al.* (2004), Malouf *et al.* (2005), Goetze *et al.* (2004), Wu C *et al.* (2004), and Wu Y *et al.* (2004).

CHAPTER 2

Artificial Heart and Cardiac Assist Devices

2.1 CARDIAC ANATOMY AND PATHOPHYSIOLOGY

As was described above in the section dealing with heart valves, the human heart consists of two ventricles responsible for ejection of blood and two atria responsible for holding and filling the ventricles. The left heart pumps blood into the systemic circulation and the right heart into the pulmonary circulation. Figure 12 depicts the heart with all four chambers.

According to the National Center for Chronic Disease Prevention, 950 000 Americans die each year due to cardiovascular disease with over 61 million Americans suffering from this disease and over 6 million hospitalizations each year as a result. Major causes of cardiac-related pathologies include atherosclerosis in either the aorta or in coronary vessels such as depicted in Figure 13.

In many cases, coronary artery blockages and/or blockages in the aorta can be alleviated via surgery to remove the blockage, either through mechanical "reaming," via laser ablation, or via balloon catheter expansion. Often, stents are placed in a coronary vessel after removal of the blockage. In those instances where the blockage cannot be removed by such means, more extensive surgery can be performed to bypass the blockage with transplanted veins or via artificial vessels composed of biomedical polymers. This procedure is termed coronary artery bypass grafts or CABGs.

Often after such surgery, the patient's repaired heart is too weak to fully support the blood pumping requirements of the systemic circulation and will require some period of reduced load in order to recover. In other cases, the repaired heart cannot ever function at a normal level of 5 L/min at 95% oxygen saturation with a mean pressure of 100 mmHg. In still other cases, the heart cannot be adequately repaired and functions at a reduced level on a permanent basis. One approach to the latter cases would be to seek a heart transplant. However, given the vast number of potential cases in need of such a transplant and the inherent need for appropriate blood typing and tissue typing, there are insufficient numbers of donor hearts. The United Network for Organ Sharing notes that donor hearts number 2000–3000 per year with the stated need at least tenfold greater.

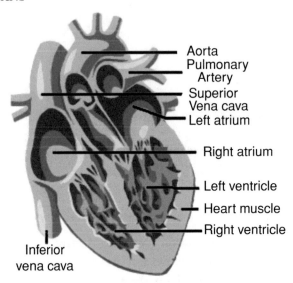

FIGURE 12: The human heart with four chambers: two atria and two ventricles.

As such, a stopgap mechanism is to employ a mechanical assist pumping system, called a ventricular assist device or VAD. This can be utilized in true "assist" mode for those patients recovering from open heart surgery and whose heart will recover full function within a few days. In those other cases where the patient's heart will not completely recover, or where surgical

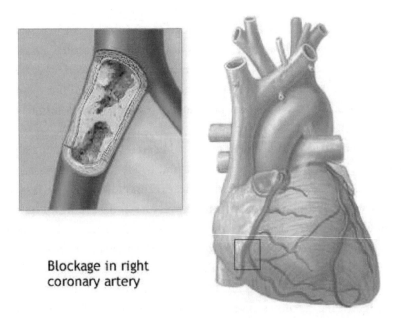

FIGURE 13: Coronary artery blockage with inset showing blockage and larger view showing location within the heart musculature.

intervention would not be beneficial, then this device can be utilized for a longer period until such time as a donor matching heart becomes available. In such cases, the device is termed as a "bridge to transplant." In the first two cases, the natural heart might pump as little as 2–3 L/min with the artificial heart pumping the remainder. When the left heart is being assisted, the device is termed an LVAD (left ventricular assist device). When the right heart is to be assisted, the device is termed an RVAD. For those cases where both hearts are being assisted, two devices are used in biventricular assist mode (BiVAD). In still other cases, when the patient's own heart cannot function even partially at the above level, it may be necessary to completely replace the failing heart with two pumps acting in tandem as a total artificial heart for end-stage heart failure.

2.2   HEART ASSIST TECHNOLOGY

Artificial hearts and heart assist devices have been in development and analysis for decades. The original pneumatically powered design, as typified by the Jarvik-3 and the Jarvik-7 pumps, consisted of a polycarbonate case surrounding a flexible polyurethane sac. The space between the two allowed high pressure air, designated as "systolic drive pressure" to collapse the blood containing sac and caused systolic ejection. A small vacuum pressure, designated as "diastolic pressure" pulled the sac toward the casing and caused diastolic filling. Artificial heart valves were employed for each artificial ventricle to prevent backflow during either portion of the cardiac cycle in a fashion similar to the natural heart. This sac type pump acted in a fashion similar to the natural heart in that it was inflow and inlet pressure sensitive (designated as preload sensitive), but not affected by the load against which the heart pumps (designated as not being afterload sensitive). The Jarvik pump is shown in Figure 14, with the air lines from the outer casing which are then connected to a drive system that provides the alternating high and low pressure to the blood sac. The two pressures, the systolic duration, and beat frequency can all be controlled and varied from a drive console.

Although pneumatically driven pumps are still utilized clinically, many original such designs have now incorporated electrically powered cam drive systems that move a piston/diaphragm arrangement to eject blood during systole and then, when the piston is pulled back toward the drive unit, initiates diastolic filling. Such blood pumps are still pulsatile and thus need valves to eliminate backflow and regurgitation. A biventricular device of this type is shown in Figure 15, with no pneumatic air lines emanating from the pump as was the case for the sac-type design.

Power for the electrically operated LVADs or BiVADs is achieved with external, rechargeable batteries as is shown in Figure 16. Transcutaneous energy transmission occurs via a pair of coils—one placed over the skin and one placed below the skin, the latter attached to the VAD. The VAD is connected to the apex (base) of the left ventricle (or to the base of both ventricles in biventricular support) with the aid of an inlet cannula. This orientation allows gravity to

FIGURE 14: The Jarvik-7 pneumatically driven artificial heart showing two ventricles and the air lines which lead to the spacing between the outer casing and the inner flexible blood sac.

assist emptying of the ventricle into the device, which offloads the ventricle in part. The VAD then pumps a portion of the cardiac output through a cannula that connects to the aorta via an end-to-side anastomosis.

Unlike the pneumatically driven blood pumps, where the air from the pneumatic drive unit is vented to the atmosphere, the electrically driven pumps are normally closed systems.

FIGURE 15: Biventricular electrically operated pusher plate design for a blood pump with four valves similar to that of the natural heart.

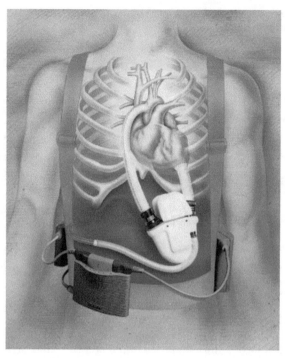

FIGURE 16: Novacor electrically powered LVAD with external battery pack and connecting cannula from apex of left ventricle and to aorta.

As such, when the diaphragm moves back toward the internal drive system, there can be a period where the air within the drive chamber is compressed, resulting in an increase in local pressure. As the diaphragm moves away from the drive unit (during systole), the air in this chamber becomes rarified and the pressure is reduced. Rather than exist with changing pressures within the drive system, early versions of electrically powered pumps contained a flexible sac, a compliance chamber, whereby the excess air in the drive system is shunted to the compliance chamber, thus maintaining a relatively constant pressure within the pump drive chamber. As the diaphragm moves forward during systole, the air is shunted from the compliance chamber into the drive chamber of the pump and vice versa during diastole. Later versions of electrically operated pumps utilize a venting system to avoid the need for a compliance chamber, which is an added component that takes up space and adds to the complications of the device operation.

The use of positive displacement pumps, such as the sac or pusher plate blood pumps, offers several advantages. These include the use of pulsatile flow similar to that of the natural heart as well as an insensitivity to afterload—i.e., the device can pump against any reasonable load, even a hypertensive load. Issues with such pumps include the sensitivity to preload and

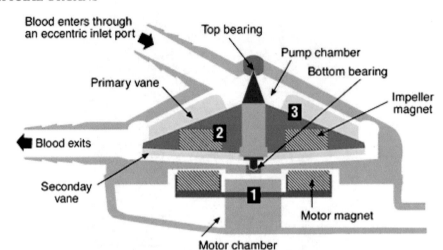

FIGURE 17: Centrifugal blood pump with inlet at top center and outlet at bottom left. Spinning rotor produces increased pressure at the periphery.

problems associated with artificial heart valves such as calcification, hemolysis, thrombosis, and mechanical failure. It was these issues that initiated the later use of rotary blood pumps as assist and/or bridge to transplant devices. Such rotary devices often require less power than their positive displacement counterparts and do not utilize valves, as they produce steady flow. Such rotary blood pumps are configured as either a centrifugal pump or as an axial pump.

The centrifugal blood pump is configured with an inlet to the pump from the top center and an outlet radially at the periphery. This arrangement is not unlike a tornado in flow design and the flow of such a device is similar—low pressure and velocity in the center with higher pressure and velocity at the periphery. Centrifugal pumps are electrically powered and utilize magnets to spin the rotor and, oftentimes, suspend the drive assembly as seen in Figure 17.

Another centrifugal design was produced by Medtronic and consists of nested cones as seen in Figure 18. Still another centrifugal design consists of a series of parallel disks spinning about an axis in unison as seen in Figure 19.

Centrifugal blood pumps have several advantages including a lack of valve-related problems, a relatively lower power drain (than positive displacement pumps), and an ability to pump at higher rates by spinning the rotor faster. Without a displacing drive chamber, there is no need

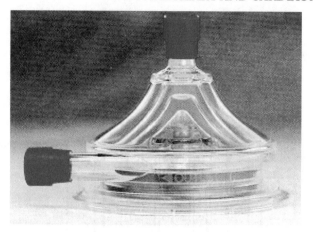

FIGURE 18: Medtronic BP-80 centrifugal blood pump with nested cones and magnetic drive.

for a compliance chamber as there is for some positive displacement pumps. The centrifugal pumps are placed below the left ventricle in a fashion similar to the positive displacement pumps with the outlet cannula extending toward the aorta as is the case for the pulsatile pump connections. Centrifugal pumps are sized similar to their pulsatile analogs and usually can produce 5–7 L/min with an increase in pressure of 100 mmHg at rotation rates of 1500–3000 rpms. Drawbacks to such pumping systems include the use of steady flow rather than pulsatile flow, as well as being significantly affected by afterload. In fact, it is possible for such pumps to spin, but produce no forward flow, if the outlet pressure generated by the pump is lower than the afterload pressure. This is known as "deadheading" and can be remedied by spinning the rotor faster to increase the outlet pressure of the pump.

The other versions of rotary pumps are the smaller axial flow pumps. Unlike the centrifugal pumps, the axial flow pumps have their inlet forward and their outlet directly behind, much as a jet engine is configured. One advantage of such an arrangement is that the flow is more stable in the flow-through arrangement. In addition, the axial pumps, being smaller than their centrifugal counterparts, can be fitted within the base of the ventricle, thus eliminating the inlet cannula. This type of pump can also be placed within the aorta to allow for flow through the device from the left ventricle, and on toward the remainder of the systemic circulation with an added pressure and flow generated from the pump. As such pumps are smaller and have lesser blood contacting area, they require higher rotation rates than their centrifugal counterparts—often reaching 15 000–25 000 rpms. It was originally thought that such rotation rates would produce harmful shearing stresses to blood elements. However, the extremely short exposure time of blood within the device allows only partial deformation of blood elements without permanent damage occurring. A typical axial flow pump is shown in Figure 20.

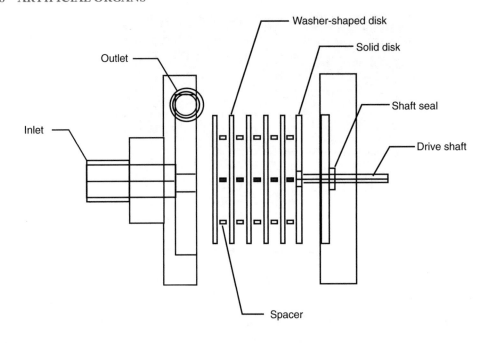

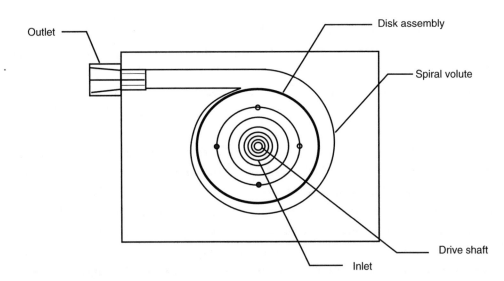

FIGURE 19: Multiple disk centrifugal blood pump based upon the Tesla Turbine design.

The axial flow pump is not only magnetically driven, but requires magnetic suspension. The rotor is magnetically suspended inside the stator, often with blood immersed bearings. A front and rear diffuser are utilized to direct blood flow toward the space between the rotor and the stator. As with their centrifugal pump counterparts, axial flow pumps are preload

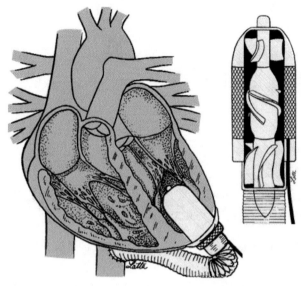

FIGURE 20: Axial flow blood pump configured within the base of the left ventricle. Flow through the device is completely axial with forward and rear diffusers to direct flow to the outer sections between the rotor and the stator.

insensitive and afterload sensitive. As such, they are relatively unaffected by inlet conditions. Axial flow pumps, being smaller in size, do not normally produce flow rates as large as centrifugal pumps, and are not normally utilized for long-term bridge-to-transplant support. However, being of a more compact size, it is far easier to configure such pumps than the larger centrifugal versions.

2.3   EVALUATION OF BLOOD PUMPS

Blood pumps are usually evaluated either experimentally on the bench (designated *in vitro*) or in animal models (*in vivo*) or in clinical studies (*in situ*). There are also computational models using finite element techniques, collectively termed "computational fluid dynamics." Experimental studies *in vitro* (on the bench) employ a mock circulatory system that mimics the human circulation in terms of the physical load against which the device must pump blood. In the bench top study, a blood analog is often employed consisting of a glycerin–water mixture. A mock circulatory system is shown in Figure 21 and consists of a compliance element simulating aortic compliance, a resistor simulating peripheral resistance, and a venous reservoir or compliance element simulating either venous inlet pressure or venous compliance.

Experimental studies of the fluid mechanics within blood pumps and inside the attached cannula (both input and output cannulae) can be typically conducted by laser Doppler anemometry (LDA) and/or particle image velocimetry. Both techniques require that the test fluid is transparent and that the associated flow field is within a transparent section. Collectively,

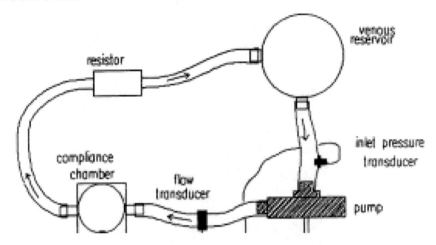

FIGURE 21: Mock circulatory system for *in vitro* testing of blood pumps.

this field is known as *flow visualization*. A particle image velocimetry (PIV) system requires the use of neutrally buoyant particles that are fluorescent so as to be easily illuminated by a dual laser system. A PIV system employs a CCD camera and dual lasers to illuminate the particles at two time intervals closely spaced (within 30 ms) with the tracks between particles calculated by the supporting computer. A typical PIV system is shown in Figure 22.

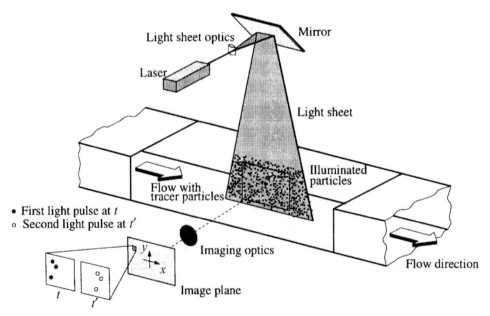

FIGURE 22: Typical PIV system with laser-illuminated particles within the flow field.

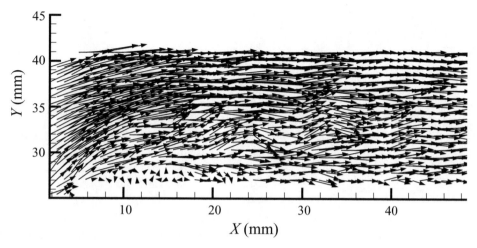

FIGURE 23: PIV vector mapping (snapshot) of particles within a flow field at one instance in time (over a 30-ms interval).

A snapshot of the fluid dynamics of the flow field generated by PIV-illuminated particles is shown in Figure 23. Such snapshots are generated 30 or more times per second and 100 of such images can be averaged to produce a flow profile indicating overall flow trends. Such an average flow profile is shown in Figure 24.

In addition to flow visualization techniques, such as LDA or PIV, it is common to measure volume flow rate with either a Doppler ultrasound flow sensor or an electromagnetic

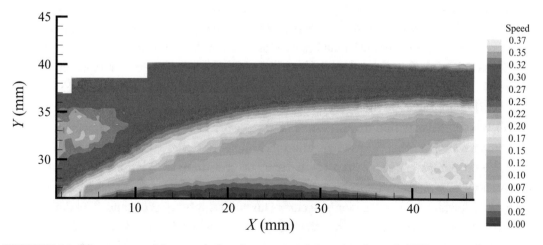

FIGURE 24: Time averaged (contour) plots from sequential vector plots of a PIV-measured flow field. Blue zone indicates low velocity (near lower wall) and red zone indicates higher velocity at outer (top) wall. Slight reversal of flow in yellow is seen at the far right.

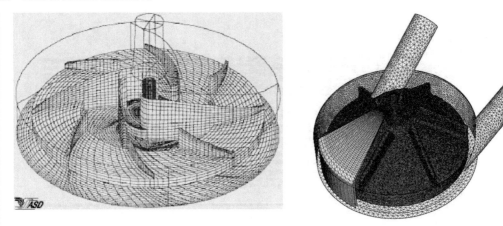

FIGURE 25: Grids generated by computational fluid mechanics algorithms to represent flow boundaries and computational accuracy steps.

flow sensor. It is also common to measure pressure within the flow field at several locations including at the inlet to the pump, at the outlet to the pump, within the compliance chamber of the mock circulatory system, and within the pumping chamber itself. Standard dome-type pressure transducers or solid-state transducers are employed. The former are less expensive, but have a limited frequency response of 100 Hz while the latter are more sensitive and accurate, can be mounted flush within a flow field conduit, but are also far more expensive and fragile.

In many cases, it may be appropriate to model the flow field by computational methods, particularly when a variety of configurations must be examined. The use of computational fluid mechanics (CFD) is often advantageous as a means of verifying experimental data as well as a method of avoiding the need to make expensive *in vitro* experiments with numerous physical configurations. Computational modeling employs the development of digital grids that depict the physical boundaries of the flow field within which the pressure–flow relationships can be analyzed. A typical grid pattern generated within a centrifugal pump is shown in Figure 25.

CFD algorithms can then generate simulated flow field information consisting of pressure and/or velocity fields within the physical grid framework as seen in Figure 26. It is possible to superimpose PIV or other flow visualization techniques onto the grid system as well.

Blood pumps are also evaluated in terms of pressure flow relations as well as for the potential for hemolysis. A typical pressure–flow relationship for a centrifugal LVAD is shown in Figure 27, which indicates a family of curves for various rpms of the pump. The data indicate that the flow is reduced at increasing afterload pressures, which is typical of centrifugal pumps. Similar performance curves of pressure versus flow are typically utilized for all LVAD designs including positive displacement pumps. Alterations in design of the pumps are often based upon the performance results noted in this figure. Hemolysis data is shown in Figure 28 and

FIGURE 26: CFD-generated flow fields within designated grid system. Left figure indicates flow field data within the device while right figure indicates stress data on walls of the device.

is generated with the pump attached to a modified mock circulatory system with whole blood used within the circulation. Hemolysis can be evaluated as either measured evidence of plasma free hemoglobin or by an index of hemolysis which is based in part on a relative scale of plasma free hemoglobin.

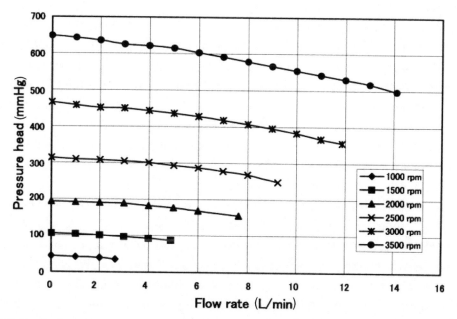

FIGURE 27: Performance curves for a centrifugal blood pump indicating pressure–flow relationships for varying pump rpms. Flow is reduced at increasing pressures (afterloads).

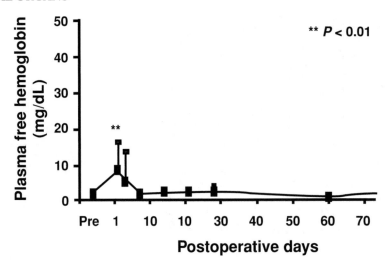

FIGURE 28: Hemolysis data for an LVAD showing levels of plasma free hemoglobin postoperative (postimplantation of pump). After an initial spike, the hemolysis data is quite low, which is typical of many LVAD designs.

Studies of the efficacies of various VAD technologies have been conducted by numerous investigators over the previous 50 years. Of late, several investigators have analyzed the role of ventricular assist devices as either a bridge to transplantation or as a long-term destination therapy including Mehra (2004), Birks *et al.* (2004), Westaby (2004), Kherani *et al.* (2004), Stevenson and Rose (2003), Matsuda and Matsumiya (2003), and Wheeldon (2003), among others. Additionally, there have been significant developments in the use of ventricular assist technology for use in pediatric applications as noted by Reinhartz *et al.* (2002), Duncan (2002), Deiwick *et al.* (2005), and Throckmorton *et al.* (2002). Biventricular support by means of mechanical devices has been studied by Samuels (2004).

The role of steady flow as generated by centrifugal and axial flow blood pumps has been debated for many years. Saito (2004), Mesana (2004), Myers *et al.* (2003), Ichikawa and Nose (2002), among others have examined this issue.

The status of using ventricular and circulatory assist devices was examined by a working group of the National, Heart, Lung, and Blood Institute as reported by Reinlib *et al.* (2003). The United Network for Organ Sharing (UNOS) has examined issues related to the use of ventricular assist technology as it pertains to the selection of patients for heart transplants. Does the use of the VAD lessen the patient's urgent need for a transplant, thus lowering their status "in line" while waiting for a transplant? Or does it increase their needs? Morgan *et al.* (2004) report on the latest findings from UNOS on that topic. Deng and Naka (2002) provide an overview of the state of the art for mechanical circulatory support as do Nemeh and Smedira (2003).

Velocity measurements within ventricular assist devices via flow visualization techniques have been reported by Yamane *et al.* (2004), Tsukiya *et al.* (2002), Manning and Miller (2002), Day *et al.* (2002), Wu *et al.* (1999), Mulder *et al.* (1997), Kerrigan *et al.* (1996), and Miller *et al.* (1995). Computational fluid dynamics (CFD) techniques have been employed by numerous investigators to analyze various design configurations of circulatory assist devices including Song *et al.* (2004a,b), Okamoto *et al.* (2003), Curtas *et al.* (2002), Anderson *et al.* (2000), and Pinotti and Rosa (1995), among others.

CHAPTER 3

Cardiac Pacemakers

3.1 CARDIAC ELECTROPHYSIOLOGY

The heart weighs between 7 and 15 ounces (200–425 g) and is a little larger than the size of your fist. By the end of a long life, a person's heart may have beat (expanded and contracted) more than 3.5 billion times. In fact, each day, the average heart beats 100 000 times, pumping about 2000 gallons (7571 L) of blood. Electrical impulses from your heart muscle (the myocardium) cause your heart to contract, and are created by the movement of ions (principally potassium) across membranes in a fashion called depolarization of cells. This depolarization is propagated along pathways as the ion transport continues. This electrical signal begins in the sinoatrial (SA) node, located at the top of the right atrium. The SA node is sometimes called the heart's "natural pacemaker" with the average heart beat occurring 72 beats/min. An electrical impulse from this natural pacemaker travels through the muscle fibers of the atria and ventricles, causing them to contract. Although the SA node sends electrical impulses at a certain rate, your heart rate may still change depending on physical demands, stress, or hormonal factors. The initiating electrical signal from the SA node travels down a preferred pathway of specialized conducting myocardial tissue toward the atrioventricular (AV) node, which is the electrical connecting point from the atria to the ventricles. After a slight transmission delay (designed to allow the atria to contract and fill the ventricles before the ventricles contract), the specialized conduction pathway continues into the ventricles. There are two bundles of fibers, called the left and right bundle branches of His, culminating in purkinje fibers, that promote rapid transmission of electrical current through both ventricles. The electrical pathways are shown in Figure 29.

3.2 THE ELECTROCARDIOGRAM

The waves of depolarization that spread through the heart during each cardiac cycle generate electrical currents, which in turn spread through the body's interstitial fluids and eventually up to the body's surface. Recording electrodes, placed on the surface of the skin on opposite sides of the heart, are used to detect such electrical potentials. These signals are filtered, amplified, and recorded as a measure of the underlying cardiac electrical activity. The record that results from this procedure is termed an electrocardiogram (ECG or EKG). An ECG is an important

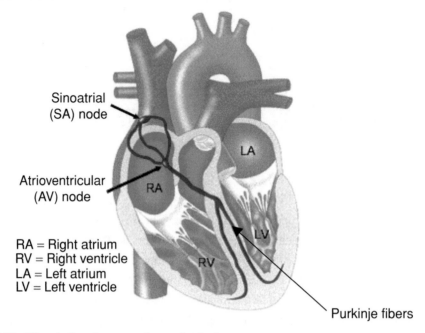

FIGURE 29: Electrical excitatory pathways in the heart.

diagnostic tool to determine whether any cardiac malfunction may have an underlying electrical reason. A typical ECG waveform is seen in Figure 30. The normal ECG consists of three basic features: a P wave, a QRS complex, and a T wave. On some patient waveforms, the QRS complex is seen as three separate waves.

The electrical currents produced as the atrial muscle cells depolarize prior to contraction generate the P wave. This is followed in time by the QRS complex and results from currents

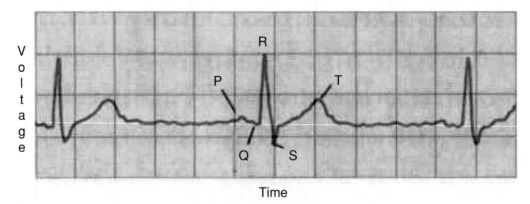

FIGURE 30: Typical ECG showing P, QRS, and T waves.

generated as the ventricles depolarize prior to their contraction. There is a slight time delay between these waves resulting from the delay produced at the AV node, as stated above. Thus, both the P wave and the QRS complex represent depolarization waves. Following ventricular contraction, the ventricle muscle cells repolarize (reverse polarity and reverse ionic transport), and these reversing electrical currents produce the T wave that typically occurs 0.25–0.35 s following ventricular depolarization.

Repolarization of the atria is masked by the QRS complex, and is thus not normally seen on the standard ECG. The intervals between these waves and the width and shape of these waves on the ECG are useful diagnostic measures of cardiac function, as is the overall frequency of the cardiac cycle. The interval between the beginning of the P wave and the beginning of the QRS complex is designated by the PQ interval. Since the Q wave is often absent, it is sometimes termed the PR interval and represents the duration of time (normally about 0.16 s) between the onset of atrial contraction and the onset of ventricular contraction. In patients with certain heart diseases including scarred and/or inflamed tissue, this may lead to a lengthening of the PR interval because more time is required for the depolarization wave to spread through the atrial myocardium and the AV node. The time required for the generation of the QRS complex is termed the QRS duration and represents the amount of time needed for ventricular depolarization. The QT interval extends from the beginning of the QRS complex to the end of the T wave and represents the time required for ventricular contraction and repolarization. Typically, the normal QT interval lasts for approximately 0.35 s. The ST interval represents the time required for the ventricles to repolarize and extends from the S wave to the end of the T wave.

The term *tachycardia* denotes an overall fast heart rate, typically at more than 100 beats/min. Tachycardia may result from fever, from stimulation of cardiac sympathetic nerves, from certain hormones or drugs, or by weakening of the heart muscle itself. When the myocardium is unable to pump blood effectively, homeostatic reflexes are activated that subsequently increase the heart rate. The term *bradycardia* denotes a slow heart rate of less than 60 beats/min and is a condition common to athletes, whose enlarged hearts pump a greater stroke volume per heart beat than that of a nonathlete. In addition, bradycardia may also result from decreased body temperature, due to certain drugs, or via stimulation of the heart by its parasympathetic nerve fibers originating from the vagus nerve. Bradycardia may also occur in patients with atherosclerotic lesions of the carotid sinus region of the carotid arteries. The term *arrythmia* refers to any electrical abnormality in the ECG including rate-related conditions as well as conduction delays or even complete blockages.

The SA node is the natural pacemaker of the heart, providing a stimulating source that initiates the electrical depolarization cycle at approximately 72 beats/min. Should there be a *block* of the excitation wave between the SA node and the AV node, then the origin of the excitation for the remainder of the heart becomes the AV node, whose normal rhythmic rate is slower than that of the SA node (approximately 55 beats/min). The remainder of the excitatory pathway (AV node to purkinje fibers) remains intact. Should there be a partial or total block between the AV node and the purkinje fibers, then alternate excitatory sources within the ventricular muscle assume the excitatory role, albeit at a much lower rate (30 beats/min). These are readily evident in the ECG waveforms as a reduced heart rate, as widening of the waves within the ECG, and/or as changes in the shape of the waves.

Premature contractions can occur within a normal rhythmic cardiac excitation (called sinus rhythm) due to altered excitatory events within an atrium or a ventricle. Premature atrial contractions (PACs or PABs for premature atrial beats) are atrial beats that occur too early due to an abnormal electrical signal. Often, things such as caffeine, alcohol, medications (especially decongestants), certain medical conditions such as hyperthyroidism, anemia, and hypertension, and stress can trigger PACs. Some people may feel a fluttering in their hearts when experiencing PACs, whereas others have no symptoms. PACs are benign and may require no treatment. Premature ventricular contractions (PVCs or PVBs for premature ventricular beats) are early ventricular contractions that occur when the ventricles contract out of sequence with normal heart rhythm. Though they are generally benign and usually do not require treatment, PVCs may result in more serious arrhythmias in those with heart disease or a history of tachycardia. For these people, antiarrhythmic drugs and an implantable cardioverter defibrillator (ICD) may be prescribed. PVCs most often occur spontaneously; however, like PACs, they can also be triggered by caffeine, alcohol, medications (especially decongestants), certain medical conditions such as hyperthyroidism, anemia, and hypertension, and stress. PACs and PVCs are readily observed as added waves within the ECG. An *ectopic* or premature atrial beat produces an added P wave within the standard ECG waveform. A PVC produces an added R wave or QRS segment within the normal ECG waveform. Repeated PAC or PCV events may indicate a serious condition that will require intervention.

Fibrillation is caused when the heart muscle begins to quiver, or fibrillate, continually and cannot contract normally. When a heart is in a state of fibrillation, there is no synchronization between the atria and the ventricles, nor any standard heart beat/contraction. This may cause the patient to experience a racing sensation—and sometimes discomfort in the chest—and/or to feel light-headed or faint. Ventricular fibrillation (VF or V Fib) is a life-threatening arrhythmia that necessitates immediate treatment with an external defibrillator, an internal defibrillator (ICD), antiarrhythmic drugs, or VT ablation. Fibrillation is easily recognized on the ECG waveform as a loss of the QRS segment replaced by a noisy, lower amplitude signal. Atrial fibrillation (AF or A Fib) is a very fast, uncontrolled atrial heart rate caused by rapidly fired signals. During an episode of AF, the atrial heart rate may exceed 350 beats/min. Not all of these beats reach the ventricles, so the ventricular rate is not this high. However, the ventricular rate is often higher than normal and can also be erratic, exceeding 100 beats/min. Sometimes an impulse will circle the atria, triggering atrial flutter, which is similar to AF. Alone, AF is rarely serious, but if a patient has high blood pressure, valvular disease, or heart muscle damage, AF can increase the risk of stroke or heart failure.

There are several treatments for AF, including medication, ablation, and an electrical therapy called cardioversion. Electrical cardioversion converts the heart rate back to normal sinus rhythm through the use of a controlled electrical shock that excites all the heart cells at once, allowing the SA node to resume its role as the heart's natural pacemaker. If medication and electrical cardioversion do not improve the AF, the physician may recommend cardiac ablation to prevent conduction of abnormal electrical impulses from the atria to the ventricles, with implantation of a permanent pacemaker to control heart rate. Ventricular fibrillation (VF) is a chaotic heart rate resulting from multiple areas of the ventricles attempting to control the heart's rhythm. Ventricular fibrillation can occur spontaneously (generally caused by heart disease) or when ventricular tachycardia has persisted too long. When the ventricles fibrillate, they cannot contract normally, hence, they cannot effectively pump blood. The instant VF begins, effective blood pumping stops. VF quickly becomes more erratic, resulting in sudden cardiac arrest or sudden cardiac death. This arrhythmia must be corrected immediately via a shock from an external defibrillator or an internal device (ICD). The defibrillator can stop the chaotic electrical activity and restores normal rhythm by depolarizing the entire heart with the idea that normal sinus rhythm will be restored if the SA node fires first and initiates the normal excitation process.

3.3   CARDIAC PACEMAKER

An artificial cardiac pacemaker is an implantable device that produces an excitatory wave at an appropriate site within the heart: a) at the SA node to replace a failing natural SA node (or correct atrial fibrillation) and generate a normal sinus rhythm, b) at the AV node where a malfunctioning AV node exists, or c) within one or both ventricles to continue excitation beyond a partial or total heart block (or to correct ventricular fibrillation). A cardiac pacemaker consists of 1) an electrical device, the pulse generator, which produces the excitatory signal, 2) the wires, called the leads, which connect the pulse generator to the cardiac tissues, 3) a battery system (housed within the pulse generator) to power the pulse generator, and 4) a programmable segment (also housed within the pulse generator) to evaluate the heart excitatory function and send the appropriate excitatory segment from the pulse generator. A schematic diagram of a cardiac pacemaker and the leads is shown in Figure 31.

The batteries within the pulse generator "can" are typically lithium iodide and have a life span of approximately 10–12 years. The pulse generator casing is typically titanium which is welded and hermetically sealed. An epoxy top is attached from which the leads are attached to the pulse generator. This allows for changes in the leads without disrupting the pulse generator case. A typical pulse generator is small in size, often less than an ounce in weight, less than two inches wide, and a quarter-inch thin. Thus, the device is roughly the size of two silver dollars stacked on top of one another. Once implanted in the upper chest, just below the skin

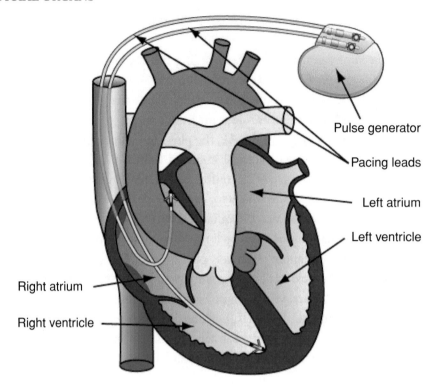

Pulse generator

Pacing leads

Left atrium

Left ventricle

Right atrium

Right ventricle

FIGURE 31: Schematic diagram of an implantable cardiac pacemaker.

near the collarbone, the pacemaker's presence is nearly invisible to the eye. When the batteries have become depleted, the entire pulse generator case is replaced. The surgical implantation procedure will be explained in more detail below. A pulse generator is seen in Figure 32, with the leads connected to the "can" seen in Figure 33.

FIGURE 32: Typical pacemaker pulse generator with epoxy top for connection to leads.

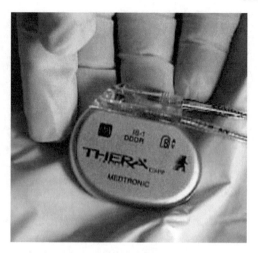

FIGURE 33: Pulse generator with leads attached.

The pacing lead, which connects to the pulse generator header, is a flexible insulated wire with an electrode tip. This tip, which can be inserted through a vein into the heart, carries impulses from the pulse generator to the heart, stimulating the heart to beat. It also carries information from the heart back to the pulse generator, which the physician accesses via a special programmer or into the pulse generator circuit embedded program. Leads that are inserted via a vein into the heart have tines on the end as shown in Figure 34 or have a screw end, which can be threaded into the cardiac muscle as shown in Figure 35.

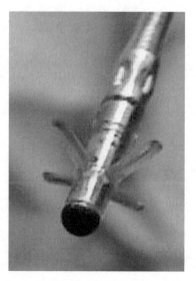

FIGURE 34: Pacemaker lead with tines on end for embedding into heart tissue.

FIGURE 35: Pacemaker lead with screw end for implantation into heart tissue.

The leads shown in Figures 34 and 35 are both designated as endocardial or transvenous leads. The tined lead shown in Figure 34 is a passive fixation device where the tines become lodged in the trabeculae (fibrous meshwork) of the heart. The helix or screw-ended lead shown in Figure 35 is an active fixation device that extends into the endocardial tissue (interior surface of the heart). Such a lead can be positioned anywhere in the heart chamber. The other category of lead is the myocardial or epicardial lead which is secured to the outer surface of the heart as shown in Figure 36. Such leads are connected to cardiac tissue via sutures or screwed into tissue as was the case with the endocardial lead.

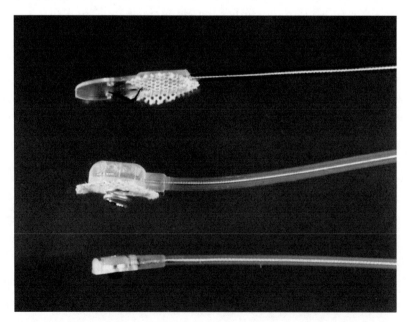

FIGURE 36: Myocardial pacemaker leads which are sewn into exterior cardiac tissue via sutures or screwed into the exterior surface.

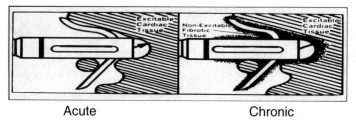

Acute   Chronic

FIGURE 37: Fibrous capsule surrounding a tined lead as a result of an inflammatory response.

Once leads have been implanted, there is often a tissue reaction that creates a fibrotic capsule around the lead, as shown in Figure 37. Such a capsule can increase the resistance to the stimulating pulse to cardiac tissue, which would require an increase in the amplitude of the original pulse from the pacemaker. This would deplete the batteries at a rate faster than normal. In order to avoid such a circumstance, some leads contain a steroid that eludes into the surrounding tissue to reduce the onset of an inflammatory response. This is seen in Figure 38.

Leads can be either unipolar or bipolar. Unipolar leads, as shown in Figure 39, are single leads inside of an insulating sheath, which send the impulse signal from the pulse generator with the returning electrical signal traveling through the tissues to the pacemaker. The unipolar lead normally has a smaller diameter than the bipolar lead and exhibits a greater tendency for pacing artifacts on the surface EKG, since the returning electrical signal travels through tissue.

The bipolar lead, as shown in Figure 40, is a coaxial lead with a larger overall diameter. The returning signal (after stimulating cardiac muscle) pathway is through the second embedded lead, rather than through tissues. As a result, the bipolar lead has a lower susceptibility to affect

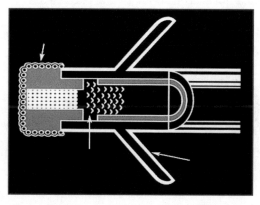

FIGURE 38: Steroid eluding lead designed to deter inflammatory response and fibrous capsule formation.

FIGURE 39: Unipolar pacemaker lead.

the EKG and is less susceptible to external interference, such as from ambient electromagnetic "noise."

The insulation on pacemaker leads is either silicone or polyurethane. Silicone is inert, biocompatible, biostable, repairable with medical adhesive, and has a long and successful history in biomedical applications. Polyurethane is also biocompatible, but additionally has a higher tear strength, lower friction coefficient, and a smaller required thickness.

Pacemaker systems that are fixed rate often utilize a single lead that is placed in the atrium (if there is an SA nodal problem) or the ventricle for AV nodal or bundle branch problems. For circumstances where there is a need for a demand style pacemaker, there must be two leads placed within the heart as seen in Figure 41.

The sensing electrode is placed within the atrium to sense any variation in normal sinus rhythm, such as what might occur during exercise of increased heart activity. The stimulating electrode might be placed within the ventricle and the rate of stimulation is dependent on the sensed SA nodal rate. Modern demand pacemakers thus have four main functions:

- Stimulate cardiac depolarization
- Sense intrinsic cardiac function
- Respond to increased metabolic demand by providing rate responsive pacing
- Provide diagnostic information stored by the pacemaker.

The output of the pacemaker, which stimulates cardiac tissue, can be visualized on the surface EKG as is seen in Figure 42.

FIGURE 40: Bipolar pacemaker lead.

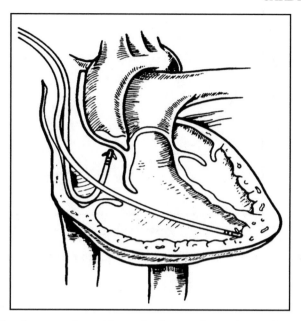

FIGURE 41: Demand style pacemaker with two leads: one for sensing and one for stimulation.

A typical stimulating pulse from the pulse generator must be large enough to cause depolarization (i.e., to "capture" the heart) and must be sufficient to provide an appropriate pacing safety margin. The pulse width must be long enough for depolarization to disperse to the surrounding tissue. As such, pulse generation waveforms are typically evaluated on a strength duration curve as noted in Figure 43.

Once a pacemaker elicits a stimulating pulse and the heart muscle depolarizes, there exists a refractory period whereby another stimulating pulse would not elicit further depolarization.

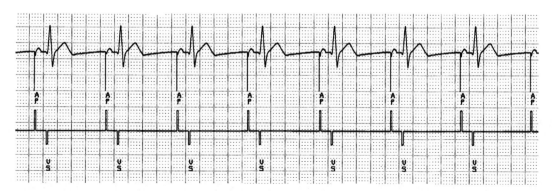

FIGURE 42: Surface EKG with pacemaker stimulation pulses noted within the EKG waveform as well as on the axis below.

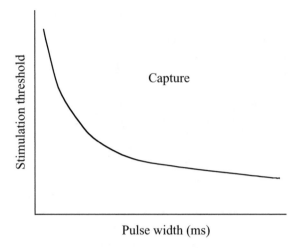

FIGURE 43: Strength–duration curve to evaluate the required necessary stimulating pulse to achieve "capture," i.e., produce a depolarization.

This same refractory period exists during normal cardiac electrical stimulation, such that any ectopic wave would not elicit a second depolarization. Repolarization would normally need to occur before a second pulse would produce another depolarizing wavefront. As was noted above, modern pacemakers often employ a sensing electrode in one heart chamber and a second stimulating electrode in another chamber. A system of classification has been established for pacemakers which is used to describe the various parameters that can be incorporated into modern pacemakers, as seen in Figure 44.

Referring to this classification system, a VOO pacemaker (see chart) stimulates the ventricle with no atrial or ventricular sensing (nor any response to sensing). This would be a fixed rate pacemaker that stimulated the ventricle regardless of the patient's intrinsic cardiac

CHAMBER PACED	CHAMBER SENSED	RESPONSE TO SENSING	PROGRAM FUNCTION	FUNCTION
V: Ventricle	V: Ventricle	T: Triggered	P: Simple	P: Pace
A: Atrium	A: Atrium	I: Inhibited	M: Multi	S: Shock
D: Dual	D: Dual	D: Dual	C: Communicating	D: Dual
O: None	O: None	O: None	R: Rate modulate	O: None
S: Single	S: Single	O: None		

FIGURE 44: Pacemaker classification system.

electrophysiology. A VVI pacemaker (see chart) senses ventricular electrical activity and paces the ventricle if there is a lack of normal intrinsic activity sensed. If normal ventricular excitatory pulses are sensed, then the pacemaker output is inhibited (the I) in the pacemaker classification. A VVIR pacemaker is identical to the VVI pacemaker except for the added "R" term indicating program function (R = rate). This pacemaker changes the stimulation rate of firing when the intrinsic ventricular stimulation warrants it. Similarly, a VAIR pacemaker senses atrial activity and stimulates the ventricles. The pacemaker pulse rate is affected by the atrial activity. If the patient's metabolic activity alters the SA nodal firing rate, then the pacemaker senses this and increases the ventricular firing rate. If there is a normal sinus rhythm, then the pacemaker output is inhibited. Similarly, a VDIR pacemaker senses intrinsic activity in both chambers with the pacemaker capable of delivering a variable rate.

Originally, pacemakers of the 1950s were external devices with the leads implanted into the heart. The first transistorized, wearable, battery powered pacemaker developed by Wilson Greatbach in 1957, and the first totally implanted pacemaker was in 1960. The battery life was approximately 18 months. In the mid-1960s, transvenous leads were developed that could be introduced through a vein, rather than require opening the chest to attach the lead. Original designs for pacemakers incorporated a fixed rate. In the mid-1960s, the first "demand" pacemakers were developed that sensed when there was a missed beat and then initiated a pulse to excite the cardiac muscle, while being inactive during normal sinus rhythm. The screw tip and tined leads were developed in the 1970s, and in 1975, the longer lasting lithium iodide battery was developed to replace the mercury–zinc battery. At that same time, the titanium case and epoxy top was developed, which better shielded the pacemaker from outside electromagnetic disturbances, such as a microwave oven. In that same time span, the first programmable pacemakers were developed as were the first dual chamber pacemakers. In the 1990s, the pacemakers became smaller with advances in batteries and in microelectronics. To date, over 2 million cardiac pacemakers have been successfully implanted.

3.4   PACEMAKER IMPLANTATION

A schematic diagram depicting the surgical implantation of a cardiac pacemaker is shown in Figure 45. The pacemaker pulse generator with its epoxy top is implanted just below one of the collarbones. The leads are then inserted through a vein leading to the heart or via minimally invasive surgery to affix them to the exterior of the heart.

The use of the easily accessible site below the collarbone allows the pulse generator with its battery pack to be easily replaced once the batteries have become depleted. By using a local anesthetic, a small opening in the skin and underlying thin tissues can be created in order to remove the pulse generator. Since the leads are connected to the epoxy top of the pulse generator by means of a force fit, then the leads can be easily removed and snapped

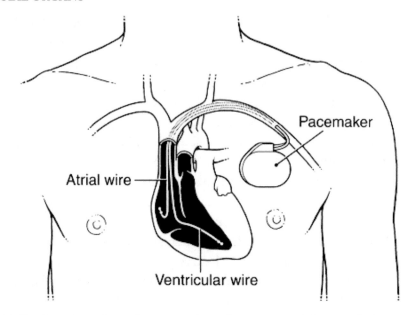

FIGURE 45: Implantation of a cardiac pacemaker pulse generator below a collarbone with the leads threaded through a nearby vein toward the heart.

into the replacement pulse generator (and battery pack) within the new epoxy top associated with the new generator. The time "off" of pacing is thus extremely short—the time it takes to pull the lead(s) out of one epoxy top and into another. The new pulse generator is then placed in the site of the old unit, and the small incision is then closed. The entire replacement procedure often takes as little as 30 min from open to close. As the modern batteries for cardiac pacemakers can last as long as 12 years, replacement of the pulse generator is infrequent.

3.5   CARDIOVERTER

If the patient is susceptible to ventricular fibrillation, then a standard cardiac pacemaker may not be able to reestablish sinus rhythm. As with the case when an individual suffers a "heart attack," where VF occurs, it is necessary to convert the patient by completely depolarizing the entire heart by means of a defibrillator. Most individuals are familiar with external defibrillators that use large paddles and a large current density to provide an overwhelming depolarization to the heart. The premise is that when the heart then completely repolarizes, that the SA node will fire before any other site within the heart and reestablish sinus rhythm. The large paddles and large current density are necessary for external defibrillation, as the skin and exterior tissues provide a large resistance to electrical current flow.

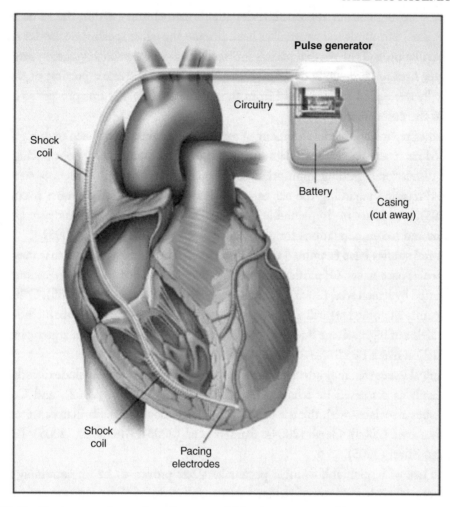

FIGURE 46: Dual pacemaker and cardioverter within the same case. Note the shocking coil which surrounds the pacing lead.

However, if the defibrillator were placed deeper within the body, then the amount of current necessary to defibrillate the heart is far less. As a result, the device is far smaller. For those patients who may require defibrillation on a regular enough basis, an implantable version can be placed in the same site as a cardiac pacemaker—below the collarbone with the leads threaded through a vein toward the heart. For those patients who require frequent cardiac pacing, but are also less frequently susceptible to ventricular fibrillation, then a combination unit of pacemaker and implantable cardioversion device (ICD) is implanted as shown in Figure 46. This device is sometimes called an implantable cardiac defibrillator (ICD).

The shocking coil for the cardioverter (defibrillator) unit (within the same housing as the pacemaker) surrounds the pacemaker lead. Should the pacer sensing electrodes detect ventricular fibrillation and the normal pacing protocol does not reestablish sinus rhythm, then the cardioverter fires to completely depolarize the heart. The pacemaker portion of the housing is electrically decoupled and insulated from the cardioverter during this process so as to avoid damage to the pacemaker circuitry.

Studies regarding the development of cardiac pacemakers have been ongoing for several decades. More recent studies include a discussion of permanent transvenous pacing by Petrie (2005), a discussion regarding appropriate selection of a pacemaker by Ellenbogen and Wood (2005), a discussion regarding the use of single chamber versus dual chamber pacing by Toff *et al.* (2005), an update on implantable pacemakers by Woodruff and Prudente (2005), and indications and recommendations for pacemaker therapy by Gregoratos (2005).

Several studies have examined the role of interfering external factors that may adversely affect a cardiac pacemaker. Of particular note has been the effect of an MRI on pacemaker activity as reported by Irnich *et al.* (2005), Vahlhaus (2005), and Del Ojo *et al.* (2005). Other external factors recently examined regarding interference with pacemaker performance included electromagnetic fields of high voltage lines by Scholten *et al.* (2005), and electromagnetic interference from induction ovens by Hirose *et al.* (2005).

Clinical issues that may adversely affect pacemaker performance include dislodged and/or fibrosed leads as discussed by Khan *et al.* (2005), Shimada *et al.* (2002), and Cutler *et al.* (1997). Issues associated with the use of implantable cardioverter defibrillators were examined by Nazarian *et al.* (2005), Gersh (2004), Sanders *et al.* (2005), Alter *et al.* (2005), Bryant *et al.* (2005), and Silver (2005).

The use of implantable cardiac pacemakers has proven to be an extremely successful "artificial organ" with millions of patients successfully treated. The later use of implantable defibrillators and dual pacemaker/defibrillators has also been successful, albeit with far less cases to date than with pacemakers alone.

CHAPTER 4

Dialysis

The human kidneys are responsible for continually cleansing blood of metabolic waste products (such as urea, uric acid, creatinine), excess ions, and excess water. The resultant filtrate is then transported to the ureter to be removed as urine. Approximately 20% of the blood supply is routed to the kidneys at any time in order to provide a continual means of removing potentially toxic materials from the blood. The water removal from the kidneys serves as a secondary means of controlling blood pressure, as total blood volume is a factor in venous return, cardiac output, and thus arterial pressure. There are two kidneys, each consisting of a network of millions of individual mass transfer elements called *nephrons*. Transport occurs across the surface of the nephron. As is the case with the lungs, which have similar elements called alveoli, the nephrons provide a means of greatly expanding the surface area available for mass transfer within a constrained volume. The theory is similar to that of a single large sphere as compared to millions of smaller spheres within the same volume. The net surface area of the large sphere is much smaller than that of the combined surface area of the numerous small spheres. Thus, mass transfer of metabolic waste products occurs by branching blood into smaller and smaller channels until the actual mass transfer occurs at the small end point—the nephron. Again, this is similar to the trachea branching into the bronchi and eventually to the alveoli in the lungs.

4.1 THE NEPHRON AND MASS TRANSFER

The kidney and its most elemental mass transfer unit, the nephron, are shown in Figure 47. Unlike most organs within the human body, the kidneys are supplied with blood directly from an arteriole, which branches from the renal artery. The use of an arteriole allows for control of the source hydrostatic pressure to each nephron (by means of vasoconstriction or vasodilation) and provides a constant source of pressure to each mass transfer unit.

The nephron consists of the glomerulus, a relatively porous membrane with pores diameters of 50 Å and pore lengths of 500 Å. These pore diameters allow a large amount of fluid to enter into the nephron tubule system, while excluding (by size) such blood elements as cells, large proteins, large sugars, etc. The glomerular filtration rate (GFR), the amount of fluid flow through all of these membranes, equals 125–150 mL/min. As the total blood flow to the kidneys

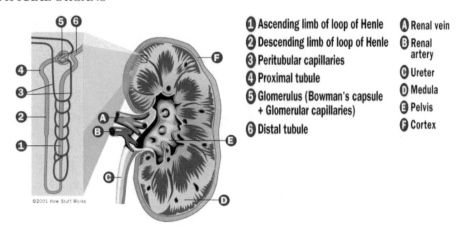

1. Ascending limb of loop of Henle
2. Descending limb of loop of Henle
3. Peritubular capillaries
4. Proximal tubule
5. Glomerulus (Bowman's capsule + Glomerular capillaries)
6. Distal tubule

A. Renal vein
B. Renal artery
C. Ureter
D. Medula
E. Pelvis
F. Cortex

©2001 How Stuff Works

FIGURE 47: The human kidney and the nephron (shown in detail to the left).

equals 1000–1200 mL/min (20–25% of total blood flow), and plasma (the liquid portion of blood) represents approximately one half of the renal blood flow (for a hematocrit of 45% plus white cells and platelets), then the GFR represents approximately 25% of plasma renal flow (150 of 600 mL/min). This means that 25% of the liquid portion of the blood is leaving the blood as it travels into the tubules of the nephron. Luckily, only about 1–2 mL/min eventually finds its way to the ureter, which indicates that the remainder is returned to the blood supply. This occurs through reabsorption of the filtrate into blood vessels that parallel each nephron, the *vasa recta* or *peritubular capillaries* as is seen in Figure 47. The glomerular filtration rate is controlled by a hydrostatic pressure gradient as well as an osmotic pressure gradient (from the blood side to the filtrate side across the glomerulus). The hydrostatic pressure gradient is a relatively large 50 mmHg (for so short of a "pore" length) and the reverse osmotic pressure gradient is 25 mmHg. The osmotic pressure gradient is reversed from the hydrostatic pressure gradient since cells and large proteins on the blood side do not travel across the pores, and exert an osmotic pressure from the filtrate side to the blood side. The net 25 mmHg pressure gradient that pushed fluid across the glomerulus is still quite large and results in a significant transport of fluid as was described above.

    The filtrate (the name of the fluid once it enters the tubules of the nephron) then follows the tubule pathway from the proximal tubule, through the loop of Henle, then through the distal tubule and the collecting duct, eventually ending in the ureter. Along the way, through a combination of concentration-driven passive diffusion as well as active transport, positive ions are transported across the tubules into the surrounding extracellular fluid. Negative ions follow the same pathway, primarily by means of a charge balance as well as passive diffusion. Water follows along the same pathway by means of a concentration-driven osmosis. All of these constituents are then absorbed by the vasa recta to return to the blood supply.

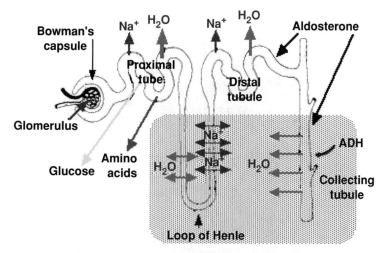

FIGURE 48: Movement of ions and water across the tubule system of the nephron as controlled by the hormones: aldosterone and ADH.

The loop of Henle operates by a countercurrent mechanism whereby the downward and upward sections are both contributing to the extracellular osmolarity, which is very high at the bottom of each loop and lessens toward the top of each loop. This occurs as ions are transported in both directions within each section of the loop, with most of the ion transport out of the loop at the bottom and into the loop at the top. The counter current mechanism allows the ascending loop to return fluid (and ions) into the descending loop and eventually into the extracellular fluid at the bottom of both loops. This fluid and ion mixture eventually finds its way back to the vasa recta.

The entire process of fluid flow and mass transfer across the glomerulus and through the tubule system is designed to transport a significant amount of fluid into the tubules with the bulk of it returning to the vasa recta. This process allows for selective retention of water and ions inside the tubules as controlled by hormones, which in turn allows for selective concentration of urine as is seen in Figure 48.

This approach—moving alot of fluid into the tubules to allow for a variable amount to be returned, cannot be duplicated by means of dialysis—the artificial cleansing of blood. Dialysis utilizes a standard transport of ions, metabolic wastes and water across a semipermeable membrane with this fluid mixture traveling across the membrane by means of simple diffusion, and not returning to the blood, as it does in the natural kidney.

4.2 DIALYSIS PROCEDURE AND THE DIALYSIS SYSTEM

Dialysis is the artificial cleansing of blood to remove the same components as those removed by the natural kidney: metabolic waste products such as urea, ureic acid, and creatinine; excess ions; and excess water. This is accomplished by means of concentration gradient-driven diffusion for

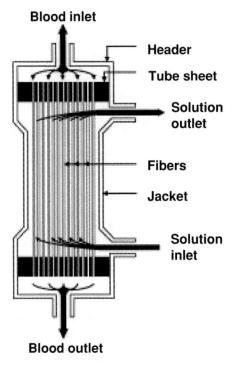

FIGURE 49: Capillary tube dialysis cartridge.

the first two components and by a pressure gradient for water. This process occurs within a capillary tube dialysis cartridge as is shown in Figure 49.

The cartridge is a polycarbonate canister containing 11 000 minute capillary tubes, each with pores small enough to allow transport of the three major blood components listed above, while too small to allow blood cells, large proteins, etc. to cross. In this fashion, the capillary tube pores are not unlike the glomerulus pores in the natural kidney nephron. The cartridge has upper and lower ports for blood to enter and exit the canister. Surrounding the capillary tubes is a fluid, called *dialysate*, which bathes the capillary tubes as serves as the recipient of the wastes, ions, and excess water which leaves the blood from within the capillary tubes. The dialysate is pumped from one side port of the cartridge and leaves from the other side port. The capillary tubes are typically composed of cellulose, which has been proven to be a biocompatible material. At the top and bottom of the capillary tube pack is a polyurethane "potting" section which appears to be a yellowish mass. In fact, this compound serves to briefly pool the incoming blood from the large inlet port so that it can more readily enter the minute capillary tubes. The reverse is true at the bottom of the capillary tubes where the blood enters the potting compound and then into the large outlet port. Dialysis cartridges come in many sizes and are matched to the size of the patient—from children to large adults as can be seen in Figure 50.

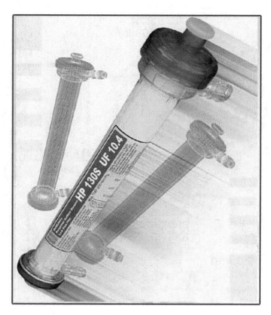

FIGURE 50: Capillary tube dialysis cartridges.

Dialysate is similar to clean blood, in that it has no waste products and has the normal level of ions. The typical dialysate mixture consists of the following normal level of blood ions:

NaCl	5.8 g/L	Na	132 meq/L
NaHO$_3$	4.5 g/L	K	2.0 meq/L
KCl	0.15 g/L	Cl	105 meq/L
CaCl$_2$	0.18 g/L	HCO$_3$	33 meq/L
MgCl$_2$	0.15 g/L	Ca	2.5 meq/L
Glucose	2.0 g/L	Mg	1.5 meq/L

The glucose is used to provide an osmotic gradient to assist in water transport from blood to dialysate. Blood with a high concentration of wastes and ions enters into the top of the dialysis cartridge with clean dialysate entering in the side port nearby. Along the length of the dialysis cartridge, simple diffusion takes place with wastes moving from the source of high concentration (blood) to the point of zero/low concentration (dialysate) across the capillary pores. Similarly, ion transport occurs from a high concentration (in the blood) to one of low concentration (in the dialysate). Water is transported from blood to dialysate by means of a concentration gradient of glucose assisted by a pressure gradient. A typical blood flow rate through the capillary tubes (as a whole) is 200 mL/min with the dialysate flow rate at 500 mL/min. The higher flow rate of dialysate ensures that dialysate with newly acquired waste products are quickly dispelled from

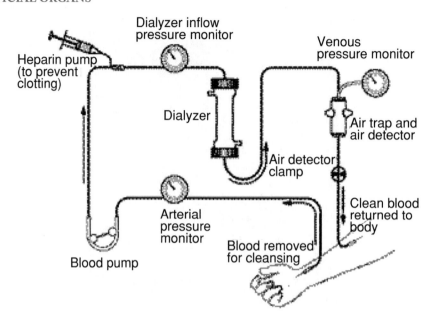

FIGURE 51: Typical dialysis system.

the cartridge and replaced with clean dialysate, thus maintaining a large concentration gradient for mass transfer.

A typical dialysis system is shown in Figure 51. Blood access from the body is connected from needles to tubing, which is routed throughout the machine and into the dialysis cartridge. Blood from the bottom of the dialysis cartridge is then routed through the remainder of the tubing and onwards to return to the body. The dialysis machine consists of various sensors and monitors along with two key elements—a roller pump which pushes blood slowly along a tubing pathway (to avoid stagnation and resultant clotting) and bubble traps to allow any ambient air from remaining in the blood to cause an air embolism.

The dialysis cartridge and tubing set are disposed of following dialysis, although in some dialysis centers, the cartridge itself is cleaned and may be reused. This latter issue will be discussed in more detail in a subsequent section.

Blood is accessed from the radial artery in the forearm and returns to the cephalic vein. Typical chronic dialysis patients undergo dialysis three times per week for 4 h/session. This results in numerous insertions of needles into the forearm. Although the skin becomes tough after time, the underlying blood vessels do not. As a result, chronic dialysis patients often undergo a minor procedure whereby an arteriovenous graft is placed below the skin connecting the radial artery to the cephalic vein as is seen in Figure 52.

The graft not only protects the blood vessels from repeated needle insertions, but also connects the high pressure artery to the low pressure vein—keeping the vein from collapsing.

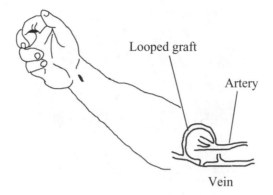

Looped graft

Artery

Vein

FIGURE 52: Arteriovenous graft in the forearm for chronic dialysis patients.

The pressure gradient from the artery to the vein also provides a gradient to propel blood through the tubing set and the dialysis cartridge. Thus, the roller pump within the dialysis system merely provides a boost for blood flow rather than having to provide the sole means of blood flow through the system.

A typical dialysis tubing set is shown in Figure 53 and includes bubble traps and sufficient tubing to connect from the artery/vein access needles to the dialysis machine, the dialysis cartridge, and back again to the forearm.

A typical dialysis machine is shown in Figure 54 with slots for the dialysis cartridge and bubble traps as well as an embedded roller pump and blood sensors.

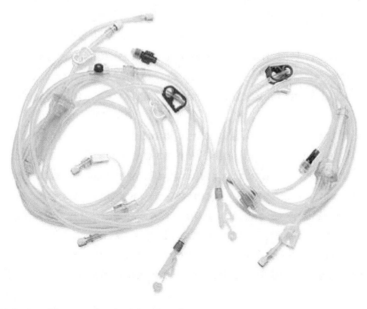

FIGURE 53: Dialysis tubing set for the blood pathway.

FIGURE 54: Dialysis machine.

The left-hand color-coded (blue and red) tubing is for the dialysate (clean and "dirty" ports). The slot to the right of this tubing is for the dialysis cartridge. To the right of the cartridge location is where the bubble traps are inserted. The roller pump is to the right of the bubble traps. At the bottom of the machine are two tubes with a tray below them. A dialysate concentrate jug is placed on the tray and one of the tubes is inserted into the jug. The machine draws dialysate concentrate from the jug, which is mixed with processed/treated water.

A typical dialysis center includes a water treatment room which converts municipal water into that which is clean enough to be used in close contact with blood. This water treatment includes a sediment filter, a water softener section, an ion exchanger section, an ultraviolet light to destroy bacteria, and a reverse osmosis unit which back-pressures water across an extremely fine (small pore) membrane to remove microscopic elements. The resulting treated water is sent to various patient treatment stations and each dialysis machine is connected to the water port via a hose located on the back of the machine. The mixture of treated water and dialysate concentrate is then transported to the dialysis cartridge through the front, color-coded tubing as was seen to the left of Figure 54.

A typical dialysis machine is approximately 60 inches high, 17 inches wide, 22 inches deep, and weights 190 pounds. It is on rollers to allow placement near the patient. The typical front-mounted screen displays the time on dialysis, time remaining, target water loss (in kg), water uptake rate (in kg/h), and blood pressure and/or heart rate.

4.3 HISTORY OF DIALYSIS

Thomas Graham, Professor of Chemistry at Anderson's University in Glasgow, coined the term dialysis in 1861. He noticed that crystalloids were able to diffuse through vegetable parchment coated with albumin (which acted as a semipermeable membrane). He called this "dialysis." Using this method, he was able to extract urea from urine. In 1913, Abel, Rowntree, Turner, and colleague constructed the first artificial kidney. They used hirudin, produced from leeches obtained from Parisian barbers, as an anticoagulant. They passed animal blood from an arterial cannula through celloidin tubes that were contained in a glass "jacket." The glass jacket was filled with saline or artificial serum. They coined the term "artificial kidney." Blood was returned into the vein of the animal via another cannula. The inventors wrote, "this apparatus might be applied to human beings suffering from certain toxic states, especially if due to kidney damage, in the hope of tiding a patient over a dangerous chemical emergency." The apparatus was never used to treat a patient (Robson JS, 1978). George Haas from Germany performed the first successful human dialysis in autumn 1924. The dialysis was performed on a patient with terminal uremia "because this was a condition against which the doctor stands otherwise powerless." The dialysis lasted for 15 min, and no complications occurred.

The first practical human hemodialysis machine was developed by WJ Kolff and H Berk from the Netherlands in 1943. This rotating drum artificial kidney consisted of 30–40 m of cellophane tubing in a stationary 100-L tank. It was Kolff who made clinicians and experimentalists interested in the treatment of uremia, and this machine delivered the effective hemodialysis treatments. This rotating drum machine is seen in Figure 55.

In 1946, Nils Alwall produced the first dialyzer with controllable ultrafiltration. It consisted of 10–11 m of cellophane tubing wrapped around a stationary, vertical drum made of a metal screen—resembling a rotating drum device stood on its end. In 1956, Kolff and Watschinger developed the principles of the Alwall machine to develop the "twin coil" artificial kidney (Figures), a modification of the "pressure cooker" dialyzer developed by Inouye and Engelberg in 1952. The first patients treated by dialysis were all believed to have acute renal failure. The methods in use for getting adequate flows of blood into the machine exhausted veins and arteries very quickly, and only a few dialysis treatments could be undertaken. The development of methods to use blood vessels repeatedly while preserving them made it possible to contemplate keeping a few patients alive for longer periods even though they had permanent

FIGURE 55: Kolff rotating drum dialysis machine—the first practical hemodialysis system.

renal failure. The arteriovenous shunt was the key development. The first substantial program for dialysis of patients with chronic renal failure began in Seattle in the same year.

Home hemodialysis was introduced to overcome the difficulties in providing adequate facilities in hospitals for the increasing number of patients being put forward for treatment. If a relative provided help for the patient, it could be carried out without the use of doctors, nurses or hospital premises, extending the number of patients that could be treated, as well as being better for the patient. However, in 1965, at the American Society of Artificial Internal Organs meeting, reports of home hemodialysis of four patients in Boston and two in Seattle were supplemented by a report of two patients treated at home in London (Shaldon). All reported success and plans to expand their programs.

Today, hundreds of thousands of chronic dialysis patients undergo routine, periodic dialysis three times per week at local dialysis centers located throughout the nation. Dialysis represents one of the most successful organ replacement systems with millions of patients treated successfully for partial or total renal failure. Diabetes and heart disease remain principal causes for renal failure with alcohol and substance abuse also accounting for numerous cases.

4.4   DIALYZER CARTRIDGE REUSE

Many local dialysis centers are privately operated facilities. Although dialysis conducted in hospitals for hospital patients (acute dialysis centers) results in single use of dialysis cartridges, many private facilities reuse cartridges. This entails the rinsing of the cartridge following patient dialysis, after which the cartridge is treated with a sterilant such as formaldehyde, gluteraldehyde, or renalin. Up to four cartridges are placed into a cleaning machine that provides several rinse steps and introduction of the sterilant. The patient's name is written on the cartridge and the cartridge is placed into a bin where it will be stored until that patient needs it for the next session.

The cartridge is then rinsed several times before used for that patient. Although there is a capital expense in purchasing the cleaning machine along with supplies and manpower needed to rinse and sterilize the cartridges, there is a net modest savings rather than purchasing a new cartridge for every patient session. The cartridges cost $10–12 each and the dialysis facility is reimbursed at a set rate for each patient session by insurance, Medicare, or Medicaid. If there is a small net savings, even after expenses associated with cleaning and storing the cartridges, then the dialysis facility would show a large profit, given the number of patient sessions conducted each year. A typical dialysis center might have 20 patient stations. Each patient undergoes dialysis three times per week for 4 h/session. If the center runs an MWF morning, MWF afternoon, T-Th-S morning and T-Th-S afternoon patient cohort, then there are a total of 240 sessions/week (3 sessions for each of 20 stations for each of four patient cohorts). As such, even a modest savings per session adds up to substantial overall savings and profit.

Dialyzer-reuse machines are shown in Figure 56.

Although cost issues are often at the forefront of the rationale to reuse dialyzer cartridges, there has been a considerable debate regarding the efficacy and safety when reusing cartridges. One health-related rationale for the reuse of cartridges is the "first use syndrome" which refers

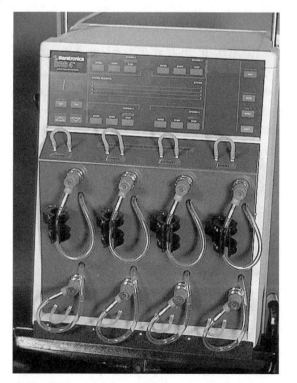

Dialyzer Reprocessing System
Model MM1000

FIGURE 56: Dialyzer-reuse machines: a four-cartridge system on the left and a single-cartridge unit on the right.

to residual manufacturing byproducts still inside the cartridge when it is unpacked from its wrapping. By reusing cartridges, the uses after the first use no longer suffer from this problem. On the other hand, the health-related arguments against reuse are that a) there is residual sterilant (formaldehyde, etc.) even after rinsing that might harm the patient over time, and b) the clearance of wastes and ions from the dialysis cartridge is reduced over time as blood byproducts clog some of the capillary tubes. Many dialysis cartridges may be reused up to 25 times per patient before the clearance levels fall below 80% of the maximum, which is the standard cutoff before the cartridge must be discarded. The Association for the Advancement of Medical Instrumentation (AAMI) sets standards for cartridge clearance levels. Numerous studies have examined the various factors regarding reuse of dialyzer cartridges including Fan *et al.* (2005), Robinson and Feldman (2005), Szathmary *et al.* (2004), Narsipur (2004), Stragier (2003), Ward and Ouseph (2003), Rahmati *et al.* (2003), and Parks (2002), among others. The issues regarding reuse continue to be debated, although the prevalence of reuse is high among the vast majority of chronic, privately operated dialysis centers.

As dialysis is utilized by hundreds of thousands of patients and the process affects blood chemistry and overall health, there have been numerous studies on the process itself, on disease states that require dialysis, on the various techniques and technologies regarding dialysis. In particular, there are issues associated with cardiovascular disease that impact hemodialysis and vice versa. Studies that have examined this link are numerous and include Familoni *et al.* (2005), Saxena and Panhotra (2005), Di Benedetto *et al.* (2005), Iorio *et al.* (2005), and Ronco and Tetta (2005), among others. Issues associated with blood access and the use of AV fistulas have been studied by Dember *et al.* (2005), Wijnen *et al.* (2005), and Peirce *et al.* (2005), among others. Issues related to blood chemistry and general health considerations have been examined by Panichi *et al.* (2005), Piccoli *et al.* (2005), Prado *et al.* (2005), Gusella *et al.* (2005), Lee *et al.* (2005), and Kiss *et al.* (2005), among others. In addition, there have been studies that have examined the effect of missed dialysis sessions in patient health as well as the desire of patients to quit dialysis including those by Gee (2005), Davison and Jhangri (2005), and Unruh *et al.* (2005), among others.

Dialysis remains a popular and cost-effective means of augmenting reduced kidney function and is a viable alternative to kidney transplantation. The latter approach can be costly and the numbers of available donor kidneys that are properly blood typed, tissue typed, and in a nearby geographic zone are relatively few in number. Although there are continuing issues related to the health of dialysis patients, particularly those with cardiovascular disease, the use of dialysis remains steady and there is no projection that it will decrease in the near future.

The future of dialysis may be in the development of a miniaturized, implantable system. Nissenson *et al.* (2005a,b) at the UCLA Medical School have developed a nanotechnology-based artificial nephron system that employs two membranes operating in series within one very small cartridge. The first membrane mimics the function of the glomerulus, allowing substances up to the size of albumin (MW 69 000). The second membrane mimics the function of the renal tubules, selectively reclaiming designated solutes, in a manner similar to that of the natural kidney. No dialysate is used in this device. As such, this miniaturized system is closer to that of an actual nephron than to currently employed dialysis systems.

References

Aagaard J. The Carbomedics aortic heart valve prosthesis: a review. *J Cardiovasc Surg (Torino)* 2004 Dec;45(6):531–4.

Alter P, Waldhans S, Plachta E, Moosdorf R, Grimm W. Complications of implantable cardioverter defibrillator therapy in 440 consecutive patients. *Pacing Clin Electrophysiol* 2005 Sep;28(9):926–32.doi:10.1111/j.1540-8159.2005.00195.x

Anderson J, Wood HG, Allaire PE, Olsen DB. Numerical analysis of blood flow in the clearance regions of a continuous flow artificial heart pump. *Artif Organs* 2000 Jun;24(6): 492–500.doi:10.1046/j.1525-1594.2000.06580.x

Arcidiacono G, Corvi A, Severi T. Functional analysis of bioprosthetic heart valves. *J Biomech* 2005 Jul;38(7):1483–90. Epub 2004 Nov 11.doi:10.1016/j.jbiomech.2004.07.007

Bach DS, Sakwa MP, Goldbach M, Petracek MR, Emery RW, Mohr FW. Hemodynamics and early clinical performance of the St. Jude Medical Regent mechanical aortic valve. *Ann Thorac Surg* 2002 Dec;74(6):2003–9; discussion 2009.doi:10.1016/S0003-4975(02)04034-1

Baldwin JT, Tarbell JM, Deutsch S, Geselowitz DB. Mean flow velocity patterns within a ventricular assist device. *ASAIO Trans* 1989 Jul–Sep;35(3):429–33.

Bech-Hanssen O, Gjertsson P, Houltz E, Wranne B, Ask P, Loyd D, Caidahl K. Net pressure gradients in aortic prosthetic valves can be estimated by Doppler. *J Am Soc Echocardiogr* 2003 Aug;16(8):858–66.doi:10.1067/S0894-7317(03)00422-X

Bernacca GM, McColl JH, Wheatley DJ. Comparison of prosthetic valve hydrodynamic function: objective testing using statistical multilevel modeling. *J Heart Valve Dis* 2004 May;13(3):467–77.

Birks EJ, Yacoub MH, Banner NR, Khaghani A. The role of bridge to transplantation: should LVAD patients be transplanted? *Curr Opin Cardiol* 2004 Mar;19(2):148–53.doi:10.1097/00001573-200403000-00015

Bolno PB, Kresh JY. Physiologic and hemodynamic basis of ventricular assist devices. *Cardiol Clin* 2003 Feb;21(1):15–27.doi:10.1016/S0733-8651(03)00002-X

Botzenhardt F, Eichinger WB, Bleiziffer S, Guenzinger R, Wagner IM, Bauernschmitt R, Lange R. Hemodynamic comparison of bioprostheses for complete supra-annular position in patients with small aortic annulus. *J Am Coll Cardiol* 2005 Jun 21;45(12):2054–60. doi:10.1016/j.jacc.2005.03.039

Botzenhardt F, Gansera B, Kemkes BM. Mid-term hemodynamic and clinical results of the stented porcine medtronic mosaic valve in aortic position. *Thorac Cardiovasc Surg* 2004 Feb;52(1):34–41.doi:10.1055/s-2004-817800

Brodell GK, Wilkoff BL. A novel approach to determining the cause of pacemaker lead failure. *Cleve Clin J Med* 1992 Jan–Feb;59(1):91–2.

Bryant J, Brodin H, Loveman E, Payne E, Clegg A. The clinical and cost-effectiveness of implantable cardioverter defibrillators: a systematic review. *Health Technol Assess* 2005 Sep;9(36):1–150, iii.

Cook WL, Jassal SV. Prevalence of falls among seniors maintained on hemodialysis. *Int Urol Nephrol* 2005;37(3):649–52.doi:10.1007/s11255-005-0396-9

Curtas AR, Wood HG, Allaire PE, McDaniel JC, Day SW, Olsen DB. Computational fluid dynamics modeling of impeller designs for the HeartQuest left ventricular assist device. *ASAIO J* 2002 Sep–Oct;48(5):552–61.doi:10.1097/00002480-200209000-00019

Cutler NG, Karpawich PP, Cavitt D, Hakimi M, Walters HL. Steroid-eluting epicardial pacing electrodes: six year experience of pacing thresholds in a growing pediatric population. *Pacing Clin Electrophysiol* 1997 Dec;20(12 Pt 1):2943–8.doi:10.1111/j.1540-8159.1997.tb05464.x

Davis PK, Pae WE Jr, Pierce WS. Toward an implantable artificial heart. Experimental and clinical experience at The Pennsylvania State University. *Invest Radiol* 1989 Jan;24(1):81–7.

Davison SN, Jhangri GS. The impact of chronic pain on depression, sleep, and the desire to withdraw from dialysis in hemodialysis patients. *J Pain Symptom Manage* 2005 Nov;30(5): 465–73.doi:10.1016/j.jpainsymman.2005.05.013

Day SW, McDaniel JC, Wood HG, Allaire PE, Landrot N, Curtas A. Particle image velocimetry measurements of blood velocity in a continuous flow ventricular assist device. *ASAIO J* 2001 Jul–Aug;47(4):406–11.doi:10.1097/00002480-200107000-00021

Day SW, McDaniel JC, Wood HG, Allaire PE, Song X, Lemire PP, Miles SD. A prototype HeartQuest ventricular assist device for particle image velocimetry measurements. *Artif Organs* 2002 Nov;26(11):1002–5.doi:10.1046/j.1525-1594.2002.07124.x

Deiwick M, Hoffmeier A, Tjan TD, Krasemann T, Schmid C, Scheld HH. Heart failure in children—mechanical assistance. *Thorac Cardiovasc Surg* 2005 Feb;53(Suppl 2):S135–40.

Del Ojo JL, Moya F, Villalba J, Sanz O, Pavon R, Garcia D, Pastor L. Is magnetic resonance imaging safe in cardiac pacemaker recipients? *Pacing Clin Electrophysiol* 2005 Apr;28(4):274–8.doi:10.1111/j.1540-8159.2005.50033.x

Dember LM, Kaufman JS, Beck GJ, Dixon BS, Gassman JJ, Grene T, Himmelfarb J, Hunsicker LG, Kusek JW, Lawson JH, Middloton JP, Radeva M, Schwab SJ, Whitting JF, Feldman HI; DAC Study Group. Design of the dialysis access consortium (DAC) clopidogrel prevention of early AV fistula thrombosis trial. *Clin Trials* 2005;2(5):413–22. doi:10.1191/1740774505cn118oa

Deng MC, Naka Y. Mechanical circulatory support devices—state of the art. *Heart Fail Monit* 2002;2(4):120–8.

Di Benedetto A, Marcelli D, D'Andrea A, Cice G, Cappabianca F, Pacchiano G, D'Amato R, Oggero AR, Bonanno D, Pergamo O, Calabro R. Risk factors and underlying cardiovascular diseases in incident ESRD patients. *J Nephrol* 2005 Sep–Oct;18(5): 592–8.

Dumont K, Vierendeels JA, Segers P, Van Nooten GJ, Verdonck PR. Predicting ATS open pivot heart valve performance with computational fluid dynamics. *J Heart Valve Dis* 2005 May;14(3):393–9.

Duncan BW. Mechanical circulatory support for infants and children with cardiac disease. *Ann Thorac Surg* 2002 May;73(5):1670–7.doi:10.1016/S0003-4975(01)03027-2

Ellenbogen KA, Wood MA. Pacemaker selection—the changing definition of physiologic pacing. *N Engl J Med* 2005 Jul 14;353(2):202–4.doi:10.1056/NEJMe058125

Familoni OB, Alebiosu CO, Avodele OE. Effects and outcome of haemodialysis on QT intervals and QT dispersion in patients with chronic kidney disease. *Cardiovasc J S. Afr* 2005 Nov 30:1–4. [Epub ahead of print]

Fan Q, Liu J, Ebben JP, Collins AJ. Reuse-associated mortality in incident hemodialysis patients in the United States, 2000 to 2001. *Am J Kidney Dis* 2005 Oct;46(4):661–8. doi:10.1053/j.ajkd.2005.07.017

Gasparini G, Curnis A, Gulizia M, Occhetta E, Corrado A, Bontempi L, Mascioli G, Maura Francese G, Bortnik M, Magnani A, Di Gregorio F, Barbetta A, Raviele A. Rate-responsive pacing regulated by cardiac haemodynamics. *Europace* 2005 May;7(3):234–41. doi:10.1016/j.eupc.2005.02.115

Gee MD. Thought and discussion among patients about stopping dialysis. *Nephrol News Issues* 2005 Nov;19(12):80–1, 83–4.

Gersh B. Do implantable cardioverter defibrillators lower mortality risk in patients with cardiomyopathy? *Nat Clin Pract Cardiovasc Med* 2004 Nov;1(1):16–17.

Goetze S, Brechtken J, Agler DA, Thomas JD, Sabik JF 3rd, Jaber WA. In vivo short-term Doppler hemodynamic profiles of 189 Carpentier-Edwards Perimount pericardial bioprosthetic valves in the mitral position. *J Am Soc Echocardiogr* 2004 Sep;17(9):981–7. doi:10.1016/j.echo.2004.05.006

Gregoratos G. Indications and recommendations for pacemaker therapy. *Am Fam Physician* 2005 Apr 15;71(8):1563–70.

Grigioni M, Daniele C, Del Gaudio C, Morbiducci U, Balducci A, D'Avenio G, Barbaro V. Three-dimensional numeric simulation of flow through an aortic bileaflet valve in a realistic model of aortic root. *ASAIO J* 2005 May–Jun;51(3):176–83. doi:10.1097/01.MAT.0000159384.36271.2C

Gura MT. Implantable cardioverter defibrillator therapy. *J Cardiovasc Nurs* 2005 Jul–Aug;20(4):276–87.

Gusella M, Rebeschini M, Cartei G, Ferrazzi E, Ferrari M, Padrini R. Effect of hemodialysis on the metabolic clearance of 5-Fluorouracil in a patient with end-stage renal failure. *Ther Drug Monit* 2005 Dec;27(6):816–18.doi:10.1097/01.ftd.0000183384.89275.f4

Hirose M, Hida M, Sato E, Kokubo K, Nie M, Kobayashi H. Electromagnetic interference of implantable unipolar cardiac pacemakers by an induction oven. *Pacing Clin Electrophysiol* 2005 Jun;28(6):540–8.doi:10.1111/j.1540-8159.2005.09565.x

Hirsch DJ, Cooper JR Jr. Cardiac failure and left ventricular assist devices. *Anesthesiol Clin North America* 2003 Sep;21(3):625–38.doi:10.1016/S0889-8537(03)00049-X

Ho TH, Van Nguyen P, Phan PK, Pham VN. Up to nine-years' experience with the Allcarbon prosthetic heart valve. *J Heart Valve Dis* 2005 Jul;14(4):512–17.

Huang CY, Tuan TC, Lee WS, Cheng CM, Lin SJ, Kong CW. Long-term efficacy and stability of atrial sensing in VDD pacing. *Clin Cardiol* 2005 Apr;28(4):203–7.

Ichikawa S, Nose Y. Centrifugal blood pumps for various clinical needs. *Artif Organs* 2002 Nov;26(11):916–18.doi:10.1046/j.1525-1594.2002.07113.x

Iorio L, Violi F, Simonelli R, Nacca RG, Caliendo A, De Santo LS. Temporary dialysis treatments for heart failure in chronic kidney disease. *Semin Nephrol* 2005 Nov;25(6):408–12. doi:10.1016/j.semnephrol.2005.05.011

Irnich W, Irnich B, Bartsch C, Stertmann WA, Gufler H, Weiler G. Do we need pacemakers resistant to magnetic resonance imaging? *Europace* 2005 Jul;7(4):353–65. doi:10.1016/j.eupc.2005.02.120

Jamieson WR, von Lipinski O, Miyagishima RT, Burr LH, Janusz MT, Ling H, Fradet GJ, Chan F, Germann E. Performance of bioprostheses and mechanical prostheses assessed by composites of valve-related complications to 15 years after mitral valve replacement. *J Thorac Cardiovasc Surg* 2005 Jun;129(6):1301–8.

Kadish A, Mehra M. Heart failure devices: implantable cardioverter-defibrillators and biventricular pacing therapy. *Circulation* 2005 Jun 21;111(24):3327–35.doi:10.1161/CIRCULATIONAHA.104.481267

Kerrigan JP, Yamazaki K, Meyer RK, Mori T, Otake Y, Outa E, Umezu M, Borovetz HS, Kormos RL, Griffith BP, Koyanagi H, Antaki JF. High-resolution fluorescent particle-tracking flow visualization within an intraventricular axial flow left ventricular assist device. *Artif Organs* 1996 Jun;20(6):534–40.

Khan EM, Voudouris AA, Hood RE, Shorofsky SR. Repositioning of a dislodged and fibrosed ventricular lead. *J Interv Card Electrophysiol* 2005 Jun;13(1):55–7.doi:10.1007/s10840-005-0729-8

Kherani AR, Maybaum S, Oz MC. Ventricular assist devices as a bridge to transplant or recovery. *Cardiology* 2004;101(1–3):93–103.doi:10.1159/000075989

Kirsch ME, Tzvetkov B, Vermes E, Pouzet B, Sauvat S, Loisance D. Clinical and hemodynamic performance of the 19-mm medtronic mosaic bioprosthesis. *J Heart Valve Dis* 2005 May;14(3):433–9.

Kiss E, Keusch G, Zanetti M, Jung T, Schwarz A, Schocke M, Jaschke W, Czermak BV. Dialysis-related amyloidosis revisited. *AJR Am J Roentgenol* 2005 Dec;185(6):1460–7. doi:10.2214/AJR.04.1309

Kleine P, Abdel-Rahman U, Klesius AA, Scherer M, Simon A, Moritz A. Comparison of hemodynamic performance of Medtronic Hall 21 mm versus St. Jude Medical 23 mm prostheses in pigs. *J Heart Valve Dis* 2002 Nov;11(6):857–63.

Koertke H, Seifert D, Drewek-Platena S, Koerfer R. Hemodynamic performance of the Medtronic Advantage prosthetic heart valve in the aortic position: echocardiographic evaluation at one year. *J Heart Valve Dis* 2003 May;12(3):348–53.

Labrosse MR, Beller CJ, Robicsek F, Thubrikar MJ. Geometric modeling of functional trileaflet aortic valves: development and clinical applications. *J Biomech* 2005 Sep 29. [Epub ahead of print]

Labrousse LM, Choukroun E, Serena D, Billes MA, Madonna F, Deville C. Prospective study of hemodynamic performances of standard ATS and AP-ATS valves. *J Heart Valve Dis* 2003 May;12(3):341–7.

Lamson TC, Ojan OS, Geselowitz DB, Tarbell JM. A two-phase fluid volume compensation chamber for an electric ventricular assist device. *Artif Organs* 1990 Aug;14(4):270–7.

Lee CT, Hsu CY, Lam KK, Lin CR, Chen JB. Inflammatory markers and hepatocyte growth factor in sustained hemodialysis hypotension. *Artif Organs* 2005 Dec;29(12):980–3. doi:10.1111/j.1525-1594.2005.00168.x

Maisel WH. Cardiovascular device development: lessons learned from pacemaker and implantable cardioverter-defibrillator therapy. *Am J Ther* 2005 Mar–Apr;12(2):183–5. doi:10.1097/01.mjt.0000155117.55919.43

Malouf JF, Ballo M, Connolly HM, Hodge DO, Herges RM, Mullany CJ, Miller FA. Doppler echocardiography of 119 normal-functioning St Jude Medical mitral valve prostheses: a comprehensive assessment including time–velocity integral ratio and prosthesis performance index. *J Am Soc Echocardiogr* 2005 Mar;18(3):252–6.

Manning KB, Miller GE. Flow through an outlet cannula of a rotary ventricular assist device. *Artif Organs* 2002 Aug;26(8):714–23.

Matsuda H, Matsumiya G. Current status of left ventricular assist devices: the role in bridge to heart transplantation and future perspectives. *J Artif Organs* 2003;6(3):157–61.

Medart D, Schmitz C, Rau G, Reul H. Design and in vitro performance of a novel bileaflet mechanical heart valve prosthesis. *Int J Artif Organs* 2005 Mar;28(3):256–63.

Mehra MR. Ventricular assist devices: destination therapy or just another stop on the road? *Curr Heart Fail Rep* 2004 Apr–May;1(1):36–40.

Mesana TG. Rotary blood pumps for cardiac assistance: a "must"? *Artif Organs* 2004 Feb;28(2):218–25.

Miller GE, Madigan M, Fink R. A preliminary flow visualization study in a multiple disk centrifugal artificial ventricle. *Artif Organs* 1995 Jul;19(7):680–4.

Miller LW. Patient selection for the use of ventricular assist devices as a bridge to transplantation. *Ann Thorac Surg* 2003 Jun;75(6 Suppl):S66–71.

Mitamura Y, Okamoto E, Hirano A, Mikami T. Development of an implantable motor-driven assist pump system. *IEEE Trans Biomed Eng* 1990 Feb;37(2):146–56.

Moffatt-Bruce SD, Jamieson WR. Long-term performance of prostheses in mitral valve replacement. *J Cardiovasc Surg (Torino)* 2004 Oct;45(5):427–47.

Morgan JA, Mazzeo PA, Flannery MR, Oz MC, Naka Y. Effects of changes in UNOS policy regarding left ventricular assist devices. *J Heart Lung Transplant* 2004 May;23(5):620–2.

Mulder MM, Hansen AC, Mohammad SF, Olsen DB. In vitro investigation of the St. Jude Medical Isoflow centrifugal pump: flow visualization and hemolysis studies. *Artif Organs* 1997 Aug;21(8):947–53.

Myers TJ, Robertson K, Pool T, Shah N, Gregoric I, Frazier OH. Continuous flow pumps and total artificial hearts: management issues. *Ann Thorac Surg* 2003 Jun;75(6 Suppl):S79–85.

Narsipur SS. Measurement of fiber bundle volume in reprocessed dialyzers. *Clin Nephrol* 2004 Feb;61(2):130–3.

Nazarian S, Maisel WH, Miles JS, Tsang S, Stevenson LW, Stevenson WG. Impact of implantable cardioverter defibrillators on survival and recurrent hospitalization in advanced heart failure. *Am Heart J* 2005 Nov;150(5):955–60.

Nemeh HW, Smedira NG. Mechanical treatment of heart failure: the growing role of LVADs and artificial hearts. *Cleve Clin J Med* 2003 Mar;70(3):223–6, 229–33.

Nissenson AR, Ronco C, Pergamit G, Edelstein M, Watts R. The human nephron filter: toward a continuously functioning, implantable artificial nephron system. *Blood Purif* 2005a;23(4):269–74. Epub 2005 May 20.

Nissenson AR, Ronco C, Pergamit G, Edelstein M, Watts, R. Continuously functioning artificial nephron system: the promise of nanotechnology. *Hemodial Int* 2005b Jul;9(3):210–17.

Noiseux N, Khairy P, Fournier A, Vobecky SJ. Thirty years of experience with epicardial pacing in children. *Cardiol Young* 2004 Oct;14(5):512–19.

Okamoto E, Hashimoto T, Mitamura Y. Design of a miniature implantable left ventricular assist device using CAD/CAM technology. *J Artif Organs* 2003;6(3):162–7.

Ostberg BN, Ostberg GN. Total artificial heart design with integrated electric motor. *Med Biol Eng Comput* 1987 May;25(3):345–6.

Padeletti L, Barold SS. Digital technology for cardiac pacing. *Am J Cardiol* 2005 Feb 15;95(4):479–82.

Panichi V, Rizza GM, Taccola D, Paoletti S, Mantuano E, Migliori M, Grangioni S, Flippi C, Carpi A. C-reactive protein in patients on chronic hemodialysis with different techniques and different membranes. *Biomed Pharmacother* 2005 Oct 25. [Epub ahead of print]

Parks MS. Practical considerations in converting to Renalin 100 for dialyzer reprocessing. *Nephrol News Issues* 2002 Nov;16(12):55, 59–62, 64.

Peirce RM, Funaki B, Van Ha TG, Lorenz JM. Percutaneous declotting of virgin femoral hemodialysis grafts. *AJR Am J Roentgenol* 2005 Dec;185(6):1615–19.

Petrie JP. Permanent transvenous cardiac pacing. *Clin Tech Small Anim Pract* 2005 Aug;20(3):164–72.

Phillips SJ. Selecting the best heart valve for your patient: mechanical or tissue. *Am Heart Hosp J* 2004 Summer;2(3):149–52.

Piccoli GB, Amgnano A, Perrotta L, Piccoli G. Daily dialysis, nocturnal dialysis and randomized controlled trials: are we asking the right questions? *Kidney Int* 2005 Dec;68(6):2913–14.

Pierce WS, Rosenberg G, Snyder AJ, Pae WE Jr, Donachy JH, Waldhausen JA. An electric artificial heart for clinical use. *Ann Surg* 1990 Sep;212(3):339–43; discussion 343–4.

Pierrakos O, Vlachos PP, Telionis DP. Time-resolved DPIV analysis of vortex dynamics in a left ventricular model through bileaflet mechanical and porcine heart valve prostheses. *J Biomech Eng* 2004 Dec;126(6):714–26.

Prado M, Roa LM, Palma A, Milan JA. Double target comparison of blood-side methods for measuring the hemodialysis dose. *Kidney Int* 2005 Dec;68(6):2863–76.

Rahmati MA, Rahmati S, Hoenich N, Ronco C, Kaysen GA, Levin R, Levin NW. On-line clearance: a useful tool for monitoring the effectiveness of the reuse procedure. *ASAIO J* 2003 Sep–Oct;49(5):543–6.

Reinhartz O, Stiller B, Eilers R, Farrar DJ. Current clinical status of pulsatile pediatric circulatory support. *ASAIO J* 2002 Sep–Oct;48(5):455–9.

Reinlib L, Abraham W; Working Group of the National, Heart, Lung, and Blood Institute. Recovery from heart failure with circulatory assist: a working group of the National, Heart, Lung, and Blood Institute. *J Card Fail* 2003 Dec;9(6):459–63.

Rezaian GR, Aghasadeghi K, Kojuri J. Evaluation of the hemodynamic performance of St. Jude mitral prostheses: a pilot study by dobutamine-stress Doppler echocardiography. *Angiology* 2005 Jan–Feb;56(1):81–6.

Richenbacher WE, Pae WE Jr, Magovern JA, Rosenberg G, Snyder AJ, Pierce WS. Roller screw electric motor ventricular assist device. *ASAIO Trans* 1986 Jul–Sep;32(1):46–8.

Roberts PR. Follow up and optimisation of cardiac pacing. *Heart* 2005 Sep;91(9):1229–34.

Robinson BM, Feldman HI. Dialyzer reuse and patient outcomes: what do we know now? *Semin Dial* 2005 May–Jun;18(3):175–9.

Ronco C, Tetta C. Dialysis patients and cardiovascular problems: can technology solve the complex equation? *Expert Rev Med Devices* 2005 Nov;2(6):681–7.

Rosenberg G, Snyder A, Weiss W, Landis DL, Geselowitz DB, Pierce WS. A cam-type electric motor-driven left ventricular assist device. *J Biomech Eng* 1982 Aug;104(3):214–20.

Rosenberg G, Snyder AJ, Landis DL, Geselowitz DB, Donachy JH, Pierce WS. An electric motor-driven total artificial heart: seven months survival in the calf. *Trans Am Soc Artif Intern Organs* 1984;30:69–74.

Saito S, Nishinaka T, Westaby S. Hemodynamics of chronic nonpulsatile flow: implications for LVAD development. *Surg Clin North Am* 2004 Feb;84(1):61–74.

Samuels L. Biventricular mechanical replacement. *Surg Clin North Am* 2004 Feb;84(1):309–21.

Sanders GD, Hlatky MA, Owens DK. Cost-effectiveness of implantable cardioverter-defibrillators. *N Engl J Med* 2005 Oct 6;353(14):1471–80.

Saxena AK, Panhotra BR. Cardiovascular mortality and dialysis access-related infections: is there a link? *Am J Kidney Dis* 2005 Dec;46(6):1149–50.

Scholten A, Joosten S, Silny J. Unipolar cardiac pacemakers in electromagnetic fields of high voltage overhead lines. *J Med Eng Technol* 2005 Jul–Aug;29(4):170–5.

Seitelberger R, Bialy J, Gottardi R, Seebacher G, Moidl R, Mittelbock M, Simon P, Wolner E. Relation between size of prosthesis and valve gradient: comparison of two aortic bioprosthesis. *Eur J Cardiothorac Surg* 2004 Mar;25(3):358–63.

Shao J, Zydney AL. Effect of bleach reprocessing upon the clearance characteristics and surface charge of polysulfone hemodialyzers. *ASAIO J* 2004 May–Jun;50(3):246–52.

Shimada Y, Yaku H, Kawata M, Oka K, Shuntoh K, Okano T, Takahashi A, Fukumoto A, Hayashida K, Kitamura N. An operative case of inferior vena cava stenosis due to fibrosis around permanent pacemaker leads. *Pacing Clin Electrophysiol* 2002 Feb;25(2):223–5.

Silver MT. Primary prevention implantable cardioverter-defibrillators: economics and ethics. *Am Heart Hosp J* 2005 Summer;3(3):205–6.

Smith WA, Hete BF, Kiraly RJ, Fujimoto LK, Jacobs GB, Ishikawa M, Butler K, Nose Y. The E4T electric powered total artificial heart (TAH). *Artif Organs* 1988 Oct;12(5):402–9.

Song X, Throckmorton AL, Wood HG, Allaire PE, Olsen DB. Transient and quasi-steady computational fluid dynamics study of a left ventricular assist device. *ASAIO J* 2004 Sep–Oct;50(5):410–17.

Song X, Wood HG, Olsen D. Computational Fluid Dynamics (CFD) study of the 4th generation prototype of a continuous flow Ventricular Assist Device (VAD). *J Biomech Eng* 2004 Apr;126(2):180–7.

Stevenson LW, Rose EA. Left ventricular assist devices: bridges to transplantation, recovery, and destination for whom? *Circulation* 2003 Dec 23;108(25):3059–63.

Stragier A. Dialyzer reuse: the debate continues. *Nephrol News Issues* 2003 Nov;17(12):97.

Sudkamp M, Schmid M, Geissler HJ, Emmel M, Gillor A, Mehlhorn U, Hekmat K. VDD-pacemaker in children—a long-term therapy? *Thorac Cardiovasc Surg* 2005 Jun;53(3):158–61.

Sung K, Park PW, Park KH, Jun TG, Lee YT, Yang JH. Comparison of transprosthetic mean pressure gradients between Medtronic Hall and ATS valves in the aortic position. *Int J Cardiol* 2005 Mar 10;99(1):29–35.

Szathmary S, Hegyi E, Amoureux MC, Rajapakse N, Chicorka L, Szalai G, Reszegi K, Derbyshire Z, Paluh J, Dodson B, Grandics P. Characterization of the DialGuard device for endotoxin removal in hemodialysis. *Blood Purif* 2004;22(5):409–15. Epub 2004 Aug 13.

Throckmorton AL, Allaire PE, Gutgesell HP, Matherne GP, Olsen DB, Wood HG, Allaire JH, Patel SM. Pediatric circulatory support systems. *ASAIO J* 2002 May–Jun;48(3):216–21.

Toff WD, Camm AJ, Skehan JD; United Kingdom Pacing and Cardiovascular Events Trial Investigators. Single-chamber versus dual-chamber pacing for high-grade atrioventricular block. *N Engl J Med* 2005 Jul 14;353(2):145–55.

Tsukiya T, Taenaka Y, Tatsumi E, Takano H. Visualization study of the transient flow in the centrifugal blood pump impeller. *ASAIO J* 2002 Jul–Aug;48(4):431–6.

Unruh ML, Evans IV, Fink NE, Powe NR, Meyer KB; Choices for Healthy Outcomes in Caring for End-Stage Renal Disease (CHOICE) Study. Skipped treatments, markers of nutritional nonadherence, and survival among incident hemodialysis patients. *Am J Kidney Dis* 2005 Dec;46(6):1107–16.

Vahlhaus C. Heating of pacemaker leads during magnetic resonance imaging. *Eur Heart J* 2005 Jun;26(12):1243; author reply 1243–4. Epub 2005 May 4.doi:10.1093/eurheartj/ehi298

Ward RA, Ouseph R. Impact of bleach cleaning on the performance of dialyzers with polysulfone membranes processed for reuse using peracetic acid. *Artif Organs* 2003 Nov;27(11):1029–34.doi:10.1046/j.1525-1594.2003.07151.x

Wernicke JT, Meier D, Mizuguchi K, Damm G, Aber G, Benkowski R, Nose Y, Noon GP, DeBakey ME. A fluid dynamic analysis using flow visualization of the Baylor/NASA implantable axial flow blood pump for design improvement. *Artif Organs* 1995 Feb;19(2):161–77.

Westaby S. Ventricular assist devices as destination therapy. *Surg Clin North Am* 2004 Feb;84(1):91–123.doi:10.1016/j.suc.2003.12.010

Wheeldon DR. Mechanical circulatory support: state of the art and future perspectives. *Perfusion* 2003 Jul;18(4):233–43.doi:10.1191/0267659103pf674oa

Wijnen E, Keuter XH, Planken NR, van der Sande FM, Tordoir JH, Leunissen KM, Kooman JP. The relation between vascular access flow and different types of vascular access with systemic hemodynamics in hemodialysis patients. *Artif Organs* 2005 Dec;29(12):960–4.doi:10.1111/j.1525-1594.2005.00165.x

Woodruff J, Prudente LA. Update on implantable pacemakers. *J Cardiovasc Nurs* 2005 Jul–Aug;20(4):261–8; quiz 269–70.

Wu C, Hwang NH, Lin YK. Statistical characteristics of mechanical heart valve cavitation in accelerated testing. *J Heart Valve Dis* 2004 Jul;13(4):659–66.

Wu Y, Gregorio R, Renzulli A, Onorati F, De Feo M, Grunkemeier G, Cotrufo M. Mechanical heart valves: are two leaflets better than one? *J Thorac Cardiovasc Surg* 2004 Apr;127(4):1171–9.doi:10.1016/j.jtcvs.2003.08.030

Wu ZJ, Antaki JF, Burgreen GW, Butler KC, Thomas DC, Griffith BP. Fluid dynamic characterization of operating conditions for continuous flow blood pumps. *ASAIO J* 1999 Sep–Oct;45(5):442–9.

Yamane T, Miyamoto Y, Tajima K, Yamazaki K. A comparative study between flow visualization and computational fluid dynamic analysis for the sun medical centrifugal blood pump. *Artif Organs* 2004 May;28(5):458–66.doi:10.1111/j.1525-1594.2004.07161.x

Yoganathan, AP, He Z, Jones SC. Fluid mechanics of heart valves. *Annu Rev Biomed Eng* 2004;6:331–62.doi:10.1146/annurev.bioeng.6.040803.140111

The Sustainable Development Goals Series is Springer Nature's inaugural cross-imprint book series that addresses and supports the United Nations' seventeen Sustainable Development Goals. The series fosters comprehensive research focused on these global targets and endeavours to address some of society's greatest grand challenges. The SDGs are inherently multidisciplinary, and they bring people working across different fields together and working towards a common goal. In this spirit, the Sustainable Development Goals series is the first at Springer Nature to publish books under both the Springer and Palgrave Macmillan imprints, bringing the strengths of our imprints together.

The Sustainable Development Goals Series is organized into eighteen subseries: one subseries based around each of the seventeen respective Sustainable Development Goals, and an eighteenth subseries, "Connecting the Goals," which serves as a home for volumes addressing multiple goals or studying the SDGs as a whole. Each subseries is guided by an expert Subseries Advisor with years or decades of experience studying and addressing core components of their respective Goal.

The SDG Series has a remit as broad as the SDGs themselves, and contributions are welcome from scientists, academics, policymakers, and researchers working in fields related to any of the seventeen goals. If you are interested in contributing a monograph or curated volume to the series, please contact the Publishers: Zachary Romano [Springer; zachary.romano@springer.com] and Rachael Ballard [Palgrave Macmillan; rachael.ballard@palgrave.com].

More information about this series at http://www.springer.com/series/15486

Robert T. Brodell
Adam C. Byrd
Cindy Firkins Smith · Vinayak K. Nahar
Editors

Dermatology in Rural Settings

Organizational, Clinical, and Socioeconomic Perspectives

 Springer

Editors
Robert T. Brodell
Department of Dermatology
University of Mississippi Medical
Center
Jackson, MS
USA

Adam C. Byrd
Department of Dermatology
University of Mississippi Medical Center
Jackson, MS
USA

Cindy Firkins Smith
Carris Health-CentraCare
University of Minnesota
Willmar, MN
USA

Vinayak K. Nahar
Department of Dermatology
University of Mississippi Medical Center
Jackson, MS
USA

The content of this publication has not been approved by the United Nations and does not reflect the views of the United Nations or its officials or Member States.

ISSN 2523-3084   ISSN 2523-3092   (electronic)
Sustainable Development Goals Series
ISBN 978-3-030-75986-5   ISBN 978-3-030-75984-1   (eBook)
https://doi.org/10.1007/978-3-030-75984-1

This Springer imprint is published by the registered company Springer Nature Switzerland AG
The registered company address is: Gewerbestrasse 11, 6330 Cham, Switzerland

This book is dedicated to dermatologists in the USA and around the world who are making a difference for the underserved by donating time and treasure to serve patients in need including: volunteer service in free clinics, university teaching service, the American Academy of Dermatology's AccessDerm™, faith-based dermatology health initiatives, performing hospital consultations, defending their patients through political action, engaging in carefully planned mission medicine on Indian reservations and around the world, and serving the many needs of their communities. Service to others is the rent you pay for your room here on Earth (Muhammad Ali).

Robert T. Brodell, MD

Adam C. Byrd, MD

Vinayak K. Nahar, MD, PhD

Cindy Firkins Smith, MD

Foreword

The happiest, most satisfied physicians work in mission-driven environments – among colleagues or staff with shared values – where they can use their skills to serve the needs of patients and their communities by improving health. The days when physicians have the privilege to make a difference in the lives of our patients *connect us* to the aspirations that first brought us to this sacred profession. Other days – when barriers get in the way of care delivery – *deprive us* of the joy of work and foster burnout. I've observed the same predictors of professional satisfaction across every practice setting and in every medical specialty including dermatology.

Despite practicing medicine in a major urban area, I have witnessed my rural colleagues experience some of the greatest joys in medicine working in tight-knit communities serving particularly vulnerable patients. Of course, they can also experience some of the greatest frustrations in medicine as they navigate unique barriers to healthcare delivery. Those challenges, however, are driving small and large practices in rural settings to become innovation engines, rethinking and reinventing care delivery.

Rural populations live at the nexus of several forces underpinning health disparities in the USA. They suffer high burdens of underlying chronic disease, including diabetes, hypertension, and opioid-use disorder. These issues complicate the care of most other health problems. Their access to care is sometimes affected by gaps in wealth, employment, and insurance coverage. Shortages of physicians and longer travel distances in many rural areas exacerbate access challenges for even those with excellent insurance. The economic challenges of practice faced by rural physicians are particularly difficult for those with high educational debt burdens. Finally, the longstanding gaps in the diversity of the physician workforce leave our profession not fully resembling the patients we serve or benefitting from a wider breadth of viewpoints and life experiences of under-represented minorities.

I'm optimistic that a cadre of energetic physicians will harness innovative solutions to advance the work of rural healthcare and the health of rural populations. Coverage for telehealth services has expanded exponentially in the wake of the COVID-19 pandemic. The broadening of insurance coverage through ACA subsidies for commercial plans and Medicaid expansion in 39 states has improved access for many patients living in rural areas. Medical schools and residency programs are recruiting trainees with a diversity of life experiences and launching new curricula focused on rural and underserved populations. National attention has highlighted the effects of racism in health, and this is driving change that is fundamental to reducing inequities.

Our profession's roots, shared values, and medical ethics will continue to drive practice improvements aimed at the betterment of the lives of our patients and public health. Health services research, iterative innovation, collaboration, and thoughtful policy debate are part of that process. I'm inspired by the efforts of many physician colleagues who are seeking opportunities for real change in rural healthcare delivery and finding joy in this meaningful work.

Conflict of Interest Disclosures: Dr Resneck serves as President-Elect of the American Medical Association, and he serves on the board of directors of the National Quality Forum. The views expressed here are those of the author and do not necessarily represent the views of the American Medical Association or the National Quality Forum.

Jack S. Resneck Jr
Professor and Vice-Chair of Dermatology
UCSF School of Medicine
San Francisco, CA, USA

Affiliated Faculty
Philip R. Lee Institute for Health Policy Studies
UCSF School of Medicine
San Francisco, CA, USA

Preface

This book was born in the COVID-19 pandemic. Suddenly, the faculty of the Department of Dermatology at the University of Mississippi Medical Center found our in-person clinics shuttered. We struggled to use teledermatology to provide access to dermatologic care for our patients. It quickly became evident that there were more barriers than the federal regulations which had largely been relaxed. The technology was sluggish, images coming through on synchronous platforms were less than sharp, and "teeing up" patients took an inordinate amount of staff time. Still, we persevered and found ways to overcome each challenge. Then, something really strange happened. Locked down in our homes, there was more time to spend with our families and extra time to spend on a project! We're Americans! Writing a book on access to care was a way to fight back, to make something good happen in a world of tragedy.

Our team had previously been assembled through a series of lectures at annual meetings of the American Academy of Dermatology. Adam Byrd, MD, was the center of the University of Mississippi Medical Center's (UMMC) efforts to provide academic dermatologic care in rural Mississippi, far from our urban medical center. Vinny Nahar, MD, PhD, lead physician in charge of research at UMMC was entrenched in writing manuscripts about efforts to utilize teledermatology and Project ECHO to address the dermatologic needs of rural patients. Cindy Firkins Smith, MD, chief executive officer and professor of dermatology at Carris Health, University of Minnesota, had spent her life working to recruit dermatologists and primary care physicians to rural areas of Minnesota. Finally, as chair of the Department of Dermatology at UMMC, Bob Brodell was ready to steer the textbook writing ship.

This book is dedicated to rural patients throughout America who struggle to find physicians for their primary and specialty care. Chapter 1 highlights the severity of this problem. The rest of the chapters consider components of the rural dermatology access to care problem and each provides potential solutions. We may not be able to solve this problem overnight, but the chapters of this book demonstrate that there is something each of us can do to help! Some of you may have additional ideas.... write to us! No idea is too small, as we work together to make big changes.

Jackson, MS, USA   Robert T. Brodell
Jackson, MS, USA   Adam C. Byrd
Minneapolis, MN, USA   Cindy Firkins Smith
Willmar, MN, USA
Jackson, MS, USA   Vinayak K. Nahar

Acknowledgments

This book was inspired by the creative ideas of dermatologists who have been motivated to improve access to dermatologic care for rural citizens around the world.

Contents

Contributors

Amelia Amon, BA University of Minnesota, Duluth, Duluth, MN, USA

Gabriel Amon, MD Department of Dermatology, University of Minnesota, Minneapolis, MN, USA

Hannah R. Badon, MD Department of Dermatology, University of Mississippi Medical Center, Jackson, MS, USA

Robert T. Brodell, MD Department of Dermatology and Pathology, University of Mississippi Medical Center, Jackson, MS, USA

Amanda S. Brown, FNP-C Department of Dermatology, University of Mississippi Medical Center, Jackson, MS, USA

Robert Hollis Burrow, BBA Department of Dermatology, University of Mississippi Medical Center, Jackson, MS, USA

Adam C. Byrd, MD Department of Dermatology, University of Mississippi Medical Center, Jackson, MS, USA

Caroline Doo, MD Department of Dermatology, University of Mississippi Medical Center, Jackson, MS, USA

Karen Dowling, DHA Children's of Mississippi, University of Mississippi Medical Center, Jackson, MS, USA

Navid Farahbakhsh, MS University of Florida College of Medicine, Gainesville, FL, USA

Ronda S. Farah, MD Department of Dermatology, University of Minnesota Health, Minneapolis, MN, USA

Maheera Farsi, DO Department of Dermatology, University of Florida College of Medicine, Gainesville, FL, USA

Hao Feng, MD, MHS Department of Dermatology, University of Connecticut Health Center, Farmington, CT, USA

Taylor Ferris, BS School of Medicine, University of Mississippi Medical Center, Jackson, MS, USA

Amy E. Flischel, MD Department of Dermatology and Pathology, University of Mississippi Medical Center, Jackson, MS, USA

Caroline P. Garraway, BA School of Medicine, University of Mississippi Medical Center, Jackson, MS, USA

Catherine Clare Gloss, BSE University of Missouri School of Medicine, Columbia, MO, USA

Eric Grisham, BS University of Missouri School of Medicine, Columbia, MO, USA

Ira D. Harber, MD Department of Dermatology, University of Mississippi Medical Center, Jackson, MS, USA

Haley Harrington, BS Louisiana State University Health Sciences Center, Shreveport, LA, USA

Lisa A. Haynie, PhD, RN, FNP-BC School of Nursing, University of Mississippi Medical Center, Jackson, MS, USA

Stephen E. Helms, MD Department of Dermatology, University of Mississippi Medical Center, Jackson, MS, USA

Hannah Hoang, BA School of Medicine, University of Mississippi Medical Center, Jackson, MS, USA

Monica Kala, BS School of Medicine, University of Mississippi Medical Center, Jackson, MS, USA

Elizabeth Kiracofe, MD Airia Comprehensive Dermatology, PLLC, Chicago, IL, USA

Lucinda L. Kohn, MD Department of Dermatology, University of Colorado Anschutz Medical Campus, Aurora, CO, USA

Carrie Kovarik, MD Dermatology at the University of Pennsylvania, Philadelphia, PA, USA

Kever A. Lewis, BS School of Medicine, University of Mississippi Medical Center, Jackson, MS, USA

Brett Macleod, BS University of North Dakota, Grand Forks, ND, USA

Tim Maglione, JD Maglione Advisors Group, LLC, Columbus, OH, USA

Spero M. Manson, PhD Centers for American Indian & Alaska Native Health, Aurora, CO, USA

Colorado Trust Chair in American Indian Health, Aurora, CO, USA

Colorado School of Public Health at the University of Colorado Anschutz Medical Campus, Aurora, CO, USA

Kari Lyn Martin, MD Department of Dermatology, University of Missouri School of Medicine, Columbia, MO, USA

Martin McCandless, BS School of Medicine, University of Mississippi Medical Center, Jackson, MS, USA

Hannah McCowan, BS School of Medicine, University of Mississippi Medical Center, Jackson, MS, USA

Nancye McCowan, MD Department of Dermatology, University of Mississippi Medical Center, Jackson, MS, USA

Ruth McTighe, MD Department of Internal Medicine and Dermatology, University of Mississippi Medical Center, Jackson, MS, USA

Chelsea S. Mockbee, MD Department of Dermatology, University of Mississippi Medical Center, Jackson, MS, USA

Anastasia Mosby, MS School of Medicine, University of Mississippi Medical Center, Jackson, MS, USA

Eliot N. Mostow, MD, MPH Dermatology Section, Northeastern Ohio Medical University, Rootstown, OH, USA

Dermatology, Case Western Reserve University, Cleveland, OH, USA

Sonal Muzumdar, BS University of Connecticut School of Medicine, Farmington, CT, USA

Vinayak K. Nahar, MD, PhD, MS Department of Dermatology, University of Mississippi Medical Center, Jackson, MS, USA

Joshua Ortego, MD Department of Dermatology, University of Mississippi Medical Center, Jackson, MS, USA

Nicholas Osborne, BS Northeastern Ohio Medical University, Rootstown, OH, USA

Leslie Partridge, FNP-C, DCNP Department of Dermatology, University of Mississippi Medical Center, Jackson, MS, USA

Ross Pearlman, MD Department of Dermatology and Pathology, University of Mississippi Medical Center, Jackson, MS, USA

Claire Petitt, BS University of Alabama School of Medicine, Birmingham, AL, USA

Curtis Petruzzelli, MS Florida State University College of Medicine, Tallahassee, FL, USA

Morgan Pfleger, BS School of Graduate Studies in the Health Sciences, University of Mississippi Medical Center, Jackson, MS, USA

James E. Roberts, MD Department of Medicine, University of Mississippi Medical Center, Jackson, MS, USA

Adam Rosenfeld, BS University of Missouri School of Medicine, Columbia, MO, USA

Erica Rusie, PharmD Scientific Affairs, TALEM Health, Trumbull, CT, USA

Manoj Sharma, MBBS, PhD Department of Environmental & Occupational Health, School of Public Health, University of Nevada, Las Vegas, NV, USA

William Taylor Sisson, DHA, MBA, CMPE Department Business Administrator, Department of Dermatology, University of Mississippi Medical Center, Jackson, MS, USA

Cindy Firkins Smith, MD, MHCI Department of Dermatology, University of Minnesota, Minneapolis, MN, USA

Rural Health, CentraCare, St. Cloud, MN, USA

Carris Health, Willmar, MN, USA

Meredith E. Thomley, BS University of Alabama at Birmingham School of Medicine, Birmingham, AL, USA

Abel Torres, MD, JD, MBA Department of Dermatology, University of Florida College of Medicine, Gainesville, FL, USA

Neha Udayakumar, BS University of Alabama School of Medicine, Birmingham, AL, USA

Laurel Wessman, MD Department of Dermatology, University of Minnesota Health, Minneapolis, MN, USA

Rural Dermatology: Statistical Measures and Epidemiology

1

Nicholas Osborne, Sonal Muzumdar,
Eliot N. Mostow, and Hao Feng

Nicholas Osborne and Sonal Muzumdar contributed equally with all other contributors.

Epidemiology: The Current State of the Dermatology Workforce

The Rural-Urban Divide

At first glance, defining rural versus urban may seem to be an easy enough task: farm vs. city; agriculture vs. service industry; small vs. large populations; or maybe sparse housing vs. lively neighborhoods. When defining these taxonomies, stereotypical distinctions like those listed above

---

The major factors that brought health to mankind were epidemiology, sanitation, vaccination, refrigeration, and screen windows.
-Former Colorado governor, Richard Lamm, 1986.

N. Osborne
Northeastern Ohio Medical University,
Rootstown, OH, USA
e-mail: nosborne@neomed.edu

S. Muzumdar
University of Connecticut School of Medicine,
Farmington, CT, USA
e-mail: muzumdar@uchc.edu

E. N. Mostow
Dermatology CWRU, Akron, OH, USA
e-mail: emostow@neomed.edu

H. Feng (✉)
Department of Dermatology, University of
Connecticut Health Center, Farmington, CT, USA

belie important considerations related to the cultural, socioeconomic, and demographic aspects of societies risking oversimplification [1]. Each method of defining rural and urban has consequences that impact the application of policy and the collection and analysis of data [1, 2]. Depending on the taxonomy used, the percentage of Americans living in rural areas ranges from 10–28% of the total population (approximately 30–90 million) [1, 3].

Significant disparities exist in health care outcomes between rural and urban residents [4, 5]. As compared to urban counterparts, rural residents are more likely to die of preventable conditions including heart disease, stroke, lower respiratory tract disease and cancer [4, 5]. Additionally, rural residents have a lower average life expectancy than their urban peers (76.7 years and 79.1 years respectively) [6]. In rural areas, there are shortages of both general practitioners and specialty physicians, including dermatologists [6] (see Table 1.1).

Dermatology

Utilizing the Area Health Resources file and American Academy of Dermatology data, there are 3.4 to 3.65 dermatologists per 100,000 people

---

R. T. Brodell et al. (eds.), *Dermatology in Rural Settings*, Sustainable Development Goals Series,
https://doi.org/10.1007/978-3-030-75984-1_1

Table 1.1 Dermatologist Density Distribution in the United States

All dermatologists	
All of US	3.4–3.65 per 100,000 individuals
Metro	4.11 per 100,000 individuals
Non-metro	1.05 per 100,000 individuals
Rural	0.085 per 100,000 individuals
Pediatric dermatologists	
All of US	1 per 385,000 children
Mohs surgeons	
All of US	0.70 per 100,000 individuals
Metro	0.78 per 100,000 individuals
Nonmetro	0.27 per 100,000 individuals
Rural	0.23 per 100,000 individuals

Nonphysician Clinicians in Dermatology: Physician Assistants and Nurse Practitioners

in the United States [7, 8]. The concentration of dermatologists is significantly lower in rural areas as compared to urban ones. In 2013, the average density of dermatologists was estimated to be 4.11 per 100,000 population in metropolitan areas as compared to 1.05 per 100,000 population in non-metropolitan areas and 0.085 per 100,000 population in rural areas [8]. Additionally, 40% of dermatologists work in the 100 densest population centers in the US [9]. Areas with the highest concentrations of dermatologists in the US include the Upper East Side of Manhattan, New York (41.8 per 100,000), Palo Alto, California (36.6 per 100,000) and Santa Monica, California (35.9 per 100,000) [9].

Pediatric Dermatology

Pediatric dermatology was recognized as a subspecialty of the American Board of Dermatology in 2000 [10]. Nationwide, there is a perceived shortage of pediatric dermatologists with wait times being the longest for any pediatric subspecialty. In the US, there is approximately 1 pediatric dermatologist for every 385,000 children; 1 pediatrician for every 1500 children; and, 1 dermatologist for every 30,000 people [10]. In surveys of pediatricians, pediatric dermatology is identified as one of the three most difficult pediatric specialties to rake referrals [10]. Wait times

average between 6 and 13 weeks nationally [10, 11].

Like general dermatologists, pediatric dermatologists are concentrated in and around large metropolitan centers, with very few practitioners in rural locales [11]. In rural areas, geographic maldistribution compounds the national shortage of pediatric dermatologists and makes accessing adequate care especially difficult.

Dermatopathology

The geographic distribution of the dermatopathology workforce has not been well-characterized. However, a recent survey of fellows of the American Society of Dermatopathology found that nearly 65% were practicing in or affiliated with an academic center [12]. Additionally, while the Northeast, Midwest and West each have approximately 20% of practicing dermatopathologists, about 30% practice in the Southern United States [12]. Given the unique characteristic that pathology samples are sent and dermatopathologists can provide professional services anywhere in the country, the geographic distribution of dermatopathologists may not impact access to care in the same manner as general and subspecialty dermatologists.

Mohs Micrographic Surgery and Procedural Dermatology

Mohs micrographic surgery (MMS) is a technique utilized to manage skin cancer located in cosmetically and functionally sensitive body areas in the United States. Compared to other skin cancer treatment methods, such as excision, MMS is associated with higher cure rates, smaller defect sizes and better aesthetic outcomes.

The MMS workforce has expanded significantly over the past few decades. From 1995 to 2016, the annual number of American College of Mohs Surgery-accredited fellowship positions increased from 25 to 84 [13]. Approximately 20% of dermatology graduates pursue training in

MMS and there are approximately 2240 practicing Mohs surgeons in the United States [13, 14]. Compared to the general dermatology workforce, the MMS workforce is more likely to be concentrated in urban areas. 94.6% of all Mohs surgeons practice in metropolitan locales while 5% practice in nonmetropolitan areas and less than 1% practice in rural areas [14]. Additionally, 98.6% of rural counties do not have a practicing Mohs surgeon [14].

Nonphysician clinicians, including nurse practitioners and physician assistants, have been employed to expand access to medical care, especially in underserved areas. Nurse practitioners are able to practice independently in 22 US states and the District of Columbia [15]. Physician assistants, in contrast, must be directly supervised by a physician [15]. The use of these physician extenders has increased significantly over time across specialties. In dermatology, membership of the Society of Dermatology Physician Assistants grew from 49 to over 2700 between 1994 and 2014 [16].

Nonphysician clinicians are used widely in dermatology, with nearly 50% of practices employing them to perform medical visits and procedures [15]. Like dermatologists, nonphysician clinicians are more likely to practice in urban areas than rural ones [15]. More than 70% of nonphysician clinicians practice in counties with a dermatologist density over 4 per 100,000 population [15]. Only 3% practice in counties without dermatologists [15].

Trends over Time

A Historical Perspective

From the late 1950s to early 1980s, the number of training positions for dermatologists increased significantly. This was largely attributable to government programs that increased the sizes of medical school classes and funding for dermatology residencies [13]. Subsequently, concerns about oversaturation of dermatologists led to a significant decrease in the expansion of new training programs in the 1980s and 1990s [13]. In the past two decades, the number of dermatologists trained annually has grown modestly, keeping pace with US population growth [13]. Similarly, the density of dermatologists in the US has increased. From 1995 to 2013, the density of dermatologists in the US increased by 21% from 3.02 per 100,000 residents to 3.65 per 100,000 residents [8]. Growth in the dermatology workforce has been disproportionately higher in urban areas than rural ones. Between 1995 and 2013, the difference between dermatologist density in rural and urban areas increased by 18% from 3.41 per 100,000 people to 4.03 per 100,000 people (rural: 0.065 in 1995 and 0.085 in 2013; urban: 3.47 in 1995 and 4.11 in 2013) [8]. Despite the increases in dermatologist density in recent years, there exists a shortage of medical dermatologists, especially in rural areas. With 20% of dermatologists training in Mohs surgery and many others performing cosmetic procedures, the availability of medical dermatologists may be constrained.

Future Projections

The demand for dermatologists in the US is projected to increase significantly over time [7]. This is partially due to a growing and aging US population. The US Census Bureau estimates that by 2060, the US population will grow by 80 million and the number of citizens over the age of 65 will double from 45 million to 95 million [17]. As a result, the prevalence and burden of skin cancer and dermatologic disease is projected to grow [7]. Additionally, the scope of dermatologists has grown in recent decades, with increased care of medically complicated and hospitalized patients further contributing to increased demand for trained dermatologists [13].

Changes in the dermatology workforce over time may have a disproportionate impact on rural residents. The dermatology workforce is aging. From 1995 to 2013, the ratio of dermatologists older than 55 to those younger than 55 increased from 0.32 to 0.57 [8]. Dermatologists in rural areas are more likely to be over the age of 55, and closer to retirement, than their urban peers [8]. Similarly, dermatologists entering the workforce are more likely to practice in urban areas [13]. As a result, disparities in access to dermatology care between rural and urban locales are expected to increase over time.

Impacts on Patients

Access to Care

Disparities in the dermatology workforce between rural and urban areas has impacted rural residents' access to dermatologic care. On a national level, rural patients experience longer wait times as compared to those in suburban and urban areas [18], although this is not shown consistently [19]. In 2007, the average wait time for new patients in rural settings was 45.6 days as compared to 31.5 days in suburban areas and 32.7 days in urban ones [18]. In addition, rural residents travel further, on average, to seek dermatologic care as compared to their urban peers [20].

Rural residents also have less access to specialized dermatological care. One study demonstrated a lower concentration of dermatology providers who prescribed injectable biologic medications, which are increasingly used to treat systemic dermatologic conditions, in rural areas as compared to urban ones [21]. As mentioned previously, access to dermatologic specialists, such as Mohs micrographic surgeons and pediatric dermatologists, is also limited in rural areas. Pediatric dermatologists are concentrated in metropolitan locales and less than 1% of all Mohs surgeons practice in rural counties.

Patient Outcomes

Impaired access to adequate dermatologic care has been demonstrated to impact patient outcomes. Dermatologist density is associated with well-defined disease-specific outcomes for patients with melanoma and Merkel cell carcinoma.

Every year, over 75,000 Americans are diagnosed with melanoma and 9000 die from the disease [22]. An increased density of dermatologists is correlated with better outcomes for melanoma. Dermatologists are more likely to diagnose melanoma at an early stage than non-dermatologist providers which leads to better patient outcomes [23]. Higher dermatologist density is associated with early melanoma detection. One study demonstrated a 39% increase in odds of early melanoma diagnosis for each additional dermatologist per 10,000 population [24]. Similarly, proximity to the nearest dermatologist is associated with decreased melanoma thickness at the time of diagnosis [25]. Another study demonstrated that increased dermatologist density, to a point, is correlated with lower melanoma mortality [26]. However, additional dermatologists over 2 per 100,000 population do not appear to impact melanoma mortality rates [26].

Merkel cell carcinoma is an aggressive neuroendocrine cancer that impacts 1600 Americans annually [27]. Dermatologist density has been associated with improved outcomes for Merkel cell carcinoma. Patients with Merkel cell carcinoma who lived in areas of higher dermatologist density had improved survival rates as compared to those living in areas with low dermatologist density [28].

Epidemiologic Perspective and Considerations

Defined epidemiologic data on dermatologist density can be used to compare rural vs urban areas, and trends over time. However, rural areas over time become developed to an extent that farms and forests merge into suburbia and the suburbs become relatively urban. We strongly advocate for developing data that provide opportunities to monitor for differences between rural and urban areas and specifically address deficiencies, when possible, to improve medical outcomes for all individuals and populations. The rest of this chapter addresses details related to issues and options for studying rural versus urban dermatology.

Statistical Measures and Urban-Rural Definitions

Introductory Framework to Data Collection and Utilization

The focus of this book is rural dermatology. The methods utilized to study rural dermatology on an epidemiologic level form the basis for assessing the impact of any solutions that may be employed to correct deficiencies in the rural health care system. Of course, as noted in the introduction, the concept of "rural" must be carefully defined [29, 30]. Each definition or classification scheme has benefits and pitfalls. In addition, it is often important to compare study results from populations that have been identified using different criteria [29, 30].

Next, there must be careful consideration as to which type of study to use [30, 31]. The study of choice has implications related to the classification scheme used to define a population with certain benefits and drawbacks. It is important to identify the risk factors that may impact health outcomes, potential confounding variables, and to carefully define the outcome measures. Models must also be considered; qualifying models are most often employed while mathematical models for study designs have been employed less often in epidemiology [30, 32].

Once the study has been chosen, data collection ensues. Obtaining data for epidemiologic studies can be obtained from public health records obtained by county, state, and federal health departments [33]. Additionally, there is an impetus to expand electronic databases for epidemiology to increase efficiency and accuracy in a large study population representative of the location of interest [34]. After the data is collected and interpreted, policies, funding, interventions, and more can be advocated.

Defining Rural Vs. Urban

Several classification schemes have been developed to account for the complex and diverse nature of what it means to be rural or urban. Some of the most commonly utilized include Metropolitan and Micropolitan Statistical Areas (Office of Management and Budget [OMB]), Rural-Urban Continuum Codes (Economic Research Service, United States Department of Agriculture [ERS, USDA]), Urban Influence Codes (ERS, USDA), Urban and rural classification (Department of Commerce, Bureau of the Census [DOC]), and Rural-Urban Commuting Area Codes (RUCA of USDA and Health Resources and Services Administration's Federal Office of Rural Health Policy [ORHP]), Metro-Centric Locale Codes (National Center for Education Statistics [NCES]), and Urban-Centric Locale Codes (NCES) [3]. Each taxonomy will be discussed in detail, including its strengthens and weakness, especially in relation to public health research. (see Table 1.2).

Table 1.2 Rural Classification Schemes [1, 29]

Defining Rural vs. Urban		
Classification Scheme	Strength	Weakness
Metropolitan and Micropolitan statistical areas	Stable county geographical unit. Useful in federal policy making	Variable sizes of counties across the nation
Rural-urban continuum codes	Stable county geographical unit. Useful in public health research	Variable sizes of counties across the nation
Urban influence codes	Stable county geographical unit. Useful in public health research	Variable sizes of counties across the nation
Urban and rural classification	Precision in representing population per area	Difficult to apply to health data, as its geographical units are not sChap
Rural-urban commuting area codes	Robust representation of rural areas	Complex and subject to change

Metropolitan and Micropolitan Statistical Areas

The OMB's taxonomy for rural v. urban is based on counties [1, 35, 36]. Rather than defining rural, this taxonomy compares metropolitan v. micropolitan v. outside core areas. The distinction between is based on the population: metropolitan with a population greater than 50,000; micropolitan (urban cluster and adjacent territory) of 10,000-50,000; and outside core areas of less than 10,000.

This taxonomy is used primarily for the establishment of federal policy [1]. These distinctions allow for various federal reimbursement, incentive, and resource allocation programs. The strength of this taxonomy scheme lies in the stability of counties as a geographic unit, the county serving as a political jurisdiction representation, and the general definitions it provides for policymakers [1, 36]. In contrast, its weaknesses revolve around the variability in county sizes across the United States, the mixture of both urban and rural areas within a single county, and subsequent over- and under-estimate of rural populations [1, 36].

Rural-Urban Continuum Codes (RUCC)

ERS, USDA distinguishes rural v. urban with the county coding: metropolitan and nonmetropolitan [35]. These two categories are further divided into three metro codes and six nonmetro codes: (a) population of one million and above, (b) 250,000-1million, (c) less than 250,000 for metro; and nonmetro further distinguish by populations (urban population of 19,999 and more; 2500–20,000 persons, or less than 2500) and its relation to adjacent metro areas [29, 35].

These codes were created to represent the heterogeneous nature of rural or non-urban areas more accurately [29]. This taxonomy is useful in public health research in tracking incidence, prevalence, mortality, and morbidity of a disease [29]. Given this taxonomy is county based, its strengths and weakness are the same as

Metropolitan and Micropolitan Statistical Areas: stable and political representation but large variability across the country.

Urban Influence Codes

The Urban Influence Code (UIC) is like RUCC in many ways [1, 35]. They are both county-based, having the same strength and weakness of stability and variability, respectively. UIC differentiates counties by the designation of metropolitan or nonmetropolitan as well. However, the metropolitan code is divided into two codes, limiting its differentiation of metropolitan areas: (a) large with a population over 1 million and (b) small with a population less than 1 million. The nonmetropolitan coding is like RUCC. It classifies each sub-code by population size and distance from large urban areas but is nine codes for OMB's micropolitan and outside core areas [35]. Like RUCC, UIC is particularly useful in public health epidemiology research. But this data set is better equipped to measure access to healthcare inequities as "the largest community within the population is more likely to reflect available health services than the total urban population in all cities and towns" [29].

Urban and Rural Classification, US Census Bureau

The United States Census Bureau's taxonomy for rural v. urban utilizes the census tract as its geographical unit [1, 35]. The bureau defines rural as those outside of urban areas and clusters (populations 2500 and above without a large commute).

Utilizing the census, this taxonomy has strength in that is more precise at representing population per area, unlike county-based taxonomies that risk over and underestimating populations [1, 36]. Although there is a definite advantage to this taxonomy, the census is difficult to apply to health data collected from zip codes or counties, has unique geography and terminology, and its unit boundaries are not stable over the years [1, 36].

Rural-Urban Committing Area Codes

The University of Washington created RUCA in conjunction with ERS [35]. This taxonomy utilizes census tract or zip code for its geographic unit [1, 35]. RUCA is divided into 33 categories, but two classification systems: four category and seven categories [35]. The four category includes: (a) urban, (b) large rural, (c) small rural, and (d) isolated. The seven category includes: (a) urban core, (b) other urban, (c) large rural core, (d) other large rural core, (e) small rural core, (f) other small rural core, and (g) isolated rural [35].

RUCA is useful in that it is a robust representation of rural areas based on their economic ties to urban and other rural areas [1]. The 33 categories allow for a more precise representation of each area. Furthermore, the zip-code unit is particularly useful in the collection of health data, as the census tract is not routinely used [1, 35]. Its weaknesses; however, lie in its complex nature and subject to change (zip codes are controlled and changed routinely by the US Postal Service) [1, 29, 36].

Statistical Measures

Pertinent Measures and Models of Study

Epidemiology is the field that accesses the various qualities of a disease, injury, or other conditions impeding upon human health and wellbeing [37]. This can lead to an understanding of etiologies, interventions and preventions, and surveillance systems. Epidemiology is a highly diverse and dynamic field of research. It can answer the *what*, *how much*, *when*, *where*, and *who* of a disease or condition under evaluation. Many variables (e.g., age, geographic location, etc.) must be considered and carefully accounted for in an intentionally and meaningfully designed study [37]. It is an essential tool in the evaluation of health disparities impacting rural communities, such as access to a dermatologist and dermatologic conditions.

In reporting epidemiologic data, several study models can be utilized to qualify a disease or condition [37]. The *what* characterizing a disease or outcome, is commonly reported in tables, organizing cases by their attributes of interest. The *how much* is a model of counts such as incidence and prevalence. Time, the *when*, is useful in representing data over a given period in graphs, diagrams, and curves. The *where* helps to graphically represent the distribution of an exposure or outcome; this model is particularly useful in identifying disparities between rural and nonrural areas. Finally, the *whom*, is a focus on personal qualities of the population of interest (e.g., age, gender, sex, education level, and more) [37, 38].

Several measurements are commonly utilized in epidemiological studies. Two key measurements are incidence and prevalence in the form of a simple count [33, 37]. Incidence is the number of new cases within a given range of time while prevalence is the total number of cases accounted for at one discrete time. These two measures uncover a wealth of information in understanding a disease or health condition's impact in our world. Incidence is useful in uncovering the etiology of the disease or condition, measuring the exposures, confounders, and outcomes within the whole population of interest [33, 37]. It is further reported as various ratios: rate, risk, and odds ratios for population or variable discrepancies [39]. Prevalence on the other hand is useful at accessing the current impact of a chronic disease or condition without the cost and length of time studying incidence would require, due to the large number of existing cases (e.g., hypertension or diabetes) [33, 37, 39]. However, given the discrete measure in time, confounding variables are present that decrease the ability of prevalence to access causative relationships of exposure and outcome. Prevalence is further reported as prevalence and prevalence odds ratios for population or variable discrepancies [39].

Other common measures are utilized to characterize the impact of a disease or condition is mortality and morbidity [30, 31]. These are especially used in surveillance systems. It is important to note that mortality is really a type of incidence where the outcome or occurrence is death rather than a disease [31].

Common Study Designs

There are many types of studies used in epidemiologic research. Three study designs will be reviewed here due to their relevance in understanding rural dermatologic disparities. They are cohort, case-control, and cross-sectional studies.

Cohort studies are those studies that compare a group of affected or at risk for a condition to a group without those without the condition or its risk factors [30]. In making this comparison, reporting subsequent health outcome incidences, risk factors are attributed to a condition or event. These types of studies are often prospective but can be retrospective. Given that incidence is measured over a length of time, these studies can last months to years, making them a costly study. There is also risk of participant dropout. However, its ability to measure incidence and point to a causative relationship between exposure and outcome makes it very useful.

Case-Control studies compares a group who already has a condition to a group who does not [30]. Whereas cohort studies ask, *"will they develop the disease"*, case-control studies ask, *"why did they develop the disease."* These studies are much more efficient in terms of time and cost; however, they are at risk of recall bias, given the assessment of past exposures.

Cross-sectional studies analyze the prevalence of a disease or condition and its association with risk factors at the time of the study [30]. It is useful in linking risk factors to conditions, but it is not useful in establishing causality. They are also cost and time efficient. Unlike case-control studies, cross-sectional occur in the present and are thus not at risk of recall bias.

Concluding Remarks

The increasing dermatologic workforce gap and healthcare disparities between urban and rural areas require a focus on accurately describing and reporting epidemiologic data using appropriate statistical measures. This is the foundation for finding solutions to minimize and/or reverse these trends. While there has been progress in better understanding and describing disparities relating to rural dermatologic care, more work is needed to devise strategies, implement interventions, and appropriately monitor the outcomes to improve the healthcare of all citizens in the United States.

Conflicts of Interest The authors have no relevant conflicts of interest.

References

1. Hart LG, Larson EH, Lishner DM. Rural definitions for health policy and research. Am J Public Health. 2005;95(7):1149–55.
2. Johnson-Webb KD, Baer LD, Gesler WM. What is rural? issues and considerations. J Rural Health. 1997;13(3):253–6.
3. Douthit N, Kiv S, Dwolatzky T, Biswas S. Exposing some important barriers to health care access in the rural USA. Public Health. 2015;129(6):611–20.
4. Richman L, Pearson J, Beasley C, Stanifer J. Addressing health inequalities in diverse, rural communities: an unmet need. SSM-population health. 2019;7:100398.
5. Cosby AG, McDoom-Echebiri MM, James W, Khandekar H, Brown W, Hanna HL. Growth and persistence of place-based mortality in the united states: the rural mortality penalty. Am J Public Health. 2019;109(1):155–62.
6. National healthcare quality and disparities report. Agency for healthcare research and quality web site. https://www.ahrq.gov/sites/default/files/wysiwyg/research/findings/nhqrdr/chartbooks/qdr-ruralhealthchartbook-update.pdf. Updated 2017. Accessed 28 May 2020.
7. Glazer AM, Farberg AS, Winkelmann RR, Rigel DS. Analysis of trends in geographic distribution and density of US dermatologists. JAMA dermatology. 2017;153(4):322–5.
8. Feng H, Berk-Krauss J, Feng PW, Stein JA. Comparison of dermatologist density between urban and rural counties in the united states. JAMA dermatology. 2018;154(11):1265–71.
9. Yoo JY, Rigel DS. Trends in dermatology: geographic density of US dermatologists. Arch Dermatol. 2010;146(7):779.
10. Prindaville B, Antaya RJ, Siegfried EC. Pediatric dermatology: past, present, and future. Pediatr Dermatol. 2015;32(1):1–12.

11. Prindaville B, Horii KA, Siegfried EC, Brandling-Bennett H. Pediatric dermatology workforce in the united states. Pediatr Dermatol. 2019;36(1):166–8.

12. Suwattee P, Cham PM, Abdollahi M, Warshaw EM. Dermatopathology workforce in the united states: a survey. J Am Acad Dermatol. 2011;65(6):1180–5.

13. Porter ML, Kimball AB. Predictions, surprises, and the future of the dermatology workforce. JAMA Dermatol. 2018;154(11):1253–5.

14. Feng H, Belkin D, Geronemus RG. Geographic distribution of US mohs micrographic surgery workforce. Dermatol Surg. 2019;45(1):160–3.

15. Adamson AS, Suarez EA, McDaniel P, Leiphart PA, Zeitany A, Kirby JS. Geographic distribution of nonphysician clinicians who independently billed medicare for common dermatologic services in 2014. JAMA Dermatol. 2018;154(1):30–6.

16. Zurfley F, Mostow EN. Association between the use of a physician extender and dermatology appointment wait times in ohio. JAMA Dermatol. 2017;153(12):1323–4.

17. Vespa J, Medina L, Armstrong DM. Demographic turning points for the united states: population projections for 2020 to 2060. United States Census Bureau Web site. https://www.census.gov/content/dam/Census/library/publications/2020/demo/p25-1144.pdf. Updated 2020. Accessed 29 May 2020.

18. Kimball AB, Resneck JS Jr. The US dermatology workforce: a specialty remains in shortage. J Am Acad Dermatol. 2008;59(5):741–5.

19. Uhlenhake E, Brodell R, Mostow E. The dermatology work force: a focus on urban versus rural wait times. J Am Acad Dermatol. 2009;61(1):17–22.

20. Rosenthal MB, Zaslavsky A, Newhouse JP. The geographic distribution of physicians revisited. Health Serv Res. 2005;40(6p1):1931–52.

21. Feng H, Cohen JM, Neimann AL. Access to injectable biologic medications by medicare beneficiaries: Geographic distribution of US dermatologist prescribers. J Dermatol Treat. 2019;30(3):237–9.

22. Melanoma incidence and Mortality, United States—2012–2016. Centers for disease control and prevention web site. https://www.cdc.gov/cancer/uscs/about/data-briefs/no9-melanoma-incidence-mortality-UnitedStates-2012-2016.htm. Published 07/01/2019. Updated 2019. Accessed 25 May 2020.

23. Pennie ML, Soon SL, Risser JB, Veledar E, Culler SD, Chen SC. Melanoma outcomes for medicare patients: association of stage and survival with detection by a dermatologist vs a nondermatologist. Arch Dermatol. 2007;143(4):488–94.

24. Roetzheim RG, Pal N, Van Durme DJ, et al. Increasing supplies of dermatologists and family physicians are associated with earlier stage of melanoma detection. J Am Acad Dermatol. 2000;43(2):211–8.

25. Stitzenberg KB, Thomas NE, Dalton K, et al. Distance to diagnosing provider as a measure of access for patients with melanoma. Arch Dermatol. 2007;143(8):991–8.

26. Aneja S, Aneja S, Bordeaux JS. Association of increased dermatologist density with lower melanoma mortality. Arch Dermatol. 2012;148(2):174–8.

27. Hughes MP, Hardee ME, Cornelius LA, Hutchins LF, Becker JC, Gao L. Merkel cell carcinoma: epidemiology, target, and therapy. Curr Dermatol Rep. 2014;3(1):46–53.

28. Criscito MC, Martires KJ, Stein JA. A population-based cohort study on the association of dermatologist density and merkel cell carcinoma survival. J Am Acad Dermatol. 2017;76(3):570–2. doi: S0190-9622(16)31019-2 [pii].

29. Hall SA, Kaufman JS, Ricketts TC. Defining urban and rural areas in US epidemiologic studies. J Urban Health. 2006;83(2):162–75.

30. Checkoway H, Pearce N, Kriebel D. Selecting appropriate study designs to address specific research questions in occupational epidemiology. Occup Environ Med. 2007;64(9):633–8. doi: 64/9/633 [pii].

31. Hernandez JBR, Kim PY. Epidemiology morbidity and mortality. NCBI StatPearls web site. https://www.ncbi.nlm.nih.gov/books/NBK547668/. Published February 2020. Updated 2020. Accessed 20 July 2020.

32. Herzog SA, Blaizot S, Hens N. Mathematical models used to inform study design or surveillance systems in infectious diseases: a systematic review. BMC Infect Dis. 2017;17(1):775.

33. Ward MM. Estimating disease prevalence and incidence using administrative data: some assembly required. J Rheumatol. 2013;40(8):1241–3. https://doi.org/10.3899/jrheum.130675.

34. Blumenberg C, Barros AJ. Electronic data collection in epidemiological research: the use of REDCap in the pelotas birth cohorts. Applied Clin Informat. 2016;7(3):672.

35. Hawley LR, Koziol NA, Bovaird JA, et al. Defining and describing rural: implications for rural special education research and policy. Rural Special Educat Quarter. 2016;35(3):3–11.

36. Coburn PhD AF, MacKinney M, Clinton A, et al. Choosing rural definitions: implications for health policy. 2007.

37. Describing epidemiologic data. In: Ruasmussen SA, Goodman Richard A, ed. The CDC field epidemiology manual. 4th 2019.

38. Lawlor DA, Chaturvedi N. Methods of measurements in epidemiology—call for a new type of paper in the IJE. Int J Epidemiol. 2010;39(5):1133–6.

39. Pearce N. Classification of epidemiological study designs. Inter J Epidemiol. 2012;41:393–7.

A Comparison of Rural and Urban Dermatology

2

Laurel Wessman, Brett Macleod, and Ronda S. Farah

Introduction

It holds true in medicine, and in dermatology, that the ability to understand, appreciate, and empathize with our patients is paramount to our success as diagnosticians and healers. Of no less significance is our respect for the subtle and not-so-subtle variations inherent in treating diverse populations, whether between or amongst patient cohorts. This certainly is the case when we consider delivering specialized dermatology care to those who are geographically privileged versus disadvantaged. We hope you will appreciate, as you make your way through this chapter, the ways in which characteristics of rural populations lead to challenges regarding optimizing and

"We are neither anti-urban nor pro-rural. We know there is a gap between urban and rural areas; we are only trying to bridge it."
H. D. Kumaraswamy

L. Wessman · R. S. Farah (✉)
Department of Dermatology, University of Minnesota Health, Minneapolis, MN, USA
e-mail: wessm018@umn.edu; rfarah@umn.edu

B. Macleod
University of North Dakota, Grand Forks, ND, USA
e-mail: brett.macleod@und.edu

delivering holistic dermatology care, as well as the distinct challenges inherent in training and maintaining an equally diverse and robust dermatology work force to meet the challenge.

Patient Demographics

Age and Geography

It is no secret, the United States population is aging [1, 2]. In fact, from 2010 to 2014, 97% of United States counties experienced increases in the population percentage of residents age 65 and older (see Fig. 2.1) [3]. Appreciation for the evolution of the population in this regard including life expectancy projections in the United States is vital to our understanding of social welfare, health care disparities, and proper and efficient allocation of resources in the medical field, including optimizing density and distribution of physicians and front line workers in all specialties [1]. From 2017 to 2060, the total life expectancy of the United States population may increase by as many as 6 years, from 79.7 to 85.6 [1]. From 2010 to 2050 the United States elderly population (ages 65 and greater) is projected to more than double, from 40.5 to 89 million [2]. The proportion of such persons reporting at least one chronic disease will also grow, as was the case from 1998 to 2008 (86.9% to 92.2%) [2]. Demand for primary care services may expand

© The Author(s), under exclusive license to Springer Nature Switzerland AG 2021
R. T. Brodell et al. (eds.), *Dermatology in Rural Settings*, Sustainable Development Goals Series, https://doi.org/10.1007/978-3-030-75984-1_2

% of country population ages 65+

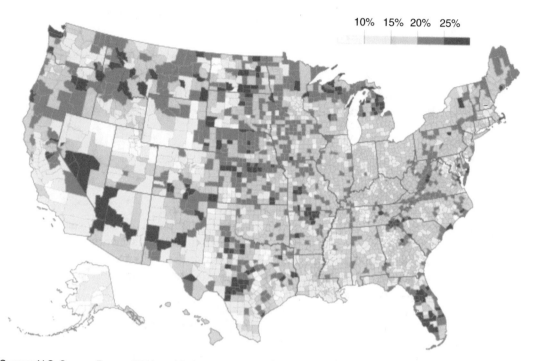

Source: U.S. Census Bureau 2014 population estimates

Fig. 2.1 Reprinted with express permission from the original authors. Percentage of county populations ages 65 and greater. https://www.pewresearch.org/fact- tank/2015/07/09/where-do-the-oldest-americans-live/. Accessed online December 2020

by 14% through the year 2025 and subspecialists will be in even higher demand, with a projected increase in dermatology visits by 16% [2]. The aging demographics of the United States population becomes increasingly relevant to the field of dermatology when one considers the increased incidence of dermatologic conditions in elderly populations, including nonmelanoma skin cancer, lentigo maligna, venous stasis ulceration, and herpes zoster among other disorders ideally managed by a board-certified dermatologist [4, 5]. While an aging population in the United States is indeed concerning enough when one considers increased need for dermatologic care, data from the Children's Hospital Association additionally highlights barriers for access to pediatric dermatology services. For instance, the average wait time for a child to see pediatric dermatology in a children's hospital was 54 days in 2012 [6].

Comparable to complexities regarding age and access to dermatologic specialty care, the distribution of the population across rural, suburban, and urban settings and relevance for access to dermatology care serves as a subject of respectable complexity. Indeed, even the definitions of rural and non-rural used by the government are imperfect and complicated. The United States Census Bureau for example, defines urbanized areas as constituting greater than or equal to 50,000 people, while an urbanized cluster refers to populations greater than or equal to 2500 people but less than 50,000 [7]. While the Bureau does not explicitly define a rural population, it is presumed to categorize populations not otherwise classified as an urban area and does not follow county or city boundaries [7]. Under this classification scheme, one in five Americans was considered rural in 2010 [7]. The Office of

Management and Budget (OMB) by contrast, designates counties rather than loosely defined areas as metropolitan (greater than or equal to 50,000 people) and micropolitan (greater than or equal to 10,000 but less than 50,000 people) with all non-metropolitan statistical sites considered rural [7]. Under this classification scheme, which associates populations more with their respective city or county boundaries, 15% of the United States population was considered rural in 2010. Therefore in essence, the census bureau may be more likely to over estimate the rurality of America while the OMB may accomplish the opposite [7]. Finally, the federal office of rural health policy (FORHP) defines rural as all non-metropolitan counties in addition to assigning Rural-Urban Commuting Area (RUCA) codes. By these standards, 18% of the population was considered rural in 2010 [7]. Although complex, understanding the significance of the aforementioned definitions of what is rural cannot be overstated, as eligible counties or census tracts located within metropolitan counties may apply for rural health grants aimed at reducing health care disparities in rural communities [7].

In general, rural communities face mounting physician shortages (including dermatologists), physicians who do practice rural medicine must do so with less support and technology- transferring more complicated patients to larger hospital systems, in turn reducing opportunities (and funding) for trainees to experience their specialty of interest in a rural site [8]. Rural Americans have a lower median household income and when broken down into age groups rural Americans earn less than their urban counterparts [9, 10]. People under 65 living in a rural county are more likely to be uninsured than their urban counterparts and even more likely to be uninsured if they lived in a county without adoption of Medicaid expansion [11].

Patients must also travel longer distances to receive specialty care, which for a variety of reasons may lead to delays in diagnosis and treatment. This has specifically been noted for deadly diseases such as melanoma, as the literature highlights greater tumor depth at diagnosis and worse outcomes for patient's who live greater distances

from a dermatologist or medical center [12–14]. For rarer cutaneous carcinomas such as Merkel Cell Carcinomas, longer distances traveled by patients was associated with later stage of diagnosis and fewer in-person clinic visits [15]. Access to newer technologies associated with improved compliance in therapy for deadly diseases may also be impacted by geography. For instance, geographic location affects access of patients who may need chemotherapy to scalp cooling technologies to reduce the likelihood of chemotherapy-induced alopecia, a reason some will refuse recommended treatments altogether [16]. In contrast to this however, not all outcomes are worse for rural residents. A study by Schram et al. demonstrated that those in urban areas were at much higher risk of atopic dermatitis, possibly related to increased stressors in daily life and exposure to pollutants [17].

Race and Ethnicity

Rural America is less diverse than its urban counterpart, with 80% of the population classified as white, compared to 58% in urban areas [18]. Hispanic and Black populations represent 9% and 8% of rural America, in contrast to 20% and 13% in urban America [18]. By contrast, the American Indian population comprises 2% and 0.5% of rural and urban America, respectively [18]. Race and ethnicity may influence morbidity and mortality associated with dermatologic conditions. For instance, blacks with mycosis fungoides may be more likely to be diagnosed at advanced stages and with increased skin involvement compared to whites, and black women may be more likely to die at a younger age from the disease [19]. In comparison, while melanoma survival has increased over the past decades it has not improved as well for minorities, particularly Hispanics [20]. Blacks also tend to present with squamous cell carcinomas at later and more aggressive stages, despite an overall lower non melanoma skin cancer incidence sin ethnic minorities [21]. One must also consider the higher rates of obesity and untreated or chronic depression in blacks and Hispanics, which may exacerbate more common dermatologic disorders such as psoriasis [22]. While the overall composition of ethnic minorities in rural communities is low, health care costs associated with cutaneous diseases presenting at late or even terminal stages are significant and resources should be allocated to addressing gaps in access to dermatology care for our patients [23]. Additionally, while we know black patients report higher satisfaction with their dermatologic care if seen in a skin of color clinic, evidence suggests that a knowledgeable dermatologist in skin of color may help to mitigate negative perceptions where such clinics currently do not exist [24].

Sexual and Gender Minorities

The 2019 Movement Advancement Project report on sexual and gender minorities in rural America revealed that 15–20% of sexual minority individuals in America live in a rural county [25]. The report highlights avoidance of healthcare in this population secondary to direct experiences or fear of discrimination from healthcare providers [25]. However, such patients may be uniquely predisposed to the development of dermatologic disease and culturally competent care delivered by a dermatologist is essential [26]. For instance those are at increased risk of alcohol and tobacco use, tanning behaviors, Human Immunodeficiency Virus (HIV) and human papilloma virus (HPV) infection and estrogen hormone therapy may be associated with keratinocyte or melanoma skin cancer [26]. Kaposi sarcoma, squamous cell carcinoma of the anus, and certain types of contact dermatitis are more common in gay and bisexual men and other men who have sex with men [27]. Reports of depression and thoughts of suicide are also more common in bisexual men and women, and gay men with acne [27]. Transgender patients additionally may require specialized gender transition services and are uniquely predisposed to adverse events from illicit cosmetic dermatologic services, such as receipt of non-FDA approved fillers [28].Additionally unique, culturally competent, and humble care is necessary for the treatment of sexual and gender minority adolescents [29, 30] in rural locations, who may have less means or ability to travel for specialized dermatology care. An action item advocated for by dermatologists [27] and encouraged herein by the authors of this text includes collection of gender identity and sexual orientation data in the clinic.

Socioeconomic Status and Healthcare Coverage

As previously alluded to, rural Americans tend to have lower median household incomes, and although the overall poverty rate is lower, the discrepancy can be at least partially explained by measures of income inequality, which tend to be higher in urban areas [9]. When broken down into age groups rural Americans still tend to earn less than their urban counterparts [10]. Census bureau data on insurance coverage rates demonstrates that individuals under 65 years of age living in a rural county are much more likely to be uninsured than their urban counterparts and even more likely to be uninsured if they lived in a county without adoption of the Medicaid expansion. For Medicare beneficiaries in 2007, 89% had supplemental coverage plus Parts A and B, the highest groups lacking additional coverage included rural residents (15%), African Americans (16%), and low income earners (between $10,000 and $20,000) [31]. The overall uninsured rates for all counties in the US have been steadily trending downward in the last few years leading up to 2016 [11].

The Dermatology Workforce

The future of medicine with regard to dermatologist distribution, density, and demographics must continuously evolve with the changing United States population in order to mitigate long wait times and improve access to empathic and compassionate care for both geographically privileged and isolated populations, enhancing our ability to reduce unnecessary morbidity and mortality from cutaneous disease.

Age and Geography

At present, while approximately 500 dermatologists enter the workforce each year, 325 exit [32]. Unfortunately for rural communities, the distribution of dermatologists is disproportionate with respect to age. From 1995 to 2013 for instance, the number of dermatologists in rural and non-metropolitan locations increased by only 6.5% [33]. In rural counties the ratio of dermatologists over 55 years of age to those under 55 years of age shows a similar discrepancy, as high as 0.93 in 2013 from 0.34 in 1995, versus 0.56 from 0.32 in metropolitan counties during the same time period [33]. Therefore, although many dermatologists are practicing longer in general, potentially alleviating an absolute shortage, rural regions are aging at a much faster rate without a sufficient supply of younger dermatologists to take over for them. Many factors likely contribute to disparities, as urbanization becomes an irrefutable trend.

The nationwide density of dermatologists has been steadily increasing in the last few decades and has risen by 21 percent from 1995 to 2013, with 3.4 dermatologists per 100,000 persons nationwide in 2017 [32, 33]. However, to adequately care for a population it is estimated that 4 dermatologists per 100,000 people are required [32]. When dermatologists per 100,000 people are broken down into 3-digit zip codes the shortage in rural America becomes devastatingly apparent: approximately70% of zip codes contain fewer than 4 dermatologists per 100,000 people while 60% have fewer than 3 [32]. As for zip codes not deemed short of dermatologists, they primarily encompass/surround large cities and academic centers [32]. Additional data reveals that even as the absolute number of dermatologists increases nationally and in non-metropolitan/rural counties there remains an ever-growing disparity in access to dermatologists for those who do not reside in metropolitan areas compared to those who do, as the percent change continues to favor the former [33]. Additionally, a recent study highlights the disproportionate numbers of dermatologists performing Mohs micrographic surgery in urban versus rural

locations, with approximately 95% practicing in metropolitan areas and 5% and 0.4% in non-metropolitan and rural areas respectively [34]. Similarly 98.1% of the approximate 300 pediatric dermatologists in the United States work in metropolitan counties, with nine states without a pediatric dermatologists at all [35].

Diversity

As one of the most competitive residencies for medical students in the United States, the race for dermatology begins early. Often dermatology residents are selected from among the top students at the top universities. Often, these students have achieved excellent scores in their board exams, dedicated years to research in dermatology, and have made the necessary connections to succeed in such a highly competitive environment. This competitive environment then creates an insurmountable obstacle to students that do not have access to the best study resources, the latest dermatology research, or faculty with professional connections. A student at a disadvantaged high school or region may experience greater barriers in forming academic foundations, which have traditionally been important for dermatology application processes. In 2018 in the United States, 52.4% of overall undergraduate students identified as White, 12.7% as Black, 20.5% as Hispanic, and 6.6% as Asian (see Fig. 2.2a) [36]. Hispanic students have seen the largest rise in enrollment with an increase of 148% since 2000 [36].

The population is filtered further through the medical school admissions process. Among medical school applicants in 2020, 43.2% identified as White, 8.2% as Black, 9.92% as Hispanic or Latino, and 21.2% as Asian (see Fig. 2.2b) [37]. Among those to matriculate, 44.7% identified as White, 8.0% as Black, 11.0% as Hispanic or Latino, and 21.56% as Asian [37]. If Asian matriculants are excluded from the data set, relatively more White applicants matriculate into medical school than remaining minorities when compared to undergraduate population demographics [37].

a   Diversity of Undergraduate Students

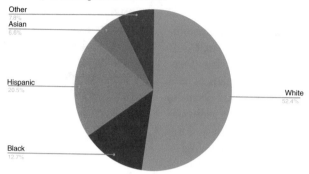

Other
7.8%
Asian
6.6%
Hispanic
20.5%
Black
12.7%
White
52.4%

Courtesy of Ronda S. Farah MD and Gretchen Bellefeille BS, University of Minnesota, Department of Dermatology[36].
Undergraduate Enrollement. National Center of Education Statistics; 2020. https://nces. ed. gov/programs/coe/indicator_cha.asp

b   Diversity of Medical School Applicants in 2020

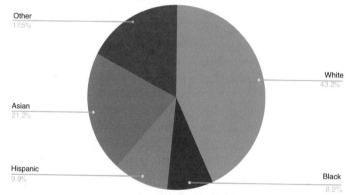

Other
17.5%
Asian
21.2%
Hispanic
9.9%
White
43.2%
Black
8.2%

Courtesy of Ronda S. Farah MD and Gretchen Bellefeille BS, University of Minnesota, Department of Dermatology[37].

2020 Facts: Applicants and Matriculants Data AAMC
https://www.aamc.org/data-reports/students-residents/interactive-data/2020-facts-applicants-and-matriculants-data

c   Diversity Among Dermatogy Applications in 2020

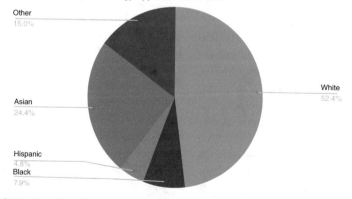

Other
15.0%
Asian
24.4%
Hispanic
4.8%
Black
7.9%
White
52.4%

Courtesy of Ronda S. Farah MD and Gretchen Bellefeille BS, University of Minnesota, Department of Dermatology[38].
Residency Application from U.S. MD-Granting Medical Schools to ACGME-Accredited Program by Specialty and Race/Ethnicity, 2020-2021. AAMC
https://www.aamc.org/media/6191/download

Fig. 2.2 (**a**) Diversity of Undergraduate Students in 2018 [36]. (**b**) Diversity of Medical School Applicants in 2020 [37]. (**c**) Diversity Among Dermatology Applicants in 2020. (Courtesy of Ronda S. Farah MD and Gretchen Bellefeuille BS, University of Minnesota, Department of Dermatology [39])

The final filter through which aspiring dermatologists must pass is the residency selection process. Among the 36,000 candidates for residency in 2015, only 3% of dermatology applicants identified as Hispanic or Latino [38]. Blacks were also less likely to apply to dermatology at a rate of 7% [38]. New data from the AAMC has shown minute interval success in attracting diverse applicants to dermatology. In 2020, Hispanic or Latino applicants encompassed 4.8% of the applicant pool, Black applicants 7.9%, and White applicants represented 47.9% (see Fig. 2.2c) [39]. These data points demonstrate moderate success with recent dermatology diversity initiatives; however, it remains that increasing diversity is significantly limited by the diversity of the applicant pool. Residency applicants in dermatology still sway from national averages for demographics. Black or African American, Latino or Hispanic, American Indian/Alaska Natives, and even White applicants are proportionally underrepresented when compared to national demographics. Asian and mixed-race are significantly overrepresented both in the overall applicant pool and among dermatology applicants [39, 40].

Sexual and gender minorities face similar struggles. From data collected in 2010, up to 30% of sexual minorities and 60% of gender minorities concealed their identity [41]. Nearly half (~43% for both groups) did so out of fear of discrimination [41]. Discrimination during medical education may portend negative impacts on performance and ultimately residency aspirations. Further, the concealment of sexual and gender minority status, as well as the issue of response bias in anonymous surveys, makes large studies of sexual and gender minorities difficult. Official data on medical students and residents unfortunately does not routinely include sexual orientation or gender beyond binary classification.

Accounting for the multitude of filters and application processes that aspiring dermatologists must pass through, it becomes apparent that the continued lack of diversity in the dermatology workforce, despite efforts to address it with holistic interview processes and diversity-driven recruitment programs, are in large part due to disparity in education at all levels. Dermatology residencies seeking to increase diversity are without question disadvantaged right from the start, as many diverse candidates have been filtered out prior to reaching this ambitious stage. Although diversity in dermatology may ideally be addressed with a systemic focus on equality in access to education and resources at high school and undergraduate levels, we applaud the continued effort by programs to train diverse cohorts of future dermatologists for the benefit of the patients we serve.

"Urbanization" of Dermatologists

Aside from the multitude of lifestyle differences that characterize urban from rural practice (to be discussed in further detail throughout this book), there are likely several factors perpetuating the shift of dermatologists, particularly new graduates, to urban centers. For one, economic stability may prove a lucrative factor, as joining an established practice or hospital system may guarantee a steady patient census and income. Those who choose to practice rurally may be more likely to operate in solo practice and may even have to establish a practice without systems support. This includes access to surgical pathology, Mohs Microgrpahic Surgery and collaboration with other specialties such as dermatology, plastics, oculoplastics, and other surgical colleagues. The idea of this may be daunting secondary to upfront costs associated with this endeavor coupled with the possibility of low financial return in settings with fewer absolute numbers of patients and a dispersed population [42]. These considerations juxtaposed with growing medical student debt make the choice to practice in an urban setting appear both compulsory and responsible. Unsurprisingly, between 2005 and 2014 the number of solo dermatologists dropped from 44% to 35% [42].

Evaluate further the familial opportunities and obligations that exist for new dermatology graduates, many of whom are women with increased childrearing and household responsibilities [33]. Graduates have been shown to be more likely to live near their hometowns, presumably with more family support, and are less likely to settle in rural locations which may offer fewer opportunities for spousal employment and lack certain desired features, such as specifically-structured school systems and athletics programs for their children [33]. Data from the AAMC demonstrates that states on average retain 1 in 2 of their residency graduates, similar also for dermatology graduates, although men were slightly less likely to remain in the state they trained (48.1%) versus their female counterparts (53.6%) [43, 44].

Presently, rural states are less likely to have both medical schools and dermatology residency programs, creating an inherent disadvantage when it comes to the recruitment and retention of new dermatology graduates to rural locations. While primary care specialties such as family medicine frequently support rural training tracks, only one accreditation council for graduate medical education (ACGME)-accredited dermatology residency supports a rural track [45]. Therefore, our current medical system as it is designed, inadvertently supports the training and retention of urban dermatologists. Efforts toward creating rural tracks and new residency positions in rural areas may help to mitigate this dilemma, as will be discussed in greater depth throughout this textbook.

Non-physician Dermatology Support

The physician shortage is a well-described phenomenon that spans multiple specialties, including dermatology. Many physicians have come to rely on physician assistant and nurse practitioner non-physician clinicians (NPCs) to help ease the burden and reduce wait times for patients seeking dermatologic care. In one study patients waited 32.3 days (increasing each year) to see a dermatologist in major urban areas and 35.1 days in mid-sized metropolitan areas [46]. An additional study found that patients in a dermatologist-sparse area will wait almost twice as long to see a dermatologist, compared to a dermatologist-dense location [47]. Longer wait times may lead to dangerous delays in diagnosis and treatment, which intern may increase both morbidity and mortality from cutaneous disease.

Although many hoped NPCs may help to alleviate the care gap between urban and rural America, an unfortunate reality shows many NPCs practicing in similar distributions as noted above for dermatologists. A survey of Medicare billing patterns showed that an overwhelming majority (70.9%) NPs and PAs billing Medicare for dermatology services were in counties that already had a dermatologist density of at least 4 per 100,000 people [48]. The same study revealed NPCs had an urban density of 0.79 per 100,000 people and a rural density of 0.36 per 100,000 people, similar to the density of dermatologists in the same areas as noted above [48]. Counties with a greater number of independently billing NPCs compared with dermatologists more likely to be rural which may indicate some success supplementing care in specific geographically isolated areas [48]. While rural patients may choose, out of necessity or preference, to see an NPC for a dermatological complaint, this may not necessarily be without risk. For instance, when melanoma staging was evaluated based on diagnostician (dermatologist versus non-dermatologist), dermatologists were consistently found to diagnose melanoma at earlier stages and at thinner depths [49].

Finally, as the United States health care system continues to adapt to medicine in a pandemic form, the expansion of teledermatology should be mentioned. Teledermatology may be a future tool to expand not only some aspects of dermatologic supervision but also medical dermatologic services and education to underserved areas. Teledermatology is not without drawbacks such as image quality and inability to perform diagnostic procedures. However, telemedicine may have some role in closing the care gap in future. A more thorough discussion regarding the benefits and drawbacks of teledermatology in urban and rural dermatology settings is beyond the scope of this chapter and will be addressed later in this text.

Closing Remarks

The intended purpose of this chapter was to enlighten our readership and emphasize differences in rural and non-rural dermatology patient populations and to emphasize how this impacts morbidity and mortality related to cutaneous disorders. We also hoped to provide a framework regarding the demographics of practicing dermatologists to set the stage for interventions which may improve our ability to provide safe and appropriate dermatologic care for our patients. The remaining chapters in this text will build on the knowledge we have gained here in an attempt to provide the highest quality care for our rural patients who both need and deserve our fullest attention and aid.

References

1. Medina LD, Sabo S, Vespa J. Living longer: historical and projected life expectancy in the United States, 1960 to 2060, P25–1145, Current population reports, P25–1145. Washington, DC: U.S. Census Bureau; 2020.
2. Dall TM, Gallo PD, Chakrabarti R, West T, Semilla AP, Storm MV. An aging population and growing disease burden will require a large and specialized health care workforce by 2025. Health Aff (Millwood). 2013 Nov;32(11):2013–20. https://doi.org/10.1377/hlthaff.2013.0714.
3. https://www.pewresearch.org/fact-tank/2015/07/09/where-do-the-oldest-americans-live/.
4. Wessman LL, Andersen LK, Davis MDP. Incidence of diseases primarily affecting the skin by age group: population-based epidemiologic study in Olmsted County, Minnesota, and comparison with age-specific incidence rates worldwide. Int J Dermatol. 2018;57(9):1021–34. https://doi.org/10.1111/ijd.13904. Epub 2018 Jan 29. PMID: 29377079; PMCID: PMC6064677.
5. American Academy of Dermatology Position Statement on Access to Specialty Care and Direct Access to Dermatologic Care.
6. Pediatric specialist physician shortages affect access to care [Internet]. Children's Hospital Association; 2012 Aug. Available from: https://www.childrenshospitals.org/%7E/media/Files/CHA/Main/Issues_and_Advocacy/Key_Issues/Graduate_Medical_Education/Fact_Sheets/Pediatric_Specialist_Physician_Shortages_Affect_Access_to_Care08012012.pdf.
7. Official web site of the U.S. Health Resources & Services Administration. 2020. Defining Rural Population. [online] Available at: https://www.hrsa.gov/rural-health/about-us/definition/index.html. Accessed 17 Oct 2020.
8. Nielsen M, D'Agostino D, Gregory P. Addressing rural health challenges head on. Mo Med. 2017;114(5):363–6. PMID: 30228634; PMCID: PMC6140198.
9. Bishaw A, Posey K. A comparison of rural and urban America: household income and poverty [Internet]. United States Census Bureau; 2016 Dec. Available from: https://www.census.gov/newsroom/blogs/random-samplings/2016/12/a_comparison_of_rura.html.
10. Cheeseman Day J, Hays D, Smith A. A glance at the age structure and labor force participation of rural America [Internet]. United States Census Bureau; 2016 Dec. Available from: https://www.census.gov/newsroom/blogs/random-samplings/2016/12/a_glance_at_the_age.html.
11. Walton T, Willyard K. Small Area Health Insurance Estimates: 2018 [Internet]. United States Census Bureau; 2020 Apr p. P30–07. Available from: https://www.census.gov/content/dam/Census/library/publications/2020/demo/p30-07.pdf.
12. Baranowski MLH, Yeung H, Chen SC, Gillespie TW, Goodman M. Factors associated with time to surgery in melanoma: an analysis of the National Cancer Database. J Am Acad Dermatol. 2019;81(4):908–16. https://doi.org/10.1016/j.jaad.2019.05.079. Epub 2019 Jun 1. PMID: 31163238; PMCID: PMC6752196.
13. Stitzenberg KB, Thomas NE, Dalton K, et al. Distance to diagnosing provider as a measure of access for patients with melanoma. Arch Dermatol. 2007;143(8):991–8. [PubMed: 17709657].
14. Snyder BM, Mounessa JS, Fazzari M, et al. Greater distance to an academic medical center is associated with poorer melanoma prognostic factors: the University of Colorado Experience. Derm Online J. 2017;23(11). PMID: 29447632.
15. Jain R, Menzin J, Lachance K, McBee P, Phatak H, Nghiem PT. Travel burden associated with rare cancers: the example of Merkel cell carcinoma. Cancer Med. 2019;8(5):2580–2586. https://doi.org/10.1002/cam4.2085. Epub 2019 Apr 5. PMID: 30950224; PMCID: PMC6536956.
16. Singer S, Tkachenko E, Sharma P, Nelson C, Mostaghimi A, LeBoeuf NR. Geographic Disparities in Access to Scalp Cooling for the Prevention of Chemotherapy-Induced Alopecia in the United States. J Am Acad Dermatol. 2020;S0190–9622(20):31171–3. https://doi.org/10.1016/j.jaad.2020.06.073. Epub ahead of print. PMID: 32610170.
17. Schram ME, Tedja AM, Spijker R, Bos JD, Williams HC, Spuls PI. Is there a rural/urban gradient in the prevalence of eczema? A systematic review. Br J Dermatol. 2010;162:964–73.

18. Cromartie J, Vilorio D. Rural population trends [Internet]. United States Department of Agriculture; 2019 Feb. Available from: https://www.ers.usda.gov/amber-waves/2019/february/rural-population-trends/#:~:text=Rural%20America%20is%20less%20racially%20and%20ethnically%20diverse,population%20%28compared%20to%2058%20percent%20in%20urban%20areas%29.

19. Huang AH, Kwatra SG, Khanna R, Semenov YR, Okoye GA, Sweren RJ. Racial disparities in the clinical presentation and prognosis of patients with mycosis fungoides. J Natl Med Assoc. 2019;111:633–9.

20. Qian Y, Johannet P, Sawyers A, Yu J, Osman I, Zhong J. The ongoing racial disparities in melanoma: an analysis of the Surveillance, Epidemiology, and End Results database (SEER) database (1975–2016). J Am Acad Dermatol. 2020;S0190–9622(20):32494. https://doi.org/10.1016/j.jaad.2020.08.097. Epub ahead of print. PMID: 32861710.

21. Gloster HM Jr, Neal K. Skin cancer in skin of color. J Am Acad Dermatol. 2006;55(5):741–60. https://doi.org/10.1016/j.jaad.2005.08.063; quiz 761–4. PMID: 17052479.

22. Jackson C, Maibach H. Ethnic and socioeconomic disparities in dermatology. J Dermatolog Treat. 2016;27(3):290–1. https://doi.org/10.3109/09546634.2015.1101409. Epub 2015 Nov 2. PMID: 26418077.

23. Ekwueme DU, Guy GP Jr, Li C, Rim SH, Parelkar P, Chen SC. The health burden and economic costs of cutaneous melanoma mortality by race/ethnicity-United States, 2000 to 2006. J Am Acad Dermatol. 2011;65(5 Suppl 1):S133–43. https://doi.org/10.1016/j.jaad.2011.04.036. PMID: 22018062.

24. Gorbatenko-Roth K, Prose N, Kundu RV, Patterson S. Assessment of black patients' perception of their dermatology care. JAMA Dermatol. 2019;155(10):1129–34. https://doi.org/10.1001/jamadermatol.2019.2063. Epub ahead of print. PMID: 31433446; PMCID: PMC6704753.

25. Where we Call Home: LGBT People in Rural America [Internet]. Movement Advancement Project; 2019 Apr p. 5–8, 38–45. Available from: https://www.lgbtmap.org/file/lgbt-rural-report.pdf.

26. Marks DH, Arron ST, Mansh M. Skin cancer and skin cancer risk factors in sexual and gender minorities. Dermatol Clin. 2020;38(2):209–18. https://doi.org/10.1016/j.det.2019.10.005. Epub 2019 Nov 23. PMID: 32115130.

27. Mansh MD, Nguyen A, Katz KA. Improving dermatologic care for sexual and gender minority patients through routine sexual orientation and gender identity data collection. JAMA Dermatol. 2019;155(2):145–6. https://doi.org/10.1001/jamadermatol.2018.3909. PMID: 30477004.

28. Hermosura Almazan T, Kabigting FD. Dermatologic care of the transgender patient. Dermatol Online J. 2016;22(10):13030/qt01j5z8ps. PMID: 28329583.

29. Boos MD, Yeung H, Inwards-Breland D. Dermatologic care of sexual and gender minority/LGBTQIA youth, Part I: An update for the dermatologist on providing inclusive care. Pediatr Dermatol. 2019;36(5):581–6. https://doi.org/10.1111/pde.13896. Epub 2019 Jul 1. PMID: 31259437; PMCID: PMC6750998.

30. Kosche C, Mansh M, Luskus M, Nguyen A, Martinez-Diaz G, Inwards-Breland D, Yeung H, Boos MD. Dermatologic care of sexual and gender minority/LGBTQIA youth, Part 2: recognition and management of the unique dermatologic needs of SGM adolescents. Pediatr Dermatol. 2019;36(5):587–93. https://doi.org/10.1111/pde.13898. Epub 2019 Jul 1. PMID: 31259441; PMCID: PMC6750974.

31. West LA, Cole S, Goodkind D, He W. 65+ in the United States: 2010. Washington (DC): US Census Bureau; 2014.

32. Glazer AM, Farberg AS, Winkelmann RR, Rigel DS. Analysis of trends in geographic distribution and density of us dermatologists. JAMA Dermatol. 2017;153:322.

33. Feng H, Berk-Krauss J, Feng PW, Stein JA. Comparison of dermatologist density between urban and rural counties in the United States. JAMA Dermatol. 2018;154:1265.

34. Feng H, Belkin D, Geronemus RG. Geographic distribution of U.S. mohs micrographic surgery workforce. Dermatol Surg. 2019;45(1):160–3.

35. Ashrafzadeh S, Peters GA, Brandling-Bennett HA, Huang JT. The geographic distribution of the US pediatric dermatologist workforce: a national cross-sectional study. Pediatr Dermatol. 2020:pde.14369.

36. Undergraduate Enrollment. National Center for Education Statistics; 2020. https://nces.ed.gov/programs/coe/indicator_cha.asp.

37. 2020 Facts: Applicants and Matriculants Data. AAMC https://www.aamc.org/data-reports/students-residents/interactive-data/2020-facts-applicants-and-matriculants-data.

38. Van Voorhees AS, Enos CW. Diversity in dermatology residency programs. J Investig Dermatol Symp Proc. 2017;18(2):S46–9. https://doi.org/10.1016/j.jisp.2017.07.001.

39. Residency Applicants from U.S. MD-Granting Medical Schools to ACGME-Accredited Programs by Specialty and Race/Ethnicity, 2020–2021. AAMC. https://www.aamc.org/media/6191/download.

40. Quick Facts United States. United States Census Bureau. https://www.census.gov/quickfacts/fact/table/US/PST045219.

41. Mansh M, White W, Gee-Tong L, et al. Sexual and gender minority identity disclosure during undergraduate medical education: "In the Closet" in Medical School: MEDSCAPE 2/18/2015. Mo Med. 2015;112(4):266.

42. Ehrlich A, Kostecki J, Olkaba H. Trends in dermatology practices and the implications for the workforce. J Am Acad Dermatol. 2017;77:746–52.

43. Physician Retention in State of Residency Training, by State [Internet]. AAMC; 2019. Available from: https://www.aamc.org/data-reports/students-residents/interactive-data/report-residents/2019/table-c6-physician-retention-state-residency-training-state.

44. Physician Retention in State of Residency Training, by Last Completed GME Specialty and Sex. 2019.
45. Department of Dermatology Highlights [Internet]. Univ. Miss. Med. Cent. Dep. Dermatol. Available from: https://www.umc.edu/som/Departments%20 and%20Offices/SOM%20Departments/Dermatology/ About%20Us/Department-Highlights.html.
46. 2017 Survey of Physician Appointment Wait Times [Internet]. Merrit Hawkins; 2017. Available from: https://www.merritthawkins.com/ news-and-insights/thought-leadership/survey/ survey-of-physician-appointment-wait-times/.
47. Xiang L, Lipner SR. Analysis of wait times for online dermatology appointments in most and least dermatologist-dense cities. J Drugs Dermatol JDD. 2020;19:562–5.
48. Adamson AS, Suarez EA, McDaniel P, Leiphart PA, Zeitany A, Kirby JS. Geographic distribution of nonphysician clinicians who independently billed medicare for common dermatologic services in 2014. JAMA Dermatol. 2018;154:30.
49. Zurfley F, Mostow EN. Association between the use of a physician extender and dermatology appointment wait times in Ohio. JAMA Dermatol. 2017;153:1323.

Making a Difference: Assessment of the Economic Viability and Impact of Rural Practice

3

Monica Kala and William Taylor Sisson

Introduction

Although not a universal truth, a population's level of formal education, access to natural resources and associated industry impact the microeconomics of a town or region and a physician's choice of a practice location. Careful research to fully understand the local or regional economics are crucial during the evaluative stages of any business venture. The concept of social norms (and the associated expectations) must also be considered, since the decision to practice in a rural area is a decision made for personal reasons by each physician/dermatologist. An overview of the factors that influence the economy of rural regions and communities is provided, herein. Specifically, this chapter presents a review of the economic impact of rural healthcare on a local population (Part 1). This explains why government programs have been developed to support rural primary healthcare. It then reviews factors effecting the supply of dermatologists in rural areas and the demand for their services (Part 2). The negative economic effects on a physician practice related to payer mix and factors that make it difficult for rural citizens to visit the dermatologist are considered (Part 3). Opportunities to overcome economic barriers are then discussed including dermatologists' partnering with Rural Health Clinics, and benefitting from low labor costs; keeping out-of-pocket patient expenses low to gain patient loyalty, and running efficient effective clinical practices (Part 4). Finally, a snapshot of a rural academic dermatology practice demonstrates that there is hope for establishing economically successful dermatology practices in rural areas (Part 5).

"No margin, no mission"
Sister Irene Kraus,
Daughters of Charity

M. Kala
School of Medicine, University of Mississippi Medical Center, Jackson, MS, USA
e-mail: mkala@umc.edu

W. T. Sisson (✉)
Department Business Administrator, Department of Dermatology, University of Mississippi Medical Center, Jackson, MS, USA
e-mail: wsisson@umc.edu

Part 1: Public Health and the Economic Impact of a Rural Practice

Local Impact of Rural Clinics

The local, state, and federal government and many private groups are motivated to improve rural healthcare because it is an important economic engine that impacts the quality of life and health of rural citizens. Several metrics demonstrate these benefits.

© The Author(s), under exclusive license to Springer Nature Switzerland AG 2021
R. T. Brodell et al. (eds.), *Dermatology in Rural Settings*, Sustainable Development Goals Series,
https://doi.org/10.1007/978-3-030-75984-1_3

On Jobs and Economic Vitality

Opening a health clinic in a rural part of the country impacts the local community outside of the realm of healthcare. New employment opportunities are created for medical doctors, nurses, pharmacists, technicians, receptionists, and many others. On average, a new physician practice near a hospital creates 24 local jobs for citizens of the community. This generates over 1.3 million dollars in income for the community including employee salaries and benefits [1]. A dermatology practice has the potential to expand and include nurse practitioners or physician assistants increasing the potential for additional jobs to be created. Additionally, as many on the practice team are likely to live in the area, their salaries stimulate the local economy allowing for an even wider impact on the community.

On Overall Community Health

When medical care is limited in a rural community, local patients are at a healthcare disadvantage for numerous reasons. There may be interruptions in receiving timely care. Additionally, patients are more likely to put off visiting their physician for fear of having to miss a day of work to travel. Others may lack an adequate mode of transportation to travel to a clinic outside of their county. The aforementioned access issues can cause illnesses to be diagnosed well after patients are symptomatic and sometimes even after they are treatable. When considering skin cancer as an example, melanoma is almost 100 percent survivable when the diagnosis is made at an early stage [2]. However, over 6,000 Americans pass away from melanoma every year [3] with the average American spending over $11,000 in the last year of life [4]. By the time melanomas are symptomatic, it can be very late in the cancer's progression [2]. Having a physician in a rural community, who has been trained to recognize benign versus malignant skin lesions reduces morbidity and mortality.

Having a primary care clinic locally is, often times, not enough. When a primary care physi-cian desires to refer a patient to a specialist, additional barriers to adequate patient care are quickly identified. An office visit in a distant urban center requires missing work and additional travel for most patients who live in rural communities. It is a widely held belief that patient outcomes are best when their primary care physician and specialists have quality communication about a patient's goals and needs [5]. However, care coordination is difficult if a primary care physician has little to no connections or affiliations with specialists [6]. Oftentimes, a lack of follow-up communication between primary care physicians and specialists creates yet another catalyst for a patient's disease to go unchecked. Additionally, the time required to schedule an appointment with a specialist may delay diagnosis and treatment [6, 7]. This delay may also increase the cost of care; a burden which falls on the patient and their community.

Part 2: Supply and Demand in the Delivery of Rural Dermatological Service

The problem of healthcare delivery in rural America is one of supply and demand. The high demand for general medical services impacts dermatologic care since primary care physicians provide the majority of skin care in rural America [8]. The meager supply or rural primary care physicians and the pressures to care for general medical problems, leaves precious little time for primary care physicians to focus on dermatology problems. The supply of dermatologists is in an even more quixotic state.

High Demand for Care in Rural America

High Demand for General Medical Services

Rural Americans are more likely to die from cancer, heart disease, and lower respiratory disease than their urban counterparts. Many of these deaths are preventable and there are numerous factors that play into these statistics. Rural health disparities are striking and significant, leading to almost 75,000 preventable deaths annually in rural communities [9]. There is certainly a substantial demand for medical services in rural areas. Table 3.1, as included as supplemental readings to this chapter, helps to further illustrate this rural health disparity in greater detail.

High Demand for Dermatologic Services in Rural America

Skin cancer is the most common cancer diagnosis in America and the incidence of skin cancer is higher in rural areas. Non-melanoma skin cancers (NMSC) are the most common diagnoses of malignancy and many experts claim that the reported numbers are an underestimate of actual cases throughout the country [10]. These NMSC routinely affect patients older than 65 years of age. Skin cancer concerns consisted of approximately 1.5 percent of skin related visits to a healthcare professional in 2013 [11]. However, the costs associated with treating skin cancers that year exceeded $6 billion. Melanoma cancer costs are also expected to rise due to increasing incidence rates, an aging population, and the development of new targeted melanoma therapies [12]. Rural citizens experience a higher incidence of many diseases including melanoma, likely due to lower socioeconomic status, decreased educational level causing them to dismiss the hazards of ultraviolet (UV) light, and failure to use sunscreen [13]. They may also have more UV exposure at work and at play. Additionally, rural citizens typically experience

Table 3.1 Health risk factors in rural versus urban communities

Risk factor	Rural prevalence	Urban prevalence
Hypertension	40%	29.4%
Obesity	39.6%	33.4% [51]
Tobacco usage	29% (men) and 25% (women)	19% (men) and 13% (women) [52]
Physical activity	19.6%	25.3% [53]

worse outcomes associated with Merkel cell carcinoma in comparison to their urban counterparts [14]. There is a pent-up demand for dermatology in rural areas. Though not well studied, it can be assumed that access to care issues may lead to other dermatologic conditions being seen at later, more severe stages in rural areas when those conditions are more difficult to treat.

Decreasing Supply of Rural Dermatologists

Too Few Rural Dermatologists

While the demand for services is increasing, there are still far too few dermatologists to meet the needs of rural communities across America. In 2013, there were 0.085 dermatologists per 100,000 people in rural counties and 4.11 dermatologists per 100,000 people in metropolitan areas; a more than 40-fold difference [14]. Having a dermatologist density of 1–2 physicians per 100,000 citizens reduces melanoma mortality by 53 percent in comparison to areas with no dermatologists. Some experts argue that increasing the number of primary care physicians is more valuable than recruiting specialists to a rural community. The imbalance is so great, however, that efforts to increase the number of dermatologists in rural areas is equally critical. This disparity has been proven to directly correlate with improved outcomes for patients with skin disease [15].

In the 1980s, there was a concern related to the oversupply of dermatologists in the country. Therefore, the number of positions was allowed to stagnate [16, 17]. In the 2000s, there were small increases in the number of dermatology residency programs at a rate of 0.9 percent per year. While this was an appropriate rise in the number of positions based on the current population, it did not account for the years when too few dermatologists were trained [17]. Additionally, the need for dermatology services has been increasing with the epidemic of skin cancers and the larger scope of dermatological practice including cosmetic dermatology; making the physician shortage even more significant.

Due to the national shortage of dermatologists, young dermatologists completing their residency have very little difficulty finding jobs. They are free to pick positions in the location of their choosing [17]. Many choose to stay in urban environments where they trained or based on personal reasons and in hopes of career progression. Shifts away from solo practice towards less demanding group practices in urban areas have also resulted in fewer dermatologists moving to rural communities. Almost half of all rural dermatological practices were owned by a single provider in 2014 [17, 18]. A physician's age also plays an interesting role in choosing the location of clinical practice. From 1995 to 2013, the number of dermatologists younger than 55 increased by 21.3 percent in urban areas but only 6.5 percent in rural areas. However, the number of dermatologists older than 55 years increased by 153 percent in rural areas in comparison to the 112.4 percent increase in urban areas [14]. As a physician ages, it is possible that the draw of a slower pace of life is more attractive. However, it is unlikely that the number of dermatologists who move from urban to rural areas without new, systematic initiatives will be enough to sustain the needs of these communities. It is likely that this maldistribution of dermatologists based on geographical location will increase in the coming years [17].

Flight of Dermatologists to Urban Areas

Rural physicians have always been in high demand. Rural dermatologists, however, are an increasingly rare breed. The maldistribution of dermatologists throughout rural and urban areas over the last few decades has steadily been worsening. Although the number of dermatologists is increasing in rural areas, the difference in dermatologist density in metropolitan versus nonmetropolitan areas has increased from a difference of 2.63 dermatologists per 100,000 people in 1995 to 3.06 dermatologists per 100,000 people in 2013, and in rural communities that difference has increased from 3.41 dermatologists per 100,000 people to 4.03 dermatologists per

100,000 people [14]. On the other hand, dermatologists in rural areas may do less cosmetic work and they are less likely to perform specialized procedures such as Mohs surgery providing more time for general dermatology services [19].

In summary, both academic and privately-held dermatology practices throughout the country are in an enviable position because there is a high demand for the services they provide. This is especially true in rural areas.

Part 3: The Economic Considerations that Negatively Impact Rural Dermatology Practice

There are a number of economic factors that dissuade physicians, including dermatologists from practicing in rural areas. Most notably, payer mix is negatively impacted because of a rural populations' low education level, high unemployment, and the high cost of private health insurance for the unemployed. In addition, distance and lack of transportation can preclude many rural patients from visiting their physicians. Finally, skilled office staff may be in shorter supply in rural areas.

Payer Mix

Education as a Factor of Payor Mix

Payor mix is the percentage of insurances held by the patients of a particular healthcare entity [20]. This correlates closely with the education gap in rural America when compared to urban centers. In fact, non-urban Americans in their prime working-age are far less likely to have a four-year degree (or higher) when compared with their urban counterparts. A recent study issued by the United States Federal Reserve indicated that 43 percent of prime working-age Americans living in urban areas had obtained a four-year college degree or higher. However, that percentage dropped to 25 percent in prime working-age Americans living in rural areas [21]. To put that into a monetary context, a 2011 study commissioned by Georgetown University's Center on Education and the Workforce found that, "a bachelor's degree is worth $2.8 million on average over a lifetime". Furthermore, "bachelor degree holders earn 31 percent more than those with an Associate's Degree and 84 percent more than those with just a high school diploma" [22].

As previously described, education level is directly related to socioeconomic status and potential medical practice earnings. Patients with higher socioeconomic status are more likely to have private health insurance which has a positive impact on the practice payor mix. Conversely, individuals within a lower socioeconomic status are much more likely to rely on lower-reimbursing, governmental payors or are forced to self-insure.

Employment as a Factor of Payor Mix

In the United States, the majority of individuals with private insurance obtain their insurance through their employer. In fact, the Kaiser Family Foundation reported that in 2018, 49 percent of all Americans were covered under an employer-sponsored health insurance plan [23]. Conversely, only 6 percent of Americans were covered under independent (non-employer-sponsored) private healthcare insurance [23]. From this, one may extrapolate that community or even regional payor mix is largely driven by the economic success of that particular locale. Additionally, large employers are more likely to provide health insurance benefits than their smaller counterparts. This "fringe benefit" offered by many larger employers is based on several factors including: overall financial margins/economies of scale, the willingness and ability to invest in human capital, as well as current federal mandates.

Signed into law in March 2010, the Patient Protection and Affordable Care Act (more commonly known as the Affordable Care Act, ACA, or colloquially as "ObamaCare,") mandates employers with more than 50 full-time employees to offer health insurance benefits [24]. In fact, "employers must offer health insurance that is affordable and provides minimum value to 95% of their full-time employees and their children up to the end of the month in which they turn age 26, or be subject to penalties" [24]. Though the politically divisiveness of the Patient Protection and Affordable Care Act leaves its future in doubt, it is currently having an impact on urban insurance coverage [24]. According to the Kaiser Family Foundation in 2019, 97.1 percent of large private sector establishments offered health insurance to their employees in the United States [25]. Conversely, only 30.8 percent of small, private sector establishments offered health insurance to their employees [25]. Congruent with the Affordable Care Act, the Kaiser Family Foundation defined large private sector establishments as those with more than 50 full-time employees while small private sector establishments were those organizations with less than 50 full-time employees [25]. The 66.3 percent difference between large and small employers is significant since the majority of large employers are associated with more urbanized areas. In summary, fewer rural patients have private insurance through their employers.

Affordability as a Factor of Payor Mix

It will come as little surprise to anyone reading this chapter that private health insurance is expensive. As health insurance premiums have risen over the last several decades, more and more Americans are searching for options. In fact since 1999, individual health insurance premiums have risen from slightly over $2,000 annually to nearly $7,000 in 2018 [26]. Over that same 20-year timeframe, family coverage has gone from just shy of $6,000 annually to nearly $20,000 annually [26]. To put this into perspective, the United States Census Bureau reported that the median household income in 2018 was $64,324 meaning that, without some type of subsidy, a family might spend up to one in three after tax dollars on their health insurance premiums [27]. Many Americans cannot afford private health insurance without employer or governmental subsidy.

Age as a Factor of Payor Mix

In 1965, then United States President Lyndon B. Johnson signed into law Titles XVIII and XIX of the Social Security Act creating the first government-issued and funded health insurance in the history of the United States. The initial legislation was quite simple when compared to today's Medicare and Medicaid systems. From 1965 through today, coverages and entitlements have expanded substantially. In fact, the 1965 budget for the initial Medicare entitlements were estimated to be around $10 billion [28]. Today, those figures far surpass $1 trillion dollars and continue to grow as the population ages, the complexity and cost of healthcare grows, and the entitlements and number of enrollees increase over time (the aforementioned figures have not been adjusted for inflation) [29].

In 1967, shortly after Medicare was signed into law, the median age of an American was 27.7 years of age [30]. According to the United State Census Bureau, the median age of an American in 2018 was 38.2 year of age [31]. That is an approximate 38 percent increase over the last 50 years. Many factors are contributing to this shift including enhanced access to nutrient rich foods, the technical automation of many physically demanding jobs, access to safe and temperature-controlled dwellings and, probably most critically, the development of better healthcare treatments – most notably with chronic healthcare conditions. As the population ages, an ever-increasing percentage of Americans will rely on government payors as their main source of access to healthcare. Based on the overall aging population and the elderly population shift associated with our non-urban regions, the propensity of governmental-insured patients in rural America is proportionately higher than in urban settings.

Payor Mix and "Allowable" Disparity

Not all payors are created equal. Anyone working within the healthcare industry understands the importance of this statement. The "allowable" rate is defined as, "the maximum amount a plan will pay for a covered health care service." [32] The allowable rates are independent of the actual charges which are set by the physician or healthcare organization. Typically, commercial (or private) insurance plans have higher allowable rates. By comparison, governmental payors (Medicare and Medicaid) have lower allowable rates. Table 3.2 illustrates the average allowable (or reimbursement) rates for several frequently used dermatologically-related Common Procedural Terminology (CPT) codes at the University of Mississippi Medical Center (see Table 3.2) Claims that were denied were excluded from the Table 3.2 dataset.

Factors Limiting Patient Access to Care

Patient Access in Rural America

Patient access is a term used within the healthcare industry to define the ease in which a patient can receive appropriate care. Patient access can

Table 3.2 Variability in reimbursement rates by payer

CPT	AETNA	BCBS	HUMANA	MEDICAID MS	MEDICARE	UNITED HEALTHCARE	VETERANS AFFAIRS	Grand Total
17000	$82	$75	$61	$41	$43	$90	$41	$54
17003	$38	$54	$26	$21	$23	$57	$25	$32
17110	$129	$138	$88	$86	$90	$155	$89	$113
88305	$74	$53	$59	$48	$52	$102	$45	$53
99213	$72	$69	$54	$58	$62	$64	$64	$66
Grand Total	$78	$74	$55	$59	$53	$81	$49	$63

Based on the reimbursement rates for the University of Mississippi Medical Center's Department of Dermatology from January 1, 2020 through June 30, 2020 include many variables but is most typically analyzed and tracked by using time as the studied variable. In many outpatient settings, clinic management will track lag time or the industry standard, third next available appointment, to better understand the ease with which a patient can access a clinic's services. Clinic management can review longitudinal trends, assess access to specific service lines, or even individual care providers in an effort to better understand opportunities for improvement. Many of these metrics are benchmarked nationally through various organizations. These benchmarks are yet another tool used to assess the current status as it relates to patient access [33].

Problems Associated with Obtaining Dependable, Skilled Labor

In many rural areas throughout the United States, access to a skilled and motivated labor force can be more difficult to obtain than in more densely populated areas. As discussed throughout this chapter, many geographic and societal factors can attribute to this shortfall. However, the ramifications of this skilled labor force shortage can significantly impact the ease in which required jobs can be performed [21]. For instance, the planning, design, and construction phase of health care facilities (clinics included) can be arduous, time consuming, and costly given numerous regulatory and logistical factors. Those individuals and teams most familiar with the various planning, design, and construction phases of building a new clinical enterprise are most often refined to a more populated, urban setting.

Part 4: Economic Incentives of a Rural Practice

Rural Health Clinics Provide Increased Medicaid Payments

As previously discussed in this chapter, patients with government-sponsored health insurance represent a greater percentage of the population in rural regions and communities. Government payors are often seen as a funding source that healthcare facilities or private practitioners may want to avoid. However, there are also benefits to governmental payors – especially in Rural America.

The Center for Medicare and Medicaid services (CMS) define Rural Health Clinics (RHCs) as, "a clinic that is located in a rural area designed as a shortage area, is not a rehabilitation agency or a facility primarily of the care and treatment of mental diseases, and meets all other requirements…" [34]. The Rural Health Clinic (RHC) is a federally-regulated designation and, although there are certainly barriers to entry, many rural practitioners and facilities have found significant success leveraging the benefits of this program.

The RHC program, originally passed by the United States Congress in 1977, was intended to: aid the aging rural population, attract younger practitioners to rural America, reduce the economic burden of providing care to rural Americans, improve patient access to care, and to promote the services of Advanced Practice Providers (nurse practitioners, physician assistants, and certified nurse midwives) in rural settings [35]. In many ways, the spirit of the 1977 legislation remains the same today.

There are several unique financial advantages surrounding the RHC program. Most notably, officially designated RHCs are typically reimbursed by federal payors at a higher level than their non-rural counterparts. This is based on a unique methodology: cost-based reimbursement [36]. Other federal healthcare entitlements, end-stage renal disease, for instance, use a similar methodology to reimburse healthcare entities for those particular services.

While typical payor reimbursement for clinic services are typically "fee-for-service" based and include an "allowable rate", RHCs are able to reconcile total services rendered, the associated overhead of those services, and periodically submit those data points to the Centers for Medicare and Medicaid Services for reimbursement [36]. The precise reimbursement rate for RHC services is dependent upon numerous variables and is driven by specific accounting practices [36]. It would be advisable for an organization or individual interested in establishing a medical practice in Rural America to investigate the effect of RHC reimbursement methods as they may impact a particular practice.

Although the financial aspects surrounding the RHC are typically advantageous to the practice, the initial enrollment process, internal accounting requirements, site surveys, and continuous scrutiny by CMS are all factors to consider when investigating the RHC designation. To better understand the RHC inclusion/exclusion criteria as set by CMS, please reference Appendix A included as a supplemental reference. Dermatologists can receive RHC benefits so long as an advanced practice provider is onsite to see patients at least 50 percent of the time the clinic is open and the main focus (at least 51 percent) of the clinic is primary care [37].

Low Cost of Labor

Although the cost of living is generally lower in rural America, many economic factors can impact employee compensation and benefits. As is true within any labor market, the economic laws of supply and demand can be even further exacerbated in areas where access to a skilled and motivated labor force is a scarce resource. Generally, however, labor is less expensive in rural areas [38].

Having access to a highly skilled and motivated workforce is essential for any clinical enterprise. The healthcare industry is largely stratified into two types of staff members – clinical and non-clinical. Clinical staff members traditionally include: nurses, technicians, medical assistants (and even advanced practice providers in some instances). Non-clinical, or clerical staff members, traditionally include: schedulers, greeters/registration staff, billers/coders, and management [39]. While it may be harder to recruit in rural areas, it may be easier to retain staff since there are fewer opportunities for jobs with similar skills in the rural market.

Gaining Patient Loyalty by Keeping Out-of-Pocket Cost to Patients Low

An estimated 31 million Americans had no health insurance in 2018. About 12.3 percent of people who live in rural counties and 11.3 percent of people in "mostly rural" counties were found to be uninsured [40]. Providing care to uninsured patients requires careful attention to the value of their services, ensuring the best care possible, while also minimizing the "out of pocket" expenses for procedures and eliminating unnecessary tests [41].

Of course, dermatologists are experts at providing high value care. This is accomplished by routinely performing surgical procedures in the office rather than at expensive outpatient surgery centers or hospitals and making an accurate diagnosis to allow for targeted treatment during the initial visit. Dermatologists can also direct and encourage patients to access primary care at local free clinics or with private physicians who will establish payment plans as needed. Additionally, rural dermatologists have learned to identify lower-priced prescription lists at specific pharmacies and even lower-priced over-the-counter options that can be used judiciously. Physicians can also utilize various virtual platforms when possible to keep patients from traveling to the clinic for unnecessary visits. Evidence-based testing practices save time for both the clinician and the patient [7]. Finally, avoiding unnecessary tests helps to further minimize healthcare costs.

Providing quality care at a low cost for patients builds trust within a community. This trust goes a long way, especially in Rural America. The physician-patient bond keeps patients within the practice, helping to ensure long-term care and ultimately resulting in better community health outcomes.

Run Efficient, Effective Practice to Offset Payer Mix with Volume

One of the major keys to any successful dermatology practice is the ability to run an efficient operation. Due to the higher prevalence of governmental payers in Rural America, the concept of operational efficiency assumes an even higher-level of importance in the rural dermatology practice.

An efficient and effective practice is only as good as the care teams that support them. Gone are the days of the "mom and pop" dermatology practice – where a small (often family) staff was able to meet the operational needs of the clinic. Payer complexity, the advent of the electronic health record, and various regulatory agencies have thrust all medical practices (dermatology included) into a much more sophisticated age of medicine.

Support staff, working collectively in teams, is often one of the most useful tools to establish and sustain an efficient practice long-term. Over time, and as practice habits and idiosyncrasies emerge through observation and repetition, support staff teams can become highly efficient when working with their dermatologist. Additionally, ensuring an appropriate overall number of clinic staff is also critically important to the efficiency of a rural dermatology practice. In a 2015 study conducted by the University of Mississippi Medical Center's Department of Dermatology, it was discovered that by adding just one additional clinical support staff member (Licensed Practical Nurse), a dermatologist was able to increase gross payments to the practice by 33 percent over an 8 month period [42]. Although exact results may vary across organizations or individual dermatologists, the concept of appropriate staffing as being a significant economic driver cannot be overstated.

To run an efficient practice, one must also consider the physical space and layout of a clinic. Although no two dermatologists practice in the exact same manner, a great deal of time and effort needs to be invested during the planning and design stages of clinic development. Too often, care providers design a clinic in haste and over-look key logistical components or worse yet – design a clinic without enough room to grow.

Finally, the advent of virtual and online learning has helped to offset most proximity-related training issues for both clinical and clerical staff. Although most rural communities have access to training opportunities for potential support staff, the overall labor pool in rural locals is much smaller; which can significantly impact the ability to recruit and retain a quality team.

Snapshot of a Rural Academic Dermatology Practice

Demographics

Located in East-Central Mississippi, Louisville is the county seat for Winston County Mississippi. In, 2019 the population of Louisville was estimated by the United States Census Bureau to be 5,983 with a Median household income of $32,131 (including 34.2% of the population considered to be living in poverty). To put that into perspective, the median household income in Mississippi is $43,567 and the median income in the United States is $64,324 [43, 44].

60.4 percent of Louisville inhabitants are African-American and 35.4 percent are Caucasian. The remaining 4.2 percent of the population consists of individuals that identify as Hispanic (3.1 percent), American Indian (0.7 percent) Asian (0.1 percent), and two or more races/other (0.3 percent). 17.5 percent of the population of Louisville is over the age of 65 and 27.9 percent are under the age of 18 [44]

Market

Given the overall lack of rural healthcare providers in Mississippi, and most notably specialists, the University of Mississippi Medical Center's Dermatology Clinic in Louisville includes a catchment area that extends well past Louisville to include numerous other counties and communities in the East-Central Mississippi area and beyond. In fact, the closest dermatology practice to the Louisville Dermatology Clinic is in Columbus, Mississippi which is approximately 60 miles from Louisville. There are eight counties (including Winston County where Louisville is located) that comprise the main catchment area for the Louisville Dermatology Clinic. As of 2018, the United States Census Bureau indicated that these eight counties included a population of nearly 167,000 people. To put this into perspective, a recent Association of American Medical Colleges (AAMC) Physician Workforce dataset indicated that the national average dermatologist to person ratio is approximately 1:30,000 [45]. As one might imagine, and based on the lag days data presented in this chapter, demand for services in Louisville are fierce.

Supply and Demand

UMMC's Louisville Dermatology Clinic has experienced a high demand for services. The data from UMMC (Table 3.3) demonstrates the demand (as illustrated by new patient access – lag days) for dermatologic services far outweighs

Table 3.3 Table 3.3 illustrates the lag time for all new patients seeking care by the various academic departments at the University of Mississippi Medical Center. Lag time, in this instance, is defined as the number of days from the date the new patient appointment was created in the electronic health record to the date of the scheduled appointment (date of service). Data accessed 10/15/2020

Academic dpartment	Average lag days by department
DERMATOLOGY	32
NEUROLOGY	31
OPHTHALMOLOGY	27
PREVENTIVE MEDICINE	26
OB-GYN	26
NEUROSURGERY	24
PEDIATRICS	21
MEDICINE	20
ORTHOPEDICS	19
SURGERY	18
PSYCHIATRY	16
ANESTHESIOLOGY	16
RADIOLOGY	13
DENTISTRY	12
OTOLARYNGOLOGY	12
GH-OB/GYN	12
GH-MEDICINE	10
GH-SURGERY	9
FAMILY MEDICINE	7
GH-PEDIATRICS	6
RADIATION ONCOLOGY	5
GH-RADIOLOGY	4
GRENADA	3
HOLMES COUNTY	1
SCHOOL OF NURSING	0
EMERGENCY MEDICINE	0

supply of dermatologic care providers. By comparison, if the Louisville Rural Dermatology Clinic lag day figure was transposed into Table 3.4 and sorted accordingly, it would remain in the top 20 percent of demanded services when compared with the other clinical services at UMMC.

Payer Mix

The University of Mississippi Medical Center's (UMMC) Department of Dermatology, uses these metrics to assess the statistical disparity between the payor mix at urban clinic locations when compared with their rural clinic location. Please note that all rural areas throughout the country are not the same and that a variety of factors can contribute to payor mix (see Figs. 3.1 and 3.2).

Building an Office

Selecting a qualified general contractor to construct a dermatology clinic in rural Mississippi was more difficult than expected when compared to similar construction projects in the capital of

Table 3.4 LOUISVILLE RURAL CLINIC is the first and only full-time rural dermatology clinic associated with UMMC

Epic department	Average lag days by epic department
UMMC BELZONI DERM	61
TUP DERMATOLOGY	48
JMM DERMATOLOGY	47
GF DERMATOLOGY	41
FSC DERMATOLOGY	38
PAV DERM	35
THE SKIN CANCER CENTER	32
LOUISVILLE RURAL CLINIC	20
UMMC LOUISVILLE DERM	17
CI DERMATOLOGY ONCOLOGY	15
STARKVILLE DERM	9
JMM PSORIASIS DERM	8
UN SOUTH DELTA HIGH	2
TH DERMATOLOGY	1
CH ELI DERMATOLOGY	0

Fig. 3.1
Figure 3.1 illustrates the cumulative payor percentages from October 2019 – September 2020 for the entire Department of Dermatology at the University of Mississippi Medical Center

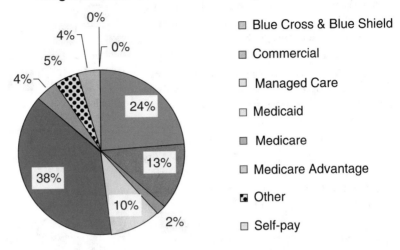

Original Insurance 12 Month Rolling

- Blue Cross & Blue Shield
- Commercial
- Managed Care
- Medicaid
- Medicare
- Medicare Advantage
- Other
- Self-pay

Fig. 3.2
Figure 3.2 illustrates the payor percentages from October 2019 – September 2020 for the Department of Dermatology's main rural practice located in Louisville, Mississippi

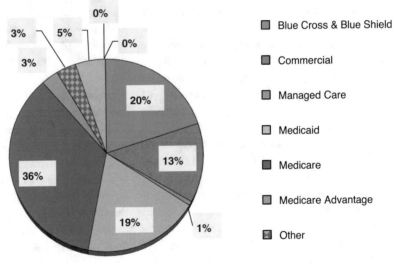

Original Insurance 12 Month Rolling

- Blue Cross & Blue Shield
- Commercial
- Managed Care
- Medicaid
- Medicare
- Medicare Advantage
- Other

Jackson. The process produced bids that were estimated to cost 30 percent more (on average) than if the construction would have taken place in a more urban Mississippi setting [46]. The cost to mobilize labor teams as well as the additional fuel, housing, and logistical costs associated with a rural construction project force general contractors to pass those increased expenses along to the end-user [46].

Financial Metrics

"Common medical and surgical dermatology procedures are typically well reimbursed" [47]. In fact, numerous articles indicate the recent desire for private equity firms to purchase dermatology practices of various sizes and complexities. A recent article published by the Journal of the American Medical Association (JAMA) as well as an article in Business Insider help to support and explain the recent trend in private equity firms seeking dermatology practices. Specifically, the JAMA article states that, "several factors draw external investors to dermatology, including high patient demand, an ongoing skin cancer epidemic, an aging population, the shortage of dermatologists, expanded insurance coverage, and perceived opportunities to expand cosmetic services, add non physician clinicians, and profit from internal redirection of ancillary laboratory services or procedural referrals" [48]. In an effort to provide a more robust picture of our current dermatology operations in Louisville, basic operational and financial metrics have been developed and included as a part of this chapter.

In 2019, the clinic in Louisville completed 4,646 office visits. The operations from this clinic generated 8,352 work relative-value units (wRVUs) which, given the practitioner's allocated effort, put him at around the 81st percentile from a productivity standpoint (2020 AAMC dataset). Table 3.5 of this chapter is a basic profit and loss statement for the University of Mississippi Medical Center's Louisville Dermatology Clinic.

Based on the metrics illustrated as part of the Louisville Dermatology Clinic, it is easy to see that being highly productive in Rural America is certainly possible given the right situation. Although payor mix, access to skilled labor, and some of the up-front costs associated with establishing a clinical practice can often be exacerbated in Rural America, the demand for services and ability to be productive and profitable are absolutely obtainable.

Part 5: Recruitment Considerations and the Advantages of a Rural Life

The most effective draw for dermatologists to move to rural communities is the individual dermatologist's connection to the area [17]. The desire for a dermatologist to return to their hometown or to an area where they once attended school is meaningful. Some programs and communities are using this emotional connection,

Table 3.5 Illustrates a basic profit/loss (P&L) statement for the University of Mississippi Medical Center's Louisville Dermatology Clinic. The P&L statement above was assembled utilizing estimations and approximations and is not intended to depict precise, GAAP required, due diligence or rigor and is to be used for educational purposes only

Louisville Dermatology Clinic	FY20 Estimates
Revenue	
Net Patient Revenue	$ 765,875
Expenses	
Salaries/Benefits	$ 662,377
Lease	$ 66,328
Other Contractual Services	$ 1,076
Commodities	$ 47,930
Total Expenses	$ 777,711
Surplus (Shortage)	$ (11,836)
Downstream Revenue Implications	
Mohs Surgery - PB NPR	$ 147,960
Dermatopathology - PB NPR	$ 52,900
Dermatopathology - HB NPR	$ 58,650
Total Downstream NPR	$ 259,510
Direct overhead	$ (155,706)
Surplus (Shortage)	$ 103,804
Total Organizational Impact	$ 91,968

along with financial incentives, to encourage dermatologists to return to their roots [17].

Living in rural America brings unexpected joys that many physicians appreciate. The overall cost of living is much lower in rural areas. When adjusted for cost of living, primary care physicians in rural America make 30 percent more than their urban counterparts [49]. Additionally, as there are very few specialists in rural communities, there is little competition for patients, and dermatologists rarely experience a shortfall in demand. In fact, wait times for patients to have an appointment can sometimes be upwards of multiple weeks. There is always a need for dermatologists in rural communities. With this demand for skin care comes the freedom to set working hours with flexibility to suit both the dermatologist and their patients.

In addition to the benefits that arise from having more control over their hours and personal lives, there is the added benefit of cheaper real estate in rural communities. In addition, the operating costs of a clinic are often lower than they would be in an urban environment. The dire need for dermatologists in rural communities also eliminates the need for advertising. Patients are likely to communicate with one another via word of mouth to family and friends in surrounding communities, who are also likely in need of specialist care.

Rural dermatologists are also able to enjoy a more relaxed lifestyle. The commute times between home and office is often measured in a few minutes. The low costs of living allow for dermatologists to more freely customize their homes as desired. It is believed that rural physician incomes allow for about 13 percent more purchasing power than urban physician incomes [49]. Rural physicians have more time for hobbies as well, especially if those include an interest in the outdoors [50]. However, while physicians can have a more relaxed lifestyle, they can also maintain a rigorous, challenging healthcare practice. Rural patients tend to present with more severe and complex medical problems than their urban counterparts; creating a mentally stimulating practice. It is also likely that patients with a wide variety of conditions will visit the facility allowing for more diversity in the day-to-day practice.

Finally, a rural physician can build a special relationship with their patients that is unmatched. The rural community has a vested interest in keeping the physician in the community and dermatologists are likely to meet a large portion of the local population. There is a great sense of job satisfaction from helping those in need and rural practitioners are likely to see citizens from surrounding communities as well. Additionally, if interested, they can diagnose other health issues and provide proper guidance to their patients, treating beyond the skin. There is no denying that a rural dermatologist is often a revered pillar of the community.

Summary: The Challenges of Recruiting and Retaining Physicians in Rural America

Living in a rural community is not for everyone, and there are some disadvantages to life away from the city. Recruiting talented staff to help run a busy rural dermatology clinic can be challenging. The same rural flight to the cities that entices dermatologists is also attractive to nurses and technicians. Therefore, physicians who hope to run their own practice must provide adequate benefits to their support staff. However, if they have family in the area, the staff also develop unmatched loyalty to the dermatologist and his/her practice. The biggest challenge is initially finding key staff with appropriate education, training, and skills.

Some physicians cite the lack of a robust social life in rural communities as a deterrent to working in a rural practice. They feel that there are fewer restaurant options and community events than city life offers. Additionally, some communities have decreased access to high speed internet and updated technologies which can make day-to-day life challenging. Dermatologists may be forced to send complicated cases and surgeries to a larger hospital as it may not be feasible to purchase the technologies necessary for these procedures in the office or at a nearby rural hospital. However, for dermatologists who are able to reconcile these challenges, the benefits of working in a rural practice are significant. (see Table 3.6).

Table 3.6 Pros and cons of living and working in Rural America

Benefits	Deterrents
Larger income[a]	Challenging to recruit qualified staff
High demand of practice	Payor mix
Flexibility of practice	Decreased nightlife
Relaxed quality of life	Decreased access to updated technology
Decreased commute	Few opportunities to Progress career
Increased purchasing power	
Variety of patients	
Deep relationship with patients	
Self-satisfying	

[a]adjusted for cost of living

Final Thoughts

As is the case with most, if not all, of the economic variables discussed in this chapter, it is important to perform one's due diligence prior to any healthcare venture. This chapter should be used merely as a starting point for your research into the economic feasibility and viability of a rural dermatology practice.

Conflicts of Interest The authors have no relevant conflicts of interest.

RHC Services RHCs Provide

- Physician services.
- Services and supplies furnished "incident to" physician services.
- NP, PA, CNM, CP and CSW services.
- Services and supplies furnished "incident to" NP, PA, CNM, CP or CSW services.
- Medicare-covered Part B-covered drugs furnished "incident to" RHC practitioner services.
- Visiting nurse services to the homebound where the Centers for Medicare & Medicaid Services (CMS) certified there is a shortage of home health agencies and certain criteria are met.
- Certain care management services.
- Certain virtual communication services Rural Health Clinic MLN Fact Sheet Page 3 of 8 ICN MLN006398.

Medicare RHC Certification

To qualify as an RHC, a clinic must be in:

- A U.S. Census Bureau-defined non-urbanized area.
- An area currently designated or certified by the Health Resources and Services Administration within the previous 4 years as one of these types of areas:
  - Primary Care Geographic Health Professional Shortage Area (HPSA) under

Section 332(a) (1)(A) of the Public Health Service (PHS) Act
  - Primary Care Population-Group HPSA under Section 332(a) (1)(B) of the PHS Act
  - Medically Underserved Area under Section 330(b) (3) of the PHS Act
  - Governor-designated and Secretary-certified shortage area under Section 6213(c) of the
  - Omnibus Budget Reconciliation Act (OBRA) of 1989

RHCs must:
- Employ an NP or PA (RHCs may contract with NPs, PAs, CNMs, CPs, and CSWs when the RHC employs at least one NP or PA).
- During operational hours, have an NP, PA, or CNM working at least 50 percent of the time.
- Directly provide routine diagnostic and laboratory services.
- Have arrangements with one or more hospitals to provide medically necessary services unavailable at the RHC.
- Have drugs and biologicals available to treat emergencies.
- Provide all these laboratory tests on site: Stick or tablet chemical urine examination or both Hemoglobin or hematocrit Blood sugar Occult blood stool specimens examination Pregnancy tests Primary culturing to send to a certified laboratory.
- Have a quality assessment and performance improvement program.
- Post operation days and hours.
- Not be primarily a mental disease treatment facility or a rehabilitation agency.
- Not be a Federally Qualified Health Center (FQHC).
- Meet all other state and Federal requirements.

RHC visits must be:
- Medically necessary.
- Face-to-face medical or mental health visits or qualified preventive visits between the beneficiary and an RHC practitioner (physician, NP, PA, CNM, CP, or CSW).
- A qualified RHC service that requires the skill level of the RHC practitioner.

RHC visits may take place:

- In the RHC.
- At the beneficiary's home (including an assisted living facility).
- In a Medicare-covered Part A Skilled Nursing Facility (SNF).
- At the scene of an accident.

RHC Visits cannot take place at:

- An inpatient or outpatient hospital (including a Critical Access Hospital).
- A facility with specific requirements that exclude RHC visits.
- More than one visit with an RHC practitioner on the same day, or multiple visits with the same RHC practitioner on the same day, counts as a single visit, except for the following:
- The patient, subsequent to the first visit, suffers an illness or injury that requires additional diagnosis or treatment on the same day (for example, a patient sees their practitioner in the morning for a medical condition and later in the day has a fall and returns to the RHC).
- A qualified medical visit and a qualified mental health visit on the same day.
- An Initial Preventive Physical Examination (IPPE) and a separate medical and/or mental health visit on the same day.

Rural Health Clinics - Materials Reference

Medicare Learning Network - Rural Health Clinic. 2019, May. Retrieved 27 Oct 2020, from https://www.cms.gov/Outreach-and-Education/Medicare-Learning-Network-MLN/MLN Products/Downloads/RuralHlthClinfctsht.pdf.

References

1. Eilrich, Doeksen, Gerald, & F. St. Clair. Estimate the economic impact of a Rural Primary Care Physician (1). 2016.
2. Livio S. Study says waiting too long to get suspected cancerous moles checked out by your doctor could be the difference between life and death. NJ.com. https://www.nj.com/gloucester-county/2011/07/study_says_waiting_to_get_susp.html. Published 19 July 2011. (9).
3. Melanoma Skin Cancer Statistics. American Cancer Society. https://www.cancer.org/cancer/melanoma-skin-cancer/about/key-statistics.html. (10).
4. Marshall S, McGarry KM, Skinner JS. The risk of out-of-pocket health care expenditure at end of life. National Bureau of Economic Research. https://www.nber.org/papers/w16170. Published 8 July 2010. (11).
5. Care Coordination. https://www.ahrq.gov/ncepcr/care/coordination.html. Published June 2014. (8).
6. Ezeonwu MC. Specialty-care access for community health clinic patients: processes and barriers. J Multidiscip Healthc. 2018;11:109–19. https://doi.org/10.2147/JMDH.S152594. Published 2018 Feb 22. (7).
7. Rothstein BE, Gonzalez J, Cunningham K, Saraiya A, Dornelles AC, Nguyen BM. Direct and indirect patient costs of dermatology clinic visits and their impact on access to care and provider preference. Cutis. 2017;100(6):405–10. (18).
8. Lowell BA, Froelich CW, Federman DG, Kirsner RS. Dermatology in primary care: prevalence and patient disposition. J Am Acad Dermatol. 2001;45(2):250–5. https://doi.org/10.1067/mjd.2001.114598. PMID: 11464187.
9. Frostenson S. The death rate gap between urban and rural America is getting wider. Vox. https://www.vox.com/science-and-health/2017/1/13/14246260/death-gap-urban-rural-america-worse. Published 13 Jan 2017. (35).
10. Rogers HW, Weinstock MA, Feldman SR, Coldiron BM. Incidence estimate of nonmelanoma skin cancer (Keratinocyte Carcinomas) in the U.S. Population, 2012. JAMA Dermatol. 2015;151(10):1081–6. https://doi.org/10.1001/jamadermatol.2015.1187. (56).
11. Lim HW, Collins SAB, Resneck JS Jr, et al. The burden of skin disease in the United States. J Am Acad Dermatol. 2017;76(5):958–72.e2. https://doi.org/10.1016/j.jaad.2016.12.043. (5).
12. Garcovich S, Colloca G, Sollena P, et al. Skin cancer epidemics in the elderly as an emerging issue in geriatric oncology. Aging Dis. 2017;8(5):643–61. Published 2017 Oct 1. https://doi.org/10.14336/AD.2017.0503. (57).
13. Azhar AF, Faheem S. Comparison of melanoma incidence in metropolitan areas versus nonmetropolitan areas in the state of Texas stratified by poverty classification. Proc (Bayl Univ Med Cent). 2019;32(3):345–7. Published 17 May 2019. https://doi.org/10.1080/08998280.2019.1593725. (55).
14. Feng H, Berk-Krauss J, Feng PW, Stein JA. Comparison of dermatologist density between urban and rural counties in the United States. JAMA Dermatol. 2018;154(11):1265–71. https://doi.org/10.1001/jamadermatol.2018.3022. (19).
15. Aneja S, Aneja S, Bordeaux JS. Association of increased dermatologist density with lower melanoma

mortality. Arch Dermatol. 2012;148(2):174–8. https://doi.org/10.1001/archdermatol.2011.345. (21).

16. Weary PE. A surplus of dermatologists. Wherefrom and whereto. Arch Dermatol. 1984;120(10):1295–97. (38).

17. Porter ML, Kimball AB. Predictions, surprises, and the future of the dermatology workforce. JAMA Dermatol. 2018;154(11):1253–5. https://doi.org/10.1001/jamadermatol.2018.2925. (39).

18. Ehrlich A, Kostecki J, Olkaba H. Trends in dermatology practices and the implications for the workforce. J Am Acad Dermatol. 2017;77(4):746–52. https://doi.org/10.1016/j.jaad.2017.06.030. (40).

19. Feng H, Belkin D, Geronemus RG. Geographic distribution of US mohs micrographic surgery workforce. Dermatologic Surgery. 2019;45(1):160–3.

20. Bush H, Gerber LH, Stepanova M, Escheik C, Younossi ZM. Impact of healthcare reform on the payer mix among young adult emergency department utilizers across the United States (2005–2015). Medicine. 2018;97(49):e13556. https://doi.org/10.1097/MD.0000000000013556. (58).

21. Bozarth A, Strifler W. Strengthening workforce development in rural areas. n.d. Retrieved 02 Oct 2020, from https://www.investinwork.org/-/media/Files/reports/strengthening-workforce-development-rural-areas.pdf?la=en. (67).

22. The College Payoff: Education, Occupations, Lifetime Earnings. n.d. Retrieved 02 Oct 2020, from https://cew.georgetown.edu/cew-reports/the-college-payoff/N. (66).

23. Health Insurance Coverage of the Total Population. 2020, April 23. Retrieved 30 Sep 2020, from https://www.kff.org/other/state-indicator/total-population/?currentTimeframe=0. (64).

24. Employer Mandate. n.d. Retrieved 02 Oct 2020, from https://www.cigna.com/employers-brokers/insights/informed-on-reform/employer-mandate. (65).

25. Percent of private sector establishments that offer health insurance to Employees, by Firm Size. 2020, September 11. Retrieved 26 Oct 2020, from https://www.kff.org/other/state-indicator/firms-offering-coverage-by-size/?currentTimeframe=0. (71).

26. Claxton GGC, Bains J, Sisko A. Health benefits In 2018: modest growth in premiums, higher worker contributions at firms with more low-wage workers. 2018, October 03. Retrieved 21 October 2020, from https://www.healthaffairs.org/doi/full/10.1377/hlthaff.2018.1001?utm_campaign=HASU%3A+10-05-18+%28Copy%29. (70).

27. Bureau U. Income and poverty in the United States: 2019. 2020, September 15. Retrieved 26 Oct 2020, from https://www.census.gov/library/publications/2020/demo/p60-270.html. (72).

28. History. n.d. Retrieved October 26, 2020, from https://www.cms.gov/About-CMS/Agency-Information/History. (73).

29. Secretary H, (OB), O. FY 2017 budget in brief - CMS – Overview. 2016, February 12. Retrieved 27 October 2020, from https://www.hhs.gov/about/budget/fy2017/budget-in-brief/cms/index.html. (74).

30. Stans MH, Eckler AR, Chartener WH. Estimates of the population of states, by age 1965–1967. 1969, April 17. Retrieved 27 Oct 2020, from https://www.census.gov/prod/1/pop/p25-420.pdf. (62).

31. Rogers L. Median age doesn't tell the whole story. 2019, July 16. Retrieved 27 Oct 2020, from https://www.census.gov/library/stories/2019/06/median-age-does-not-tell-the-whole-story.html. (69).

32. Allowed Amount - HealthCare.gov Glossary. n.d. Retrieved 30 Sep 2020, from https://www.healthcare.gov/glossary/allowed-amount/. (63).

33. Polaha J, Sunderji N. Patient access: how do we measure it? Famil Syst Health J Collaborat Family Healthcare. 2019;37(3):191–4. https://doi.org/10.1037/fsh0000442. (75).

34. Rural health clinics. n.d. Retrieved October 19, 2020, from https://www.cms.gov/Medicare/Provider-Enrollment-and-Certification/CertificationandComplianc/RHCs. (68).

35. Duke EM. Starting a rural health clinic - a how-to manual. 2004. Retrieved 27 Oct 2020, from https://www.hrsa.gov/sites/default/files/ruralhealth/pdf/rhc-manual1.pdf. (76).

36. Medicare learning network - rural health clinic. 2019, May. Retrieved 27 Oct 2020, from https://www.cms.gov/Outreach-and-Education/Medicare-Learning-Network-MLN/MLNProducts/Downloads/RuralHlthClinfctsht.pdf. (77).

37. Rural health information hub. n.d. Retrieved 30 Oct 2020, from https://www.ruralhealthinfo.org/topics/rural-health-clinics. (86).

38. Rural America at a Glance, 2017 Edition. 2017, November. Retrieved 09 Nov 2020, from https://www.ers.usda.gov/webdocs/publications/85740/eib182_brochure%20format.pdf?v=2736.

39. Sisson WT. An investigation of the disparate motivational factors of staff members within an academic medical clinic (unpublished doctoral dissertation). The University of Mississippi Medical Center, Jackson, Mississippi. 2017. (79).

40. Bureau USC. Rates of uninsured fall in rural counties, remain higher than Urban Counties. The United States Census Bureau. https://www.census.gov/library/stories/2019/04/health-insurance-rural-america.html. Published September 24, 2019. (16).

41. Terhune C. The $200 billion perils of unnecessary medical tests. PBS. https://www.pbs.org/newshour/health/200-billion-perils-unnecessary-medical-tests. Published May 24, 2017. (17).

42. Kindley KJ, Jackson JD, Sisson WT, Brodell R. Improving dermatology clinical efficiency in academic medical centers. Inter J Health Sci. 2015;9(3):351–4.

43. U.S. Census Bureau QuickFacts: Mississippi. n.d. Retrieved 30 Oct 2020, from https://www.census.gov/quickfacts/fact/table/MS/BZA115218. (81).

44. U.S. Census Bureau QuickFacts: Louisville city, Mississippi. n.d. Retrieved 30 Oct 30 2020, from https://www.census.gov/quickfacts/fact/table/louisvillecitymississippi/IPE120219. (82).

45. Data and Reports - Workforce - Data and Analysis - AAMC. 2015. Retrieved 29 Nov 2017, from https://www.aamc.org/data/workforce/reports/458490/1-2-chart.html. (84).
46. Moore B. Construction cost in rural mississippi [Telephone interview]. 2020 15 Oct. (78).
47. Krause P. Private equity firms are suddenly buying dermatology practices - here's why. 2016, August 22. Retrieved 30 Oct 2020, from http://www.businessinsider.com/why-private-equity-firms-buy-dermatology-practices-2016-8. (80).
48. Resneck JS. Dermatology practice consolidation fueled by private equity investment: potential consequences for the specialty and patients. JAMA Dermatol. 2018;154(1):13–4. https://doi.org/10.1001/jamadermatol.2017.5558. (30).
49. Reschovsky JD, Staiti AB. Physician incomes in rural and urban America. Issue Brief Cent Stud Health Syst Change. 2005;92:1–4. (6).
50. Physicians offer insights on practicing rural medicine. HealthLeaders media. https://www.healthleadersmedia.com/welcome-ad?toURL=%2Fstrategy%2Fphysicians-offer-insights-practicing-rural-medicine. Published 2 April 2008. (28).
51. Befort CA, Nazir N, Perri MG. Prevalence of obesity among adults from rural and urban areas of the United States: findings From NHANES (2005–2008). https://onlinelibrary.wiley.com/doi/pdf/10.1111/j.1748-0361.2012.00411.x. Published 31 May 2012. (32).
52. Buettner-Schmidt K, Miller DR, Maack B. Disparities in rural tobacco use, smoke-free policies, and tobacco taxes. West J Nurs Res. 2019;41(8):1184–202. https://doi.org/10.1177/0193945919828061. (33).
53. Whitfield GP, Carlson SA, Ussery EN, Fulton JE, Galuska DA, Petersen R. Trends in meeting physical activity guidelines among urban and rural dwelling adults — United States, 2008–2017. MMWR Morb Mortal Wkly Rep. 2019;68:513–8. https://doi.org/10.15585/mmwr.mm6823a1. (34).

Government and Private Efforts to Incentivize Rural Practice

4

Robert Hollis Burrow and Joshua Ortego

As outlined in the previous chapter, the economic impact of a dermatology practice in a rural area can be substantial and the potential exists for rural practices to be profitable. In this chapter, we will look at both public and private programs designed to incentivize the establishment of business and healthcare in rural areas. This is motivated by the desire to promote economic development and population health. The content is divided into two major categories of support (public and private) and two sub-categories related to the type of support being provided (monetary and non-monetary).

Public Incentives

Monetary

The United States Department of Agriculture Rural Development is a mission area within the USDA "committed to helping improve the economy and quality of life in rural America" by using "loans, grants and loan guarantees to help create jobs and support economic development and essential services such as housing; health care; first responder services and equipment; and water, electric and communications infrastructure" [1]. This governmental organization provides one of the major connections to federal dollars intended for rural development in all its forms. The two programs most pertinent to our discussion are the Rural Business Development Grants and the Business & Industry Loan Guarantees.

The Rural Business Development Grants program (RBDG) was established to provide grant money for rural projects that leads to new or further development of rural businesses, as well as funding distance learning networks and employment related adult education programs. To this end, RBDGs may cover a broad range of development activities. RBDG are competitive grants, and are evaluated at the level of state offices. Possible applicants include public entities such as towns, communities, state agencies, nonprofits, and institutions of higher education. A dermatologist partnered with an institute of higher learning would therefore be eligible to apply. It would also be possible to contact a rural community directly to see whether they would be interested in applying on behalf of a private practice physician. The program lists criteria utilized to

Show me an incentive and I will show you an outcome. Charlie Munger.

R. H. Burrow · J. Ortego (✉)
Department of Dermatology, University of Mississippi Medical Center, Jackson, MS, USA
e-mail: rburrow@umc.edu; jortego@umc.edu

© The Author(s), under exclusive license to Springer Nature Switzerland AG 2021
R. T. Brodell et al. (eds.), *Dermatology in Rural Settings*, Sustainable Development Goals Series,
https://doi.org/10.1007/978-3-030-75984-1_4

evaluate applications, including job creation, percent of nonfederal funding committed, economic need, consistency with local development priorities, and experience of the grantee with similar efforts. In order to qualify for consideration, the financing must benefit a "small and emerging" rural business defined as being "outside the urbanized periphery of any city with a population of 50,000 or more" and as having "fewer than 50 employees and less than $1 million in gross revenues" [2]. While there is no maximum amount of financing through the grant, smaller requests are preferred. The general range of grant money is $10,000 to $500,000 with no cost-sharing. The use of funds is limited to the following: training and technical assistance, acquisition or development of land, construction, conversion, or renovation of buildings; equipment and utilities, pollution control, capitalization of revolving loans, distance adult job training, and rural transportation improvement [2].

The Business & Industry Guaranteed Loan Program (BIGLP) was also established by the United States Department of Agriculture (USDA) to "improve the economic health of rural communities by increasing access to business capital through loan guarantees" [3]. Commercial lenders utilize this program to provide affordable financing for rural businesses. For-profit businesses, nonprofits, cooperatives, federally-recognized tribes, public bodies, and individuals can all qualify for these loan guarantees, as long as they are United States citizens or permanent residents and can demonstrate the funds will create or save jobs for rural residents. As above, rural areas are defined as being outside the urban periphery of a town of fewer than 50,000 people. The lender and borrower's headquarters may be located anywhere in the United States, as long as the project is in one of these eligible rural areas. There are, of course, restrictions on the use of the borrowed funds. They are not permitted to be used for lines of credit; owner-occupied or rental housing; golf courses; racetracks or gambling facilities; churches, church-controlled organizations, or charities; fraternal organizations; lending, investment, or insurance companies; projects

involving more than 1 million dollars and the relocation of 50 or more jobs; agricultural production; payment to another lender; or any project where the lender will retain an ownership interest in the borrower. The maximum amount of the loan guarantee provided by this program varies based upon the total loan amount (Table 4.1). Maximum loan terms vary, with 30 years for real estate, 15 years or useful life for machinery and equipment, and 7 years for working capital. Full amortization is required, without balloon payments. Paying interest only is allowed for the first 3 years, with fixed or variable rates negotiated between the lender and borrower. Fees consist of an initial 3% of the guaranteed amount, followed by an annual 0.5% of the outstanding principal [3].

Another area for primary care physicians (and, potentially dermatologists aligned with a group of primary care physicians) to explore is the Rural Health Clinic program which provides financial incentives for rural practice. In 1977, Congress passed the Rural Health Clinic Services Act with the intention of improving access to primary healthcare in rural communities and promoting collaboration between physicians, nurse practitioners, and physician assistants. Due to a low number of practitioners serving Medicare patients in rural areas, the law created special payment mechanisms for Federally-certified Rural Health Clinics (RHC). This provides enhanced Medicare reimbursement utilizing cost-based methodology to pay an "all-inclusive rate (AIR) for medically necessary, face-to-face, primary health services and qualified preventive health services furnished by an RHC practitioner"

Table 4.1 Maximum loan guarantee provided per total loan amount

Total Loan Amount	Maximum Amount of Guarantee
Up to $5 million	80%
Between $5 million and $10 million	70%
Between $10 million up to $25 million	60%

[a]Information obtained from the US Department of Agriculture Rural Development Program (www.rd.usda.gov)

[4]. RHC practitioners include physicians, nurse practitioners, physician assistants, certified nurse-midwives, clinical psychologists, and clinical social workers. In order to qualify, a clinic must meet several specific criteria (Table 4.2). There are also specific criteria that must be met for a patient visit to be reimbursed using the enhanced Medicare rate. The encounter must be a medically necessary, face-to-face visit between the Medicare beneficiary and an RHC practitioner. These can be medical, mental health, or qualified preventive visits, and they must require the skill level of the RHC provider. The location of the visit may be in the RHC itself, the patient's home, a Medicare-covered skilled nursing facility, or at the scene of an accident. They do not qualify if they take place in an inpatient or outpatient hospital [4]. In fact, a dermatologist could benefit from this program as well! If an RHC consisted of two or three primary care physicians, a primary care NP or PA, and a single dermatologist, the clinical visit percentage for primary care could remain well above 50 percent. This would then enable the clinic to remain a Federally-certified RHC, and provide the derma-

tologist an excellent incentive to practice in a rural, underserved area.

Non-monetary

There are also non-monetary benefits provided by government organizations that can serve the needs of physicians choosing to practice in a rural area. Created by Congress in 2000, the Delta Regional Authority (DRA) is one such organization. The DRA "works to improve regional economic opportunity by helping to create jobs, build communities, and improve the lives of the 10 million people who reside in the 252 counties and parishes of the eight-state Delta region" [5]. The states included in this region are Alabama, Arkansas, Illinois, Kentucky, Louisiana, Mississippi, Missouri, and Tennessee. The DRA seeks to address social challenges within the Delta, including increasing access to quality healthcare. There are two special programs that can be of great benefit to dermatologists considering a rural practice in this region [5].

The first is the Delta Region Community Health Systems Development Technical Assistance Program designed to "enhance healthcare delivery in the Delta Region through intensive technical assistance to providers in select rural communities, including critical access hospitals, small rural hospitals, rural health clinics and other healthcare organizations" [6]. The program focuses their assistance into several areas (Table 4.3).

Eligibility for this program is very simple, with applicants only needing to be a critical access hospital, small rural hospital, rural health clinic, or other healthcare organization located within the 252 counties and parishes that make up the Delta Region. Applications are accepted on a continuous basis [6].

The second program specifically benefits non-domestic physicians who have been trained in the United States. The Delta Doctors program was designed by the DRA to help further address access to healthcare through allowing "foreign physicians who are trained in this country to

Table 4.2 Rural health clinic certification qualifications

Located within a Census Bureau-defined non-urbanized area

Designated within the last 4 years by the HRSA as either a health professional shortage area or medically underserved area

Employ either a nurse practitioner or physician assistant who works at least 50% of operational hours

Directly provide routine diagnostic and lab services including:
    Chemical urine exam
    Hemoglobin or hematocrit
    Blood glucose
    Fecal occult blood
    Pregnancy tests
    Primary culturing for send off

Arrangements with at least one hospital for medically necessary services unavailable

A stock of drugs and biologicals for emergencies

A quality assessment and quality improvement program

Cannot primarily be a mental disease or rehabilitation facility, and cannot be a federally qualified health center

ᵃInformation obtained from the Centers for Medicare & Medicaid Services (www.cms.gov)

work in medically underserved areas or health professional shortage areas for three years through a J-1 visa waiver" [7]. Normally, non-domestic medical school graduates would be sent back to their home country for a minimum of 2 years and not be allowed to immediately practice in the United States. Through the Delta Doctors program, the DRA recommends J-1 visa waivers to the State Department allowing graduates to stay in exchange for 3 years of service in a medically under-served or health professional shortage area. This increases the number of providers available within the Delta Region, which in turn positively impacts both the health and economy of the area. Both primary care and specialist physicians may apply for the waiver. Applicants must agree to provide direct patient care for 3 years in the 252 counties and parishes of the Delta Region, working at least 40 hours per week or 160 hours per month. They must also agree to not discriminate against a patient's inability to pay for services or payment using Medicaid, Medicare, or an equivalent indigent healthcare program. In order to protect the interests of American jobs, the facility seeking to hire someone through the Delta Doctors program must have made "a good

Table 4.3 Focus areas of the delta region community health systems development technical assistance program

Improving hospital or clinic financial operations

Implementing quality improvement activities to promote the development of an evidence-based culture leading to improved health outcomes

Increasing use of telehealth to address gaps in clinical service delivery and improve access to care

Enhancing coordination of care

Strengthening the local health care system to improve population health

Providing social services to address broader socio-economic challenges faced by patients (e.g., housing, child care, energy assistance, access to healthy food, elderly support services, job training, etc.)

Ensuring access to and availability of emergency medical services (EMS)

Identifying workforce recruitment and retention resources targeted to rural communities

Other areas to be determined in consultation with the health resources & services administration and the Delta Regional Authority

[a]Information obtained from the Delta Regional Authority (www.dra.gov)

faith effort to recruit an American physician for the opportunity in the same salary range, without success for at least 45 days" [7, 8]. This program is an incredible opportunity for foreign-born physicians with residency training in the United States and will hopefully result in many choosing to stay and continue their work in an underserved rural area.

Private Incentives

Monetary Incentives

As opposed to publicly sponsored opportunities, resources available to rural clinics from the private sector tend to be more varied and individualized. These are usually loan opportunities that are made available by banks, savings and loans or credit unions in the form of grants and specialized loans and may rely on guarantees or incentives offered by the governing municipality. For these reasons, private resources are not often listed or marketed but businesses seeking to qualify for these opportunities can individually approach lenders to identify these opportunities. In some rural communities, qualifying businesses, including physicians, can apply for support in the form of reduced rates of utilities including energy, telecommunications, information technology support, and other resources necessary for the functioning of a successful clinic.

Non-Monetary Incentives

While the Accreditation Council of Graduate Medical Education (ACGME) does not allow for individuals to pay for their own residency education, a rural healthcare organization such as hospitals and medical centers can sponsor residency positions in exchange for a contractual agreement to serve in a rural area. The Department of Dermatology at the University of Mississippi Medical center recently helped a rural medical center in northeast Mississippi by training a resident who agreed to return and practice there upon completion of residency. This win-win situation represents another way private interests can recruit specialists to underserved areas. This opportunity is explained in further detail in Chap. 7.

Disclosures The authors have no relevant disclosures.

References

1. United States Department of Agriculture Rural Development. About RD [Internet]. Washington, D.C.: United States Department of Agriculture; date unknown [cited 2020 Nov 3]. Available from: https://www.rd.usda.gov/about-rd.
2. United States Department of Agriculture Rural Development. Rural Business Development Grants [Internet]. Washington, D.C.: United States Department of Agriculture; date unknown [updated 2019 Dec; cited 2020 Nov 3]. Available from: https://www.rd.usda.gov/sites/default/files/fact-sheet/508_RD_FS_RBS_RBDG.pdf.
3. United States Department of Agriculture Rural Development. Business & Industry Loan Guarantees [Internet]. Washington, D.C.: United States Department of Agriculture; date unknown [cited 2020 Nov 3]. Available from: https://www.rd.usda.gov/programs-services/business-industry-loan-guarantees.
4. Centers for Medicare & Medicaid Services. Rural Health Clinic Fact Sheet [Internet]. Woodlawn, Maryland: Centers for Medicare & Medicaid Services; 2019 May [cited 2020 Nov 3]. Available from: https://www.cms.gov/Outreach-and-Education/Medicare-Learning-Network-MLN/MLNProducts/Downloads/RuralHlthClinfctsht.pdf.
5. Delta Regional Authority. About Delta Regional Authority [Internet]. Clarksdale, MS: Delta Regional Authority; 2020 [cited 2020 Nov 3]. Available from: https://dra.gov/about-dra/about-delta-regional-authority/.
6. Delta Regional Authority. Delta Region Community Health Systems Development Program [Internet]. Clarksdale, MS: Delta Regional Authority; 2020 [cited 2020 Nov 3]. Available from: https://dra.gov/initiatives/promoting-a-healthy-delta/delta-community-health/.
7. Delta Regional Authority. Delta Doctors Fact Sheet [Internet]. Clarksdale, MS: Delta Regional Authority; 2020 [cited 2020 Nov 3]. Available from: https://dra.gov/images/uploads/content_files/Delta_Doctors.pdf.
8. Delta Regional Authority. Delta Doctors [Internet]. Clarksdale, MS: Delta Regional Authority; 2020 [cited 2020 Nov 3]. Available from: https://dra.gov/initiatives/promoting-a-healthy-delta/delta-doctors-how-to-apply/.

The Practice of Austere (Resource-Limited) Dermatology

Ross Pearlman, Martin McCandless,
Amy E. Flischel, and Carrie Kovarik

Ross Pearlman and Martin McCandless contributed equally with all other contributors.

Introduction

The term "austere" originally referred to medical treatment provided in severely resource limited settings. These settings included high-altitude mountains, diving, and wilderness environments [1]. In addition to these settings, the contemporary use of this term includes limitations in access to care in developing countries, rural areas of developed countries, and even underserved urban environments. The opening chapter of this book discusses the differences between rural and urban life. There are millions of people who live "in between" these extremes all over the world including those communities at the most resource-limited end of the spectrum. Areas without a well-established healthcare system, which includes proper trained professionals, adequate facilities, and affordable medications, can benefit from ethical interventions aimed at improving population health with respect to dermatologic care.

Access to dermatologic care in resource-limited settings is demanding and requires sustainable and affordable interventions. Fortunately, previous successful attempts to increase access to dermatologic care in austere settings have provided a framework for future interventions. Effective training of local health care professionals provides the foundation for treating the disease burden in the regions of interest. The general approach to building sustainable dermatologic services should include consultation with local health authorities, planning and delineation of targeted interventions, and careful implementation to establish a patient base. Recent technological advances have led to telehealth services, which offer an important link to expert consultation that is often lacking in austere community

"The World Is Big and I Want to Have a Good Look at it before it Gets Dark." – John Muir

R. Pearlman (✉) · A. E. Flischel
Department of Dermatology and Pathology,
University of Mississippi Medical Center,
Jackson, MS, USA
e-mail: rpearlman@umc.edu; aflischel@umc.edu

M. McCandless
School of Medicine, University of Mississippi
Medical Center, Jackson, MS, USA
e-mail: mmccandless@umc.edu

C. Kovarik
Dermatology at the University of Pennsylvania,
Philadelphia, PA, USA
e-mail: Carrie.Kovarik@pennmedicine.upenn.edu

© The Author(s), under exclusive license to Springer Nature Switzerland AG 2021
R. T. Brodell et al. (eds.), *Dermatology in Rural Settings*, Sustainable Development Goals Series,
https://doi.org/10.1007/978-3-030-75984-1_5

47

environments. Special techniques in procedural dermatology and dermatopathology may expand disease intervention opportunities in resource-limited environments. Successful management of common dermatologic diseases requires availability of basic, affordable medications for the management of these skin conditions. This chapter reviews the basic concepts and techniques focused on sustainable and affordable dermatologic care in austere settings.

Sustainable Implementation and Use of Dermatology Services in Austere Settings

Access to dermatological care is limited and maldistributed in austere settings. On a global scale, the International Foundation for Dermatology reported that nearly 3 billion people in 345 developing rural communities lacked adequate dermatological care [2]. This problem is evident in many developing countries that have little to no access to dermatology specialist care, even in urban areas, and are austere environments in toto with respect to dermatology. In Africa, the supply of dermatologists varies by country from around 49 per million residents to less than 1 per million residents. Less than half of African countries have any type of formal post-graduate dermatology training available to local physicians [3]. Lack of a stable dermatology physician pipeline ensures continued lack of adequate access to dermatology care in these regions.

To address the shortage of dermatology access in resource-limited settings, strategies have been developed to implement supplemental training for existing medical personnel. Additionally, an improved understanding of the local resources and disease burden, whether in the United States or internationally, has enhanced the sustainability and affordability of care in these environments. Utilizing the available primary health care resources plays a key role in augmenting the access to quality dermatological care.

Public health solutions should be proportionate to the disease burden of a population [4]. Dermatological diseases are not always a top pri-

ority when compared to other prevalent and severe diseases such as hypertension, diabetes, malaria or other local infectious diseases. Therefore, any community intervention should focus on preserving availability of local healthcare resources to optimize population health as well as expand access to specialty care. Due to the limited availability of dermatologists in resource-limited settings, primary care physicians and other local healthcare professionals typically serve as the central providers of dermatology services. Ideally, these providers should have a solid foundational knowledge of dermatologic disease, an appreciation for the breadth of the field, and ready access to a dermatology specialist for at least remote consultation and questions.

Any attempt to construct programs increasing access to dermatologic services should also focus on enhancing quality of care. In some instances, the lowest-income countries had worse health outcomes despite the increased use of health care in the past decade [5]. A Global Burden of Diseases study of maternal mortality in Ethiopia found that a total of 16,740 maternal deaths occurred in 1990 whereas 15,234 maternal deaths occurred in 2013. Even with additional governmental health programs such as the Health Development Army, the annual mortality rate declined by 1.6%, significantly lower than the targeted 5.5% set by the Millennium Development Group [6]. The paradox between poor outcomes and increased health care expenditure is multifactorial including but not limited to a lack of quality, integration, use, and longitude of care [7]. The availability of care does not necessarily equate to quality care. For example, poor quality of care was demonstrated in the Indian incentive-based fertility program, Janani Suraksha Yojana, where an increase in in-hospital facility births did not significantly reduce maternal or newborn mortality [8]. High-quality care includes introspective evaluation, integration of cultural factors, recognition of co-existing and asymptomatic conditions, accurate diagnosis, timely and relevant treatments, appropriate referral, and follow-up with the patient to adjust treatment options as needed. Similarly, the approach to providing dermatology care in resource-limited settings should be carefully constructed to avoid increasing "availability" of care without consideration of patient outcomes and population health. Local health system reform should always respect community values, preferences, and autonomy.

Figure 5.1 shows our approach to planning interventions in austere environments to improve long-term, sustainable access and quality of care. This conceptual approach is supported by global health solutions from multiple public health campaigns that have pioneered intervention implementation and learned a number of crucial lessons to enable successful execution.

This entire process will not be successful unless the local health authorities recognize the need to improve dermatologic care in their area and extend an invitation for assistance. Consultation with local health authorities serves as the first step in understanding the organizational structure of available dermatologic and health services in resource-limited areas. This first step reveals the relationship between the healthcare infrastructure and the performance of health care professionals, the perceptions held about the expertise of the local health care workers, and the user experience of the patients. Any change suggested to local healthcare infrastructure should be directed by local health authorities. Every effort must be made to avoid paternalistic approaches to intervention. Successful integration with local health systems depends on functional policy-making to maneuver political, cultural, and social pressures that may serve as barriers to increased access to quality care. Effective implementation of local dermatologic interventions starting via consultation with local health authorities was demonstrated by Mahé et al. in Mali. In this circumstance, consultation with local health authorities revealed which diseases were important to target in the region

[9]. Using well-established communication techniques can help facilitate these high-quality discussions for planning interventions [10]. The dermatologist can ask community and healthcare leaders to explain in their own words information presented to them, which then enables them to address any confusion or misunderstanding.

The establishment of targeted interventions is the second step to expand quality dermatology services in austere environments. Multiple intervention modalities have been described to expand access to dermatology services including training global dermatology specialists, increasing primary prevention education of the public, placing dermatologists part-time on the ground in clinics, and expanding teledermatology and telepathology access [11]. Programs focused on teaching dermatology diagnoses and treatments to local health professionals have demonstrated efficacy [12]. Depending on the geographical location, the dermatologist should identify the scope and significance of skin diseases associated with the community. Customizing the scope of dermatologic diagnoses may aid in enhancing appropriate interventions relevant to the area of interest. Baseline data regarding global geographic and temporal variation in skin disease impacts can be found through the Global Burden of Disease Study 2010. This database contains country-level mortality from 1990 to 2010 which included benign and malignant cutaneous and subcutaneous diseases. In 2010, developing countries exhibited significantly greater ($P < 0.001$) skin disease-mortality than developed countries except for melanoma, basal cell carcinoma, and squamous cell carcinoma with little variation in the previous two decades [13]. The common skin diseases should be hierarchically categorized to inform health care workers of relevant treatment modalities. Categorization

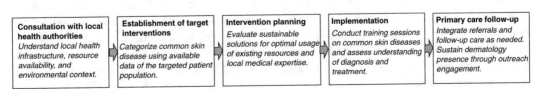

Fig. 5.1 The general approach to planning dermatology interventions in austere environments. (Adapted from Mahe 2005)

of skin diseases should be based on the prevalence of disease and disease burden on the patient population. Identifying relevant skin diseases may provide the basis to focus planning of sustainable interventions.

Most of the time, sustainable quality care does not involve putting a dermatologist on the ground full-time due to the inherent scarcity of specialist availability. Additionally, one specialist may not be able to sufficiently support a large patient population. The most practical solution is to recruit local medical expertise including primary care physicians, nurses, or volunteers. The implementation of digestible "algorithmic" approaches to diagnoses may seem to be a simple solution but poses risks to the population. When conducting educational or training sessions on diagnosis and treatment, the information must be conveyed in a way that makes sense to the providers who are using them.

Previous outreach efforts to provide dermatology in austere settings taught local health care providers about skin diseases, differential diagnoses, and corresponding treatments as an illustrated booklet or slide presentation using graphics and common language. The health care workers were then asked a set of standardized questions to assess retention of the material. Further evaluations from this study found that after 2 years post-training, participants exhibited long-term retention [12]. This study demonstrates that effective training of local healthcare resources yields sustainable, long-term improvements in local dermatology services. In addition to teaching the most common skin diseases, local medical staff should be taught to recognize malignant patterns and familiarize themselves with elements of dangerous and life-threatening diseases. Such cases should be referred to a dermatologist located at a nearby healthcare facility or accessed through teledermatology. Establishing a partnership with a dermatologist who is cognizant of the local health needs can be invaluable to the treatment team on the ground. In the United States, a local partner may serve as a location for in-person care. Internationally, this care can be coordinated with local Ministries of Health or established partnerships.

Many global health programs have delivered sustainable health services abroad while expanding global health training activities. When programs involve the immediate delivery of care, trainees may be limited to the treatment of acute problems. Educational experiences should also address the broad determinants that affect community-centered care and public health [14]. In addition, it is critical for any program with rotating providers to ensure proper follow-up care for patients. These considerations should be incorporated into the curriculum of these programs to build an effective model of global health involvement through sustainable practices. In Mexico, the Community Dermatology Civil Association has established a series of two-day programs with rotating dermatologists to expand access to dermatologic services in austere environments of Mexico, including in the Sierra Madre del Sur mountain range, home to several groups of isolated indigenous peoples. This program was structured with an education program of Day 1 for local health care providers and a "jornada" or free clinic on Day 2. This program treated 4677 patients and taught hundreds of general practitioners in only 33 sessions [15]. A similar Community Dermatology program in Argentina has replicated these results using this same model of training plus patient care [16].

Programs targeting dermatology residents in the United States offer global health opportunities to broaden their exposure to dermatology worldwide and facilitate cross-cultural exchange. The American Academy of Dermatology Residents' International Grant (AAD RIG) offers the archetype for such programs, providing international educational opportunities for senior dermatology residents to attend elective rotations in various austere environments in developing countries. Its mission is to expose young dermatologists to austere settings while developing broad public health improvement. Through this broad global health model, previous AAD RIG participants continue to engage in service-oriented medical and academic activities both locally and globally [17]. The AAD RIG program currently offers rotations at four international austere sites including Nepal, South Africa,

Botswana, and Peru. The AAD also facilitates opportunities for residents to rotate in austere environments within the United States. The Native American Health Service Resident Rotation is a two-week program in rural Chinle, Arizona at an Indian Health Services clinic.

Telemedicine is an ever-expanding method for advancing the availability of specialty consultation using remote communication. While most dermatological services in developing countries involve long waitlists and long-distance travel, teledermatology offers a solution for diagnosing, triaging, and treating disease through remote communication. This method of two-way and mobile communication can serve as an extension for further research and evaluation studies. A telemedicine study reported that 11 locations in Botswana successfully trained a total of 24 clinicians and contributed to the management of 643 patients [18]. Although teledermatology provides patients and providers with improved access to care, these pilot studies revealed technical and social challenges that limit the benefit of teledermatology services including high costs, technical infrastructure malfunction, damage to the devices, lack of connectivity to high-speed internet, and misalignment with the culture of providers and patients.

One dermatologist in Botswana demonstrated a simple and sustainable solution to these common technical issues: the use of a dedicated WhatsApp number for clinicians and patients to send store-and-forward consults. WhatsApp is a mobile messaging application that allows users to send free, secure text messages and images using a wireless internet connection. To simplify matters, this application is already the predominant electronic communication platform in Botswana and used by over 1 billion people worldwide [19]. The WhatsApp platform facilitated simple exchange of clinical photographs for diagnosis and treatment recommendations. A natural consequence of this increased access to dermatology specialist consultation was a trend of expanding knowledge by referring providers with respect to diagnosis and management [20]. The creative use of technology will continue to present new opportunities for interventions to increase access to dermatology services in austere environments in the future.

The final steps are sustainable implementation of the intervention and ensuring subsequent follow-up of dermatology. The group in Mali gathered 400 suitable healthcare workers from 112 surrounding primary centers to be trained to identify common skin diseases [12]. The high attendance and support from local health authorities for simplified approaches in training for skin diseases facilitated the assimilation of intervention into an existing health system. However, it is important to note that the "amount" of skin disease care does not necessarily reflect the quality of care, leading to false complacency about the progress of skin health.

Healthcare workers in local communities are often responsible for follow-up of patient diagnosis and treatment plans. This role will be overseen frequently by the patient's primary care provider. They should communicate any results of labs or pathology to the patient and ensure that management plans are adjusted as needed. Seeing the patient again via in-person consultation or a telehealth appointment to assess progress allows an opportunity to measure the quality and effectiveness of the healthcare intervention.

Special Techniques for Resource-Limited Settings

Austere clinical environments present several challenges in addition to a dearth of clinical personnel. These settings often suffer from a shortage of critical medical supplies, medications, and lab support. At some sites, rotating dermatologists are tasked with bringing their own surgical tools, drugs, and sanitary clothing such as scrubs [21]. Practicing in resource-limited settings may require creative approaches to common clinical situations. The following techniques and drugs are ideal for successful operation clinics in resource-limited settings.

Specific cutaneous biopsy techniques enable clinicians to collect tissue samples for the diagnosis and management of skin diseases. Use of local anesthetics such as lidocaine with epinephrine is standard practice at most medical centers [22]. Shave and punch biopsies are essential procedures for physicians to diagnose and manage some skin conditions including possible malignancies. Shave biopsies are one of the quickest, simplest, and most cost-effective procedures. Generally, a scalpel or Dermablade is used to remove a targeted lesion such as papillomas, or neoplasms such as basal or squamous cell carcinomas. Aluminum chloride 20% solution, silver nitrate, or Monsel's solution can be used to achieve hemostasis. Punch biopsies are used for excisional or incisional purposes depending on the tissue needed. This technique extends through the dermal and subcutaneous layers to obtain a narrow, cylindrical specimen. A small circular excision can be made, alternatively, with a scalpel blade if a punch tool is not available. Wound closure in resource-limited environments may be achieved with traditional suturing techniques if available. Alternatively, straight needle stitching or cyanoacrylate adhesives can be used depending on available resources and wound size.

With a shortage of trained dermatologists and dermatopathologists in resource-limited settings, skin biopsy specimens in difficult cases have been submitted for teledermatopathology for evaluation. Partnership with or assisting local labs with training in tissue processing and staining (i.e. hematoxylin and eosin, Giemsa, Gram, Ziehl-Neelsen, etc) may help better manage representative specimens, proper tissue processing, and reliable reporting. The transport of fixed tissue to an experienced pathology laboratory may be a feasible option through partnerships formed between Ministries of Health and local organizations or academic institutions [23].

The use of cryosurgery is a minimally invasive technique to destroy benign, pre-malignant, or malignant skin lesions using a cryogen such as liquid nitrogen. The mechanism of cellular damage is multifaceted with the subzero temperature resulting in vascular damage leading to tissue ischemia. Each freeze-thaw cycle induces osmotic gradients which disrupt the integrity of the cell membrane [24]. Most benign lesions including skin tags, solar lentigo, seborrheic keratosis, and verrucous papules are responsive to $-20\,^{\circ}C$ treatment. Small lesions typically regress after the first treatment; however, larger benign lesions may need an additional 2 to 6 treatments at 3 to 4-week intervals [25]. Precancerous lesions (actinic keratoses) and low-grade malignancies such as Bowen's disease (squamous cell carcinoma in-situ) and superficial basal cell carcinoma can be treated using cryosurgery. This technique offers a low-cost and quickly-administered intervention versus excision. Since malignant lesion margins are ill-defined, there is a possibility of recurrence after treatment. Therefore, it may be recommended to err on the side of overtreatment in some cases and to provide regular follow-ups [26]. Thermal conduction is an important factor when considering cryotherapy. Poor thermal conductors such as hyperkeratotic lesions may need prior debulking with curettage and a focused-spray tip for effective treatment [27].

Although cryosurgery greatly expands the practitioner's arsenal for intervention, it is important to consider the logistic challenges associated with availability of cryosurgery. Proper sustainable storage of liquid nitrogen may be challenging in resource-limited areas. Gun-style cans store up to 500 mL of liquid nitrogen allowing for easy transportation. Long-term storage of liquid nitrogen in heavy, well-insulated containers called dewars requires continual replenishing as the largest 50 L dewars may only store liquid nitrogen for up to 2 months. A comparative cost model of liquid nitrogen-stored malaria vaccines found that the availability and cost of liquid nitrogen varied widely across African countries. The recurrent costs were for purchasing liquid nitrogen (assuming US $1.00/L), storage, transportation, labor, and energy usage at the facilities with liquid nitrogen storage containers located in the national stores rather than regional or local-level stores [28]. Continual shipment partner stores or suppliers of liquid nitrogen to rural communities are needed if resource-limited settings lack proper storage capabilities. Partnership with other providers who may have liquid nitrogen, such as gynecologists performing cervical cryosurgyery, may make this treatment more feasible.

Dermatological pharmaceuticals and therapeutics are often in low supply or absent in resource-limited settings. Ideal medications allow for local pharmacies to stock or compound agents in-house. However, obtaining a steady supply of raw materials can be challenging and sustainability is never guaranteed. Multiple partnerships across different health and industry sectors are necessary to maintain medicinal stocks and financial stability. Therefore, feasible treatment options must consider the regulations, cost, availability, and ability to follow-up [29]. A list of fundamental dermatology medications should be made depending on the stock at the in-house pharmacy or partner availability. Table 5.1 shows a list of dermatology drugs adapted from the World Health Organization List of Essential Medicines with addition of several essential agents for the practice of dermatology [30]. All of these are available as generic drugs in the United States and most are available at a reasonable cost.

Table 5.1 Essential drugs and their preparations for use in dermatology, adapted from the WHO Model List of Essential Medications. Drugs that are not on the original WHO Model List are noted with an (*) [30]

Anesthetics and Analgesics	
Lidocaine + epinephrine	Injection
Bupivacaine	Injections
Ibuprofen	Tablet
Acetylsalicylic acid	Suppository, tablet
Anti-Staphylococcals	
Cefalexin	Powder for reconstitution with water, oral
Clindamycin	Lotion, capsule, oral liquid
Dicloxacillin	Capsule, powder for oral liquid
Doxycycline	Capsule, tablet
Sulfamethoxazole + trimethoprim	Table, oral liquid
Other antibiotics	
Amoxicillin/clavulanic acid	Oral liquid, tablet

Table 5.1 (continued)

Anesthetics and Analgesics	
Lidocaine + epinephrine	Injection
Ciprofloxacin	Capsule, oral liquid, solution for infusion
Levofloxacin	Tablet
Topical antiseptics	
Chlorhexidine	Solution
Ethanol	Solution
Povidone iodine	Solution
Antivirals	
Acyclovir	Tablet, oral liquid, powder for injection
Valacyclovir*	Tablet
Valganciclovir	Tablet
Immunosuppressants	
Azathioprine	Powder for injection, tablet
Cyclosporine	Capsule, concentrate for injection
Methotrexate	Powder for injection
Methylprednisolone	Injection
Prednisone, prednisolone	Tablet, oral solution
Antifungals	
Amphotericin B	Powder for injection
Fluconazole	Capsule, injection, oral liquid
Griseofulvin	Oral liquid
Ketoconazole*	Cream, shampoo
Miconazole	Cream or ointment
Selenium sulfide	Detergent-based suspension
Terbinafine	Cream, ointment
Topical anti-infective medicines	
Mupirocin	Cream, ointment
Gentamicin*	Ointment
Silver sulfadiazine	Cream
Topical steroids	
Betamethasone	Cream, ointment
Triamcinolone	Cream, ointment, injection
Hydrocortisone	Cream, ointment
Agents affecting skin differentiation and proliferation	
Acitretin*	Capsule
Benzoyl peroxide	Cream, lotion, wash
Coal tar	Solution
Fluorouracil	Ointment
Podophyllum resin	Solution
Salicylic acid	Solution
Urea	Cream, ointment
Scabicides, pediculicides, antihelminthics	
Benzyl benzoate	Lotion
Ivermectin	Tablet
Albendazole	Tablet

Research

Establishing an evidence-based approach to these dermatologic access interventions in austere environments facilitates their continued growth, success, and ethical implementation. Involving local physicians in research planning, assessment of data, organization and writing of will lead to the most successful outcomes. Including them as authors of the submitted publication is, of course, an ethical obligation. However, relationships with communities in austere settings can be tenuous. It is critical that education and patient care remain the primary foci of any dermatologic healthcare intervention.

Medical Service Trips: Ethical Considerations

Medical service trips (MST) or "mission medicine" should be distinguished from the processes presented in this chapter, which are a blueprint for establishing long-term access to dermatologic care. In contrast, MSTs aim to provide treatment in a short-term or limited timeframe. The traditional MST model introduces several issues that detract from the value of medical volunteer efforts in resource-limited settings.

In 2000, Bishop and Litch described the phenomenon of medical tourism and the effects it had on healthcare in the austere mountainous regions surrounding Sagarmatha National Park and Mount Everest in Nepal. Healthcare providers on MSTs established adhoc clinics along heavily trafficked trails. Physicians often saw patients from single consultations with no knowledge of local maladies, no plan for adequate follow-up, and no communications with the patient's other local providers. Bishop and Litch argue that many physicians they encountered assumed that their assessment held greater value than assessment by local healthcare. The authors often observed treatments that were at odds of interfered with management of patient's chronic conditions. Medical tourists rarely consulted with local healthcare workers or sought to integrate their skillset into the existing health-

care infrastructure [31]. This fragmentation of care can decrease quality of care as physicians evaluate patient complaints with little to no clinical context. Some critics have suggested that medical student MSTs focus on improving student resumes, practicing unpolished clinical skills, and providing travel opportunities at the expense of patient quality of care [32]. Qualitative results of a 2016 survey of healthcare students visiting clinic sites in rural Mississippi and the Dominican Republic demonstrated that some students understood that they were likely the primary beneficiaries of their volunteer trip, rather than the local population. However, students had not considered impacts of their volunteer activities on local healthcare efforts [33].

Practice of medicine by insufficiently supervised or qualified volunteers can result in significant harm to already vulnerable populations. For-profit tourist outfits have been anecdotally reported to recruit "medical" volunteers as young as 16, some of whom do not even have a complete high school degree [34]. Often times, diagnosis and treatment of local variants of disease in the developing world are beyond the scope of Western trained specialists. For example, vesicovaginal fistula repair surgery in developing countries versus the developed world is distinct due to the rarity of prolonged obstructed labor in the West. This pathology results in a completely different clinical scenario than most Western surgeons have encountered and requires specialized training with appropriate case load. In the 2000's, many women were harmed in West Africa due to incompetent surgical management by surgeons visiting on short-term "fistula trip" [35]. It is certainly possible that subtle variations in dermatologic disease presentation could result in misdiagnosis and mistreatment, which is why it is critical to work in appropriately supervised settings.

The Health Volunteers Overseas (HVO) model is an attempt to structure sustainable, ethical rotations by qualified health professionals overseas [36]. The goal of the model is to make incremental contributions to long-term goals that are overseen and monitored by local host institutions.

This model establishes their core attributes of short-term volunteer programs that may make them sustainable, ethical, and productive: (1) focus on education, (2) equitable partnerships with host institutions, and (3) structured management of services. This model is currently being used to extend access to dermatology services in Cambodia, Costa Rica, Nepal, Uganda, and Vietnam [36]. A similar foundational philosophy is used in the structuring of programs for dermatology residents by the AAD Resident International Grant and Native American Health Service Residents Rotation Program. All rotations are housed within existing local host academic teaching programs. Residents have the opportunity to learn and provide dermatology services under the guidance of public sector dermatologists and with access to teledermatology consultation if needed. A special emphasis is placed on providing and arranging appropriate follow-up for patients [37].

Conclusions

Providing dermatology services in resource-limited settings poses a significant challenge even to the most prepared experts. A stepwise approach to implementing a dermatology public health intervention gives the clinician the best chance of sustainable and ethical improvements to care for the local populations. Key considerations include the public health needs of the local populations and the availability of on-the-ground healthcare support personnel. Local supply chains will influence the availability of critical clinical materials and drugs and are essential element to consider in the planning process. All in all, there are nearly unlimited opportunities to improve dermatology care in resource-limited environments in the United States and worldwide with a careful stepwise approach and ethical intervention.

Conflicts of Interest The authors have no relevant conflicts of interest.

References

1. Imray CH, Grocott MP, Wilson MH, Hughes A, Auerbach PS. Extreme, expedition, and wilderness medicine. Lancet. 2015;386(10012):2520–5.
2. Coustasse A, Sarkar R, Abodunde B, Metzger BJ, Slater CM. Use of Teledermatology to Improve Dermatological Access in Rural Areas. Telemed J E Health. 2019;25(11):1022–32.
3. Mosam A, Todd G. Dermatology Training in Africa: Successes and Challenges. Dermatol Clin. 2021;39(1):57–71.
4. Mahé A, Faye O. Response to 'Providing dermatological care in resource-limited settings: barriers and potential solutions'. Br J Dermatol. 2018;178(2):574–5.
5. Kruk ME, Gage AD, Arsenault C, Jordan K, Leslie HH, Roder-DeWan S, Adeyi O, Barker P, Daelmans B, Doubova SV, English M, Elorrio EG, Guanais F, Gureje O, Hirschhorn LR, Jiang L, Kelley E, Lemango ET, Liljestrand J, Malata A, Marchant T, Matsoso MP, Meara JG, Mohanan M, Ndiaye Y, Norheim OF, Reddy KS, Rowe AK, Salomon JA, Thapa G, Twum-Danso NAY, Pate M. High-quality health systems in the Sustainable Development Goals era: time for a revolution. Lancet Glob Health. 2018;6(11):e1196–252.
6. Tessema GA, Laurence CO, Melaku YA, Misganaw A, Woldie SA, Hiruye A, Amare AT, Lakew Y, Zeleke BM, Deribew A. Trends and causes of maternal mortality in Ethiopia during 1990–2013: findings from the Global Burden of Diseases study 2013. BMC Public Health. 2017;17(1):160.
7. Roro MA, Hassen EM, Lemma AM, Gebreyesus SH, Afework MF. Why do women not deliver in health facilities: a qualitative study of the community perspectives in south central Ethiopia? BMC Res Notes. 2014;7:556.
8. Ng M, Misra A, Diwan V, Agnani M, Levin-Rector A, De Costa A. An assessment of the impact of the JSY cash transfer program on maternal mortality reduction in Madhya Pradesh. India Glob Health Action. 2014;7:24939.
9. Mahé A, Faye O, N'Diaye HT, Ly F, Konaré H, Kéita S, Traoré AK, Hay R. Definition of an algorithm for the management of common skin diseases at primary health care level in sub-Saharan Africa. Trans R Soc Trop Med Hyg. 2005;99(1):39–47.
10. Ubel PA, Scherr KA, Fagerlin A. Empowerment failure: how shortcomings in physician communication unwittingly undermine patient autonomy. Am J Bioeth. 2017;17(11):31–9.
11. Fuller LC, Hay RJ. Global health dermatology: building community, gaining momentum. Br J Dermatol. 2019;180(6):1279–80.

12. Mahé A, Faye O, N'Diaye HT, Konaré HD, Coulibaly I, Kéita S, Traoré AK, Hay RJ. Integration of basic dermatological care into primary health care services in Mali. Bull World Health Organ. 2005;83(12): 935–41.

13. Boyers LN, Karimkhani C, Naghavi M, Sherwood D, Margolis DJ, Hay RJ, Williams HC, Naldi L, Coffeng LE, Weinstock MA, Dunnick CA, Pederson H, Vos T, Dellavalle RP. Global mortality from conditions with skin manifestations. J Am Acad Dermatol. 2014;71(6):1137–1143.e17.

14. Evert J, Bazemore A, Hixon A, Withy K. Going global: considerations for introducing global health into family medicine training programs. Fam Med. 2007;39(9):659–65.

15. Estrada R, Chavez-Lopez G, Estrada-Chavez G, Paredes-Solis S. Specialized dermatological care for marginalized populations and education at the primary care level: is community dermatology a feasible proposal? Int J Dermatol. 2012;51(11): 1345–50.

16. Casas I. Community dermatology in Argentina. Dermatol Clin. 2021;39(1):43–55.

17. Lonowski SL, Rodriguez O, Carlos CA, Forrestel AK, Kovarik CL, Wanat KA, Williams VL. Looking back on 10 years of the American Academy of Dermatology's Resident International Grant Experience in Botswana. J Am Acad Dermatol. 2019;

18. Littman-Quinn R, Mibenge C, Antwi C, Chandra A, Kovarik CL. Implementation of m-health applications in Botswana: telemedicine and education on mobile devices in a low resource setting. J Telemed Telecare. 2013;19(2):120–5.

19. Yueng K. WhatsApp passes 1 billion monthly active users. 2016 [Last Accessed July 3, 2017]; Available from: https://venturebeat.com/2016/02/01/ whatsapp-passes-1-billion- monthly-active-users.

20. Williams V, Kovarik C. WhatsApp: an innovative tool for dermatology care in limited resource settings. Telemed J E Health. 2018;24(6):464–8.

21. Kovarik C, Williams V, Forrestel A. Resident International Grant Botswana rotation pre-departure handbook. 2019 [Accessed 12 Dec 2020]; Available from: https://www.aad.org/member/career/awards/ resident-international.

22. Frank SG, Lalonde DH. How acidic is the lidocaine we are injecting, and how much bicarbonate should we add? Can J Plast Surg. 2012;20(2):71–3.

23. Tsang MW, Kovarik CL. The role of dermatopathology in conjunction with teledermatology in resource-limited settings: lessons from the African Teledermatology Project. Int J Dermatol. 2011;50(2):150–6.

24. Bischof JC. Quantitative measurement and prediction of biophysical response during freezing in tissues. Annu Rev Biomed Eng. 2000;2:257–88.

25. Farhangian ME, Snyder A, Huang KE, Doerfler L, Huang WW, Feldman SR. Cutaneous cryosurgery in the United States. J Dermatolog Treat. 2016;27(1):91–4.

26. Kuflik EG. Cryosurgery for skin cancer: 30-year experience and cure rates. Dermatol Surg. 2004;30(2 Pt 2):297–300.

27. Goldberg LH, Kaplan B, Vergilis-Kalner I, Landau J. Liquid nitrogen: temperature control in the treatment of actinic keratosis. Dermatol Surg. 2010;36(12):1956–61.

28. Garcia CR, Manzi F, Tediosi F, Hoffman SL, James ER. Comparative cost models of a liquid nitrogen vapor phase (LNVP) cold chain-distributed cryopreserved malaria vaccine vs. a conventional vaccine. Vaccine. 2013;31(2):380–6.

29. Chang AY, Kiprono SK, Maurer TA. Providing dermatological care in resource-limited settings: barriers and potential solutions. Br J Dermatol. 2017;177(1):247–8.

30. World Health Orgnaization (WHO) WHO Model List of Essential Medicines - 21st Edition. 2019 [Accessed 9 Dec 2020]; Available from: https://apps.who.int/ iris/rest/bitstreams/1237479/retrieve.

31. Bishop R, Litch JA. Medical tourism can do harm. BMJ. 2000;320(7240):1017.

32. Snyder J, Dharamsi S, Crooks VA. Fly-By medical care: Conceptualizing the global and local social responsibilities of medical tourists and physician voluntourists. Global Health. 2011;7:6.

33. Rovers J, Japs K, Truong E, Shah Y. Motivations, barriers and ethical understandings of healthcare student volunteers on a medical service trip: a mixed methods study. BMC Med Educ. 2016;16:94.

34. Sullivan N. The Trouble with Medical "Voluntourism" 2017 [Accessed 28 Dec 2020]; Available from: https://blogs.scientificamerican.com/observations/ the-trouble-with-medical-voluntourism/.

35. Wall LL, Arrowsmith SD, Lassey AT, Danso K. Humanitarian ventures or 'fistula tourism?': the ethical perils of pelvic surgery in the developing world. Int Urogynecol J Pelvic Floor Dysfunct. 2006;17(6):559–62.

36. MacNairn E. Health volunteers overseas: a model for ethical and effective short-term global health training in low-resource countries. Glob Health Sci Pract. 2019;7(3):344–54.

37. Introcaso CE, Kovarik CL. Dermatology in Botswana: the American Academy of Dermatology's Resident International Grant. Dermatol Clin. 2011;29(1):63–7.

Rural Dermatology Residency Slots: Priming the Pump

6

Maheera Farsi, Navid Farahbakhsh, and Abel Torres

Introduction

Patients in rural, poor, and highly underrepresented minority areas often do not have adequate access to dermatologists [1]. Currently, less than 10% of dermatologists practice in a rural setting, while 40% practice in the 100 densest US areas [2]. Improving the shortage of rural dermatology physicians requires adequate recruitment of incoming residents with an interest in rural dermatology and retention of these graduates in these rural settings. While there are unique challenges with recruitment and retention, many of these challenges may be surmountable.

Rural residency training tracks, in some medical specialties, have been demonstrated to improve rural physician recruitment and retention [3]. The factors, which influence retention,

Living in a rural setting exposes you to so many marvelous things –
the natural world and the particular texture of small-town life, and
The Exhilarating Experience of Open Space.
— Susan Orlean

M. Farsi (✉) · A. Torres
Department of Dermatology, University of Florida College of Medicine, Gainesville, FL, USA
e-mail: Abeltorres@ufl.edu

N. Farahbakhsh
University of Florida College of Medicine, Gainesville, FL, USA
e-mail: Nfarahbakhsh@ufl.edu

have been studied in Family Medicine residency programs, but not in Dermatology residency training [4]. Applying these factors as part of the dermatology residency selection process would likely improve success in post-residency retention.

One of the first rural dermatology residency programs was started at the Loma Linda University Department of Dermatology in 2014. This program was designed to recruit one incoming dermatology resident each year to train for 2 years in an urban or suburban setting and spend the final year of training in a rural setting. The rural setting was initially designated in Michigan, but later expanded to include rural areas in Georgia and Tennessee. This program had a plethora of applicants and was successful in recruiting four residents. However, the effort proved to be unsuccessful in retaining them to work in a rural setting, and thus, the program was closed in 2020 with the last two participants being allowed to opt out of the rural track. The exact reasons why there was success in recruiting, but not retaining, are still being sorted out. This will be discussed later in this chapter.

A second rural dermatology residency track was created at the University of Mississippi Medical Center (UMMC) in 2018. To date, recruitment and retention of this program has proven to be successful perhaps due to the specific criteria used to recruit and retain incoming residents by the UMMC Dermatology

Department. In 2021, this program has four residents training in their rural track.

Possible Factors Influencing Recruitment of Rural-Minded Practitioners

The variety of factors have been demonstrated to enhance rural recruitment in Family Medicine residency programs - the most important factors (from a scale (1) little importance to (5) for extremely important) include: significant other's wishes (4.50); meaningful work (4.38); local community (4.25); and, medical community/work environment (4.20) [4]. Table 6.1 displays these recruitment factors from the most to least important. Practice setting does not necessarily change these responses. When physicians were surveyed in both rural and urban practice settings, it was found that their 'spouse' had the highest mean score (3.19 for rural and 3.34 for urban setting) for influencing their practice location [5]. The mean score for spouse was also statistically significant ($p < 0.0002$) when compared to other factors

Table 6.1 Factors influencing rural retention

Rank	Factor	Mean (sd)
1	Significant other's wishes	4.50 (0.52)
2	Meaningful work	4.38 (0.81)
3	Local community	4.25 (0.58)
4	Medical community/work environment	4.20 (0.94)
5	Work/life balance	4.06 (0.85)
6	Broad scope of practice	4.06 (1.12)
7	Job security	3.81 (0.91)
8	Need for health care in the community	3.73 (0.80)
9	Proximity to family and friends	3.63 (1.09)
10	Income/benefits	3.50 (1.03)
11	Local school system	3.50 (1.15)
12	Professional development	3.38 (1.15)
13	Teaching opportunities	2.88 (1.09)
14	Loan repayment opportunities	2.00 (1.41)

With permission from [4]

Scale: Not at all important (1), A little important (2), somewhat important (3), very important (4), extremely important (5)

Abbreviation: *Sd* standard deviation

such as desire to live in their hometown, location of residency, and mentors [5].

Whether these same factors are as strong for Dermatology, only time will tell. Logically, it makes sense, but there is a difference. Dermatology is a highly competitive specialty with many more qualified applicants than positions [6]. In fact, applicants to dermatology are often willing to relocate anywhere, apply multiple times after failing to match, and participate in research training or other activities as they desperately attempt to secure a coveted residency slot. For the residency selection committee, it may be difficult to accurately discern the applicant's commitment to rural life. In fact, the applicant and their families themselves may temporarily convince themselves that selection criteria, including rural practice requirements, meet their life goals when, in fact, they do not. Thus, for dermatology, these factors may be more important in predicting retention more than recruitment.

The UMMC Dermatology program uses many of these factors to help recruit potential residents to their rural track. One of their most important recruitment methods is identifying medical students early in their medical school training who may be interested in practicing rural dermatology. The residency program uses the medical school's Dermatology Interest Group (DIG) to identify these students and then pairs them with a faculty mentor to work on rural dermatology research projects, outreach events, and rural medical school electives to assess their true interest in practice outside an urban area.

A holistic approach is taken by the UMMC Department of Dermatology to evaluate potential students for this track. Aside from medical school grades, United States Medical Licensing Examination (USMLE) scores, personal statements, and interviews, other factors are taken into account. These include the residents and spouses' hometown, lifestyle choices, students' preference on training in a rural area, and rural medicine experiences. Ultimately, the most significant factor in selection includes identifying interested students early on during medical school training, long-term mentorship with

students, and identifying ties or a sensible rationale for the student to pursue a career in rural
dermatology.

The Loma Linda University Department of
Dermatology program used similar criteria in
evaluating potential applicants but differed in that
an emphasis was not placed on identifying medical students early and establishing those rural
track mentoring relationships. In addition, there
were significant differences in the program that
likely had a significant impact on retention.

Factors Influencing· Retention

The factors which influence recruitment may or
may not overlap with those for retention. Rural
upbringing has been reported to be the strongest
predictor of retention and rural practice choice

[4, 5, 7, 8]. The residents' familiarity with their
upbringing and surroundings can improve happiness, trust, and comfort with the program.
Figure 6.1 displays an integrated understanding
of the potential for rural exposure to improve
rural retention [8]. The factors highlighted represent major criterion utilized by the UMMC rural
dermatology track to select residents.

While this can be considered a "pre-
determined" factor for retention, there are other
practice-related and lifestyle factors that may
play an even more important role [8–10]. These
"flexible factors" include parenting of children,
compatibility with the medical community,
workload, financial sustainability, sociocultural
integration, and transition to a new practice environment [8]. Spousal satisfaction and integration
into the local community are key retention factors
that can play a major role in long-term success of

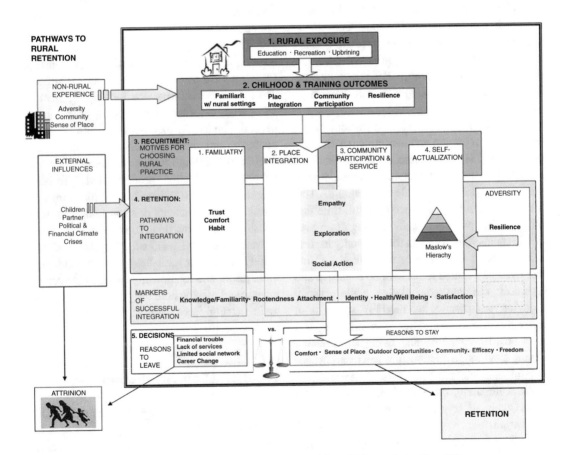

Fig. 6.1 Conceptual map of how rural exposure affects rural retention. (With permission from [8])

maintaining rural physicians [11]. Another factor which is frequently overlooked is the concept of sense of place and self-actualization. The resident's emotional ties to a rural community and motivation to lead a happy and fulfilling life play important roles in retention [8]. Interestingly, these same factors impact the retention of physicians in urban settings [8]. Residency programs interested in improving access to care for regional rural populations can use these factors to select residents in their program or consider establishing a rural track residency track to bring more focus to the problem. In its most rudimentary form, simply asking candidates about their upbringing, motivations for pursuing medicine, and relationship with rural medicine may have a positive effect or rural practice patterns.

The Loma Linda program attempted to use rural exposure and familiarity with a region as a selection criterion, but perhaps the definition of "rural" was less clearly stated to applicants. Although comfort in a Michigan, Georgia, or Tennessee setting was broadly assessed, perhaps narrowing the focus to a more specific region, such as a hometown in rural Mississippi, or specifying rural roots of a spouse may have been more beneficial.

As previously mentioned, a factor that is essential for retention and recruitment is mentorship. In the UMMC program, rural residents have early and continued mentorship throughout their medical school and residency training. This includes continuing guidance by a full-time faculty member practicing and teaching at an academic center 90 miles from the university in rural Mississippi. A rural mentor who takes the time and makes a genuine effort to hone cognitive and technical dermatologic skills, while clearly enjoying rural practice and rural life, serves as an impactful role model. They also offer support with board preparation and living adjustments. Strong mentorship relationships can result in successful implementation of a rural dermatology track. Ultimately, these rural residents will be more likely to stay and ultimately practice in rural areas.

In the Loma Linda program, early mentorship was not emphasized, and the structure of the program was such that it was assumed that mentorship would take place during the resident's third year at the rural site. In addition, formal mentorship training of the mentors and mentees was not implemented, and data now demonstrates that formal mentorship training for any mentor relationship is critical [8, 12, 13]. Similarly, mentorship relationships usually thrive when the mentor is not in a supervisory role over the mentee. This was not possible with how the third year was structured [8, 12, 13].

The UMMC program was able to establish successful mentorship by identifying interested medical students early during their medical school training and pair them with a dermatology faculty member practicing at the central urban clinics and contact, early and often, with a rural faculty member. The Dermatology Department invests early with specific medical students. These efforts include, but are not limited to: (1) financial support for summer research in rural communities that leads to publications; (2) support for expenses involved in medical student training at a rural office including free housing; (3) aggressive mentoring; (4) emotional support and efforts to instill passion for helping those in need; and, (5) entry into tutoring programs, if needed. The faculty mentors work with these students throughout their medical training, assist them in navigating through the competitive dermatology application process, and ultimately become a trusted advocate for their mentee. These medical students develop long-term relationships that allow them to solidify their desire to train in a rural residency program and ultimately work in a rural field. The investment of time, effort, and relationship formed between the mentor and mentee hold the promise of being the main driving force behind the success of the UMMC program.

Successful Implementation

The UMMC Dermatology Department selects one resident per application cycle and uses the above described selection criteria when recruiting incoming medical residents into their rural track. Rural track residents work alongside their co-residents for 9 months throughout the year and for three consecutive months every year they live in rural Mississippi where they work with a full-time academic dermatology faculty member at a rural clinic site. This three-month rotation substitutes for their general dermatology rotation that traditional residents perform at urban dermatology clinics in Jackson, Mississippi. Housing is provided for residents during these 3 months of training. The first rural site was an academic dermatology clinic established in Louisville, Mississippi approximately 95 miles from the main Dermatology Department. The rural residents are allowed to bring their spouse with them during this time, though they usually choose to come back to Jackson, Mississippi on the weekends. They attend didactic teaching rounds on Monday at the main campus before returning to the rural clinic after their conferences. Thus, during the rural general dermatology rotation, residents have clinics from Tuesday to Friday. The residents enjoy being home during the weekend, especially if they have children. They also have the opportunity to be involved in weekly didactics with non-rural track residents and continue to interact and form relationships with one another, fostering comradeship amongst the group.

By contrast, although the Loma Linda program used similar selection criteria as noted previously, it involved the relocation of the residents and their families to the rural training site for most of their third and final year. No gradual introduction to the rural setting was possible due to the nature of the program funding. In fact, an out of state third year was the only option. In retrospect, having the resident spend 3–4 months in the rural setting each year as opposed to the entire third year could have been more effective, but this still could have put a significant stress on the resident's family life. Although long-distance learning was implemented to assure continuation of didactic and other teaching, this resulted in physical separation from their colleagues, except for elective time to participate in some selected program activities. Thus, there was no gradual adjustment to the rural setting, co-resident interactions suffered, and on-going mentor relationships suffered. Furthermore, the relocation impact on the families may have been more acute in nature, and with the sites being out of state, the isolation impact may have been more severe.

The residents serving in the UMMC rural track are required to serve a minimum of 3 years at a rural location upon graduation. This location is agreed upon by the residents and the hospital site prior to their enrollment in the program. During the mentorship process, the faculty mentor identifies the site where the student is interested in establishing a life-long practice. This allows for open communication and understanding of the student's expectations upon graduation of residency. The medical student's willingness to work in this area and commit to the program, outside of the Residency Match, has been the bedrock of the program's success. At this time, 50% of the rural dermatology residents went to medical school in Mississippi and about 50% attended medical school in other nearby southern states. The value of educational diversity is important, and medical students from other states typically have ties to rural Mississippi or have expressed an interest in returning back to their hometowns in rural Mississippi. This three-year commitment allows for residents to have job security in their preferred rural location and for residency programs to improve their retention of rural residents and meet the demands of their underserved population. The residents also uniformly express interest in being rural "academic" physicians with research interests. They look forward to the benefits of working "long-term" for the university and enjoying teaching residents and medical students who will be rotating through their future practice.

Residents have a contractual agreement to complete a minimum of 3 years post-residency in their selected rural city. This three-year period correlates with the 3 years of residency training the rural track resident completes. If residents fail to meet the requirements of their contract, they would be required to pay back - in total - the cost of the 3 years of their residency training.

The Loma Linda program participants were also contractually required to serve a minimum amount of time in a rural setting post-residency completion and had a similar "pay-back" clause if they failed to do so. Most accepted applicants even had a chance to visit the sites before being selected. However, two of the sites were added later in the program, and the exact location and future employment contract were not finalized prior to enrollment in the program. The lack of post graduate practice location finalization, was implemented in order to allow residents more flexibility in building future plans as they progressed in their training. Unfortunately, this led to real and perceived misunderstandings between the residents and program sponsors, which negatively impacted the entire program.

Conclusions

One of the major risks associated with a rural dermatology track is that if residents do not follow through with their commitment of working in a rural setting. When a resident disengages from the rural track, it is impossible to replace them to sustain the program. There is substantial time, resources, and finances allocated to train a rural dermatology resident, and unsuccessful retention increases this burden. This highlights the importance of appropriate selection criteria, gradual transition, and the importance of clear employment contractual terms and post-residency agreements related to site and type of clinical practice.

Camaraderie amongst the co-residents, faculty, mentors, and employers is also an important factor to consider. The rural track residents have a unique schedule that may require them to be away from their co-residents and significant others for a period of time. This adds to the considerable stresses experienced by all residents. The rural track residents also have unique experiences which their co-residents, on the traditional track, may not truly understand. These factors hold the potential to alter the resident relationship dynamic. The UMMC program philosophy helped to alleviate these strains. Attendance is expected at all group residency settings, such as didactics and social gatherings. The UMMC pro-

gram also offers non-rural track residents the opportunity to enjoy rural elective rotations. All residents can, therefore, share the special benefits of rural training. Although similar adjustments were discussed for the Loma Linda program, they were late in the course of the program and never fully implemented.

Of all of these factors, mentorship is probably the most important. Lack of mentorship and teaching will negatively impact any program. The student and resident should feel supported, mentored, and have a professional outlet to express any concerns that arise. The stress of moving to a new rural town, along with the social and cultural adjustments and new professional identity, can play a significant role in the residents' and their families' overall happiness. This affects their commitment to stay in a rural dermatology setting [14]. Perhaps if mentorship development and a well-planned mentorship experience had been aggressively pursued for the rural faculty participants, the Loma Linda program would have had a different outcome.

The six-year Loma Linda experiment was not a success. It was started with good intentions and alumni support but unfortunately failed to meet its objectives. Whether it could have been ultimately rehabilitated will never be known. There was a glimmer of hope in that one of the two residents that actually completed the training at the rural teaching sites voluntarily stayed on at the rural site. Yet, even that resident eventually did not stay in the rural setting. The remaining half of the four total residents were provided the choice of opting out of the program and did so before participating in the rural component. Perhaps the lack of success was because of some or all of the factors discussed above including contract issues, or the personalities of all the parties involved, or perhaps because the structure of the out of state component doomed it from the start. The UMMC experiment is still developing but seems to be garnering momentum and goodwill. It remains to be seen which of the factors described above will be key to its success, but we cannot help but think that mentorship will be a major factor.

References

1. Vaidya T, Zubritsky L, Alikhan A, Housholder A. Socioeconomic and geographic barriers to dermatology care in urban and rural US populations. J Am Acad Dermatol. 2018;78(2):406–8. https://doi.org/10.1016/j.jaad.2017.07.050.

2. Yoo JY, Rigel DS. Trends in dermatology: geographic density of US dermatologists. Arch Dermatol. 2010;146(7):779. https://doi.org/10.1001/archdermatol.2010.127.

3. Patterson DGLR, Schmitz D, Phillips R, Skillman SM, Doescher MP. Rural residency training for family medicine physicians: graduate early-career outcomes, 2008–2012. Seattle: University of Washington; 2013.

4. Morken C, Bruksch-Meck K, Crouse B, Traxler K. Factors influencing rural physician retention following completion of a rural training track family medicine residency program. WMJ. 2018;117(5): 208–10.

5. Kazanjian A, Pagliccia N. Key factors in physicians' choice of practice location: findings from a survey of practitioners and their spouses. Health and Place. 1996;2(1):27–34.

6. Alikhan, A, Sivamani RK, Mutizwa MM, Felsten LM. Advice for fourth year medical students beginning the dermatology residency application process: Perspectives from Interns who matched. Dermatol Online J. 2009;15(10).

7. Geyman JP, Hart LG, Norris TE, Coombs JB, Lishner DM. Educating generalist physicians for rural practice: how are we doing? J Rural Health. 2000;16(1):56–80. https://doi.org/10.1111/j.1748-0361.2000.tb00436.x.

8. Hancock C, Steinbach A, Nesbitt TS, Adler SR, Auerswald CL. Why doctors choose small towns: a developmental model of rural physician recruitment and retention. Soc Sci Med. 69(9):1368–76.

9. Laven G, Wilkinson D. Rural doctors and rural backgrounds: how strong is the evidence? A systematic review. Aust J Rural Health. 2003;11(6):277–84. https://doi.org/10.1111/j.1440-1584.2003.00534.x.

10. Mayo E, Mathews M. Spousal perspectives on factors influencing recruitment and retention of rural family physicians. Can J Rural Med. 2006;11(4):271–6.

11. Cameron PJ, Este DC, Worthington CA. Professional, personal and community: 3 domains of physician retention in rural communities. Can J Rural Med. 2012;17(2):47–55.

12. Roots RK, Li LC. Recruitment and retention of occupational therapists and physiotherapists in rural regions: a meta-synthesis. BMC Health Serv Res. 2013;13:59. Published 2013 Feb 12. https://doi.org/10.1186/1472-6963-13-59.

13. Gillham S, Ristevski E. Where do I go from here: we've got enough seniors? Aust J Rural Heal. 2007;15(5):313–20. https://doi.org/10.1111/j.1440-1584.2007.00900.

14. Law M, Lam M, Wu D, Veinot P, Mylopoulos M. Changes in personal relationships during residency and their effects on resident wellness: A qualitative study. Acad Med. 2017;92(11):1601–6. https://doi.org/10.1097/ACM.0000000000001711.

Training Medical Students in a Rural Dermatology Clinic

7

Haley Harrington, Ross Pearlman, and Amy E. Flischel

Medical students in the United States have been primarily trained at university-based medical schools since the early twentieth century following the Flexner Report in 1910 which eschewed training based solely on an apprenticeship model in practitioner's clinics. Flexner's approach favored 2 years of didactic training in the basic sciences followed by clinical training, in large part, at university hospitals and associated clinics. This led to significant improvements in the quality of physicians and a Renaissance in American Medicine in the areas of education, research, and patient care. Eliminating training in the hinterland, unfortunately is, to a large degree, responsible for the migration of medical care to the cities and away from rural areas. This has contributed to a critical access to care problem for primary care and medical specialties, including dermatology, in most parts of the country away from urban centers.

This chapter reviews the nature of medical education in the field of dermatology, issues related to the supply and demand of the dermatology workforce and considers the benefits of rural training and the barriers that prevent widespread adoption of rural medical student education. Identifying solutions to training medical students in rural areas may be the most important step in solving the rural access to care problem in dermatology.

Medical Student Education in the Field of Dermatology

Due to the prevalence of skin disorders, dermatological education is an important part of every medical students' training. Skin-related complaints are common in the outpatient setting among all specialties, most notably primary care. Approximately 6% of all visits to physicians involve a primary problem with the skin, hair, or nails, but only 40% of those are seen by dermatologists [2]. Medical students, unfortunately, rarely receive thorough training in dermatologic disease [3]. Surprisingly, approximately 50 percent of U.S. medical schools do not offer electives in dermatology. Many schools offer only a

---

"Medical education is not just a program for building knowledge and skills in its recipients… it is also an experience which creates attitudes and expectations."
-Abraham Flexner

H. Harrington
Louisiana State University Health Sciences Center, Shreveport, LA, USA
e-mail: hharr3@lsuhsc.edu

R. Pearlman · A. E. Flischel (✉)
Department of Dermatology and Pathology, University of Mississippi Medical Center, Jackson, MS, USA
e-mail: rpearlman@umc.edu; aflischel@umc.edu

short, introductory clinical dermatology elective to a limited number of their students. A median of 10 hours of dermatology instruction over the 4-year period of medical education represents less than 0.3% of a medical student's total education [4]. The limited curriculum time devoted to dermatology makes it impossible to properly educate medical students in the field of dermatology. Limited exposure also makes it difficult for these students to discover the potential a career in dermatology might hold for their future [5].

Summing up Supply and Demand

The difficulties of training medical students in a rural setting are amplified when considering the need for training dermatologists in general. The influx of dermatology nurse practitioners is one way the general need for dermatologists has been satiated. There are even more significant shortages in the dermatologist workforce in rural areas and these disparities have continued to increase over time. The number of metropolitan dermatologists has grown while the nonmetropolitan and rural areas continue to fall behind with regard to access to dermatologic care. While the density of dermatologists in rural areas increased from 0.84 to 1.05 per 100,000 people between 1995 and 2013, the density of dermatologists in urban areas increased from 3.47 to 4.11 per 100,000 people during that time. Although the total number of rural dermatologists is increasing, the gap between rural and urban access to dermatologists has continued to grow and merely maintains the 4:1 ratio discrepancy. From this data, it is not surprising to learn that younger dermatologists favor practicing in urban communities, perhaps because they have never been exposed to the benefits of a rural practice. In 2013, the percentage of older dermatologists was two-fold greater in nonurban than urban communities. Data also indicated that the number of rural dermatologists younger than 55 years was diminishing. These findings suggest that the urban-rural workforce disparities will continue as dermatologists retiring in the upcoming years affect supply in rural areas even more severely [6].

Location of graduate medical education training often predicts the area where physicians establish their practice. Unfortunately, many rural states have limited residency training positions [7]. This fact not only contributes to the workforce density gap with greater wait times for patients to see the dermatologist, but it also affects the exposure that medical students are able to receive during training [8]. In other words, as the number of rural dermatologists diminishes, it is increasingly difficult for medical schools to develop opportunities for their students to receive dermatology training in a non-urban setting.

The Benefits of Training Medical Students in Rural Clinics

Most medical schools have physicians in private practice, group single-specialty practice, or large multispecialty groups who serve the educational mission of the university as affiliate faculty volunteers [9, 10]. The University of Mississippi Medical Center has attacked this problem head on. They require all medical students to participate in a rural primary care clerkship during their fourth year of medical school [11]. This setting offers special benefits to students when implemented effectively. Students often have less competition with classmates, encounter greater opportunities to evaluate patients at varying stages of disease, experience more continuity of care, and play a more hands-on role in the evaluation and treatment of patients when compared to the typical academic medical student training experience.

A comprehensive review of 72 studies indicated that medical students who train in rural clinics see a greater patient volume and have opportunities to do more procedures [12]. While there were no substantial negative effects on USMLE scores observed between students trained in a rural versus urban setting, preceptor evaluations demonstrated significant positive effects on rural students' clinical and professional skill development. This observation was noted in a study evaluating a 16-week preceptorship during the fourth year of medical school where the student lived in a rural community, worked with a primary care physician, and completed a community-oriented project [13]. A survey of 96 medical students before and after completing this preceptorship reported that medical students had a greater understanding of health systems and the community after completion of the rotation. The students also self-reported improved confidence in patient education skills and greater ability to handle undifferentiated and acute problems [14]. Shorter duration 8-week clinical rotations in rural settings have also been highly valued among medical students. Third-year rural clerkship medical students took patient histories, performed physical examinations, educated patients, and performed necessary clinical testing. The students that participated in the rural clinic rated the experience highly because of the reasonable autonomy, collaboration opportunities with community providers, and understanding of the needs of the underserved populations that it provided [15]. Overall, data indicates that longer rotations are superior to shorter rotations and are more likely to build future rural workforce capacity. When comparing a 6-week rural rotation to an urban home-based rotation with only a few days spent in the rural setting at a time, students were more able to make meaningful local connections to the people and community [16]. Unfortunately, dermatologists working in rural clinics have remained a largely untapped educational resource for clinical training of medical students.

Benefits of a Rural Dermatology Clinical Experience

If training in primary care rural clinics is beneficial for medical students, it would not be surprising that rural dermatology experiences might produce similar advantages. It has been shown that physician assistants trained in a rural setting are more comfortable diagnosing and treating dermatological conditions than their counterparts trained in urban settings [17]. This is likely the result of experience gained from the high number of dermatologic patients seen in busy rural practices [17]. Although there is little literature describing the training benefits for medical students specifically in a rural dermatology clinic, the potential advantages are clear. Rural medical students report higher patient loads and greater opportunities to perform procedures than those not trained in a rural setting, leading to a significant improvement in clinical skills [12]. Rural experiences also increase student self-confidence, as they are able to work one-on-one with a dedicated preceptor [18]. Students get to know their teachers well and have increased opportunities for personalized teaching and mentoring [19]. In addition, rural experiences cultivate resilience and competence in clinical skills, as rural physicians often operate with fewer resources and referral options than urban counterparts. Rural physicians are, in fact, special because they develop the ability to recognize the limits of their environments and adapt their clinical practice to meet the needs of their community [20]. Each of these factors contribute to the strong clinical training experience received by medical students in a rural setting and reiterate the importance of creating more opportunities for students to take advantage of rural opportunities.

Barriers to the Development of Programs to Train Medical Students in Rural Clinics

Challenges arise when considering the consistency of rural training experiences and faculty development. The university must harness its system for evaluating affiliate faculty while providing guidelines, adequate supervision, and faculty development to ensure quality improvement. Initial steps would require each rural dermatologist apply for a faculty position, and the university would need to review qualifications such as board certification before accepting rural faculty to participate in the program. Successful implementation may also require the development of guidelines to appropriately delineate expectations to guarantee appropriate and adequate supervision [21]. (see Chap. 11).

An additional consideration must be given to providing living quarters for medical students rotating at sites far from the university. It would be difficult for many students to maintain their apartments in the city and afford the additional expense of a hotel during the period of the preceptorship.

Perhaps most importantly, medical students need to be convinced of the value of a rural experience. Many medical students are not aware of the opportunities and the adventure associated with a period away from urban life. Students training in a metropolitan area are attracted by the close contact with peers and colleagues, entertainment and night life, and access to restaurants and numerous cafes. Students report concerns regarding the quality of rural teachers, distance from larger libraries, clinical exposure to complex patients, and limited opportunities to pursue esteemed internships, residencies, or fellowships [19].

A medical student's background is an additional factor that contributes to the dilemma associated with sparking interest in rural training. There is a direct relationship between a student's rural background and their inclination to pursue training in a rural setting. When 100 first-year medical students were surveyed, 86% of students from a rural background versus 30% of students from an urban background expressed the desire to train in a rural setting [22]. Moreover, studies indicate that rural origin seems to be the greatest predicting factor for a physician to consider rural practice [23]. In 2017, a survey of 508 medical graduates was performed regarding practice location, which showed that a rural background was a substantial predictor of rural practice with an odds ratio of 3.91 (p-value of <0.001) [24]. Likewise, physicians who are raised in rural communities are two to four times more likely to practice in rural areas [25]. Medical students with spouses and children also need to overcome the hardships imposed by a period of separation if the rural clerkship is distant from the university. Ultimately, student-reported barriers must be considered, in the context of a particular student's background, and addressed to potentially increase student interest in training in a rural setting in a sustainable manner.

Overcoming the Difficulties of Rural Dermatology Training for Medical Students

A variety of strategies can be utilized to increase medical student and physician recruitment to underserved areas, such as loan repayment, higher reimbursement for medical care in rural areas, funding rural graduate medical education training positions, and greater recruitment of students with rural origins and diverse backgrounds [26]. Highlighting the benefits of a rural lifestyle, clean air, ease of transport, and the associated supportive environment may increase recruitment [19]. Developing new partnerships between rural clinics and residency programs may provide another creative alternative to increase residency training positions and meet the rural health care needs within the field of dermatology.

The University of Mississippi Medical Center (UMMC) is taking an aggressive approach. They created a project entitled IMPACT the RACE, which stands for Improved Primary Care for the Rural Community through Medical Education. This project was created to enhance the education of medical students and encourage them to pursue careers in rural communities. It involves community-based activities throughout all four years of medical school that include incorporating tele-health to address unmet needs, outreach to middle school, high school and college students to increase rural recruitment, and partnerships with rural physicians to expand clinical "shadowing" opportunities.

The Department of Dermatology at UMMC also established a free standing academic rural residency dermatology practice staffed by a full-time faculty member. Dermatology, family practice, and plastic surgery residents rotate along with and medical students in this office located in Louisville, Mississippi, about 90 miles from the University Hospital in Jackson. A two-bedroom home is available for lodging to allow rotators to be fully engaged in the rural community. This program attracts a variety of medical students from both rural and urban areas across North America, including Minnesota, Idaho, Mississippi, Tennessee, North Carolina, and

Canada. There are so few academic rural dermatology rotations that students must travel from around the country, and even from outside of the country, to be a part of this experience.

To help create a well-rounded learning experience in rural areas, a number of resources are available. Search engines provide rapid, updated, inexpensive access to medical information that will be valuable throughout the physicians' career. In addition, innovative journal clubs, online community groups, and virtual grand rounds are even more frequent in the post-COVID-19 era. These low-cost, interactive, remotely accessible online resources allow for self-directed learning in a rural setting [25]. The American Academy of Dermatology (AAD) has created a standardized, online curriculum for medical student dermatology education. When surveyed, students found the AAD modules beneficial and time efficient. Additionally, and importantly, the modules have shown to significantly improve a student's dermatology knowledge. Use of this curriculum in supplement with clinical exposure in rural dermatology can help standardize dermatology learning for medical students [27].

Summary

It is essential to recognize the advantages of training more students in rural clinics and the potential for this training impacting the rural physician shortage. While many barriers exist, there is an impetus to overcome the obstacles associated with training medical students in a rural dermatology clinic [8]. If the percentage of medical students choosing careers in non-urban areas is increased by only a small amount, rural healthcare disparities will be reduced. This growth could develop into a positive feedback cycle: increasing the number of rural dermatologists leads to increased medical student exposure to rural dermatology, leading to more clinically prepared students who may be more inclined to practice in rural areas [23]. In addition, a cadre of volunteer affiliate faculty or fulltime academic faculty in rural offices will enhance the limited

clinical education offerings provided by medical schools in the field of dermatology. These developments will improve the availability of hands-on activities in the clinic that many medical students crave and provide enhanced learning experiences with increased opportunities to participate in patient care and treatment plans and greater one-on-one time with faculty. The time for developing these rural teaching programs in dermatology is now.

Disclosure The authors have no relevant conflicts of interest related to the content of this chapter.

References

1. Douthit N, Kiv S, Dwolatzky T, Biswas S. Exposing some important barriers to health care access in the rural USA. Public Health. 2015;129(6):611–20. https://doi.org/10.1016/j.puhe.2015.04.001.
2. Federman DG, Reid M, Feldman SR, Greenhoe J, Kirsner RS. The primary care provider and the care of skin disease: the patient's perspective. Arch Dermatol. 2001;137(1):25–9. https://doi.org/10.1001/archderm.137.1.25.
3. Murase JE. Understanding the importance of dermatology training in undergraduate medical education. Dermatol Pract Concept. 2015;5(2):95–6. https://doi.org/10.5826/dpc.0502a18. Published 2015 Apr 30.
4. McCleskey PE, Gilson RT, DeVillez RL. Medical student core curriculum in dermatology survey. J Am Acad Dermatol. 2009;61(1):30–35.e4. https://doi.org/10.1016/j.jaad.2008.10.066.
5. Philips RC, Dhingra N, Uchida T, Wagner RF Jr. The "away" dermatology elective for visiting medical students: Educational opportunities and barriers. Dermatol Online J. 2009;15(10):1. Published 2009 Oct 15.
6. Feng H, Berk-Krauss J, Feng PW, Stein JA. Comparison of Dermatologist Density Between Urban and Rural Counties in the United States. JAMA Dermatol. 2018;154(11):1265–71. https://doi.org/10.1001/jamadermatol.2018.3022.
7. Resneck JS Jr, Kostecki J. An analysis of dermatologist migration patterns after residency training. Arch Dermatol. 2011;147(9):1065–70. https://doi.org/10.1001/archdermatol.2011.228.
8. Uhlenhake E, Brodell R, Mostow E. The dermatology work force: a focus on urban versus rural wait times. J Am Acad Dermatol. 2009;61(1):17–22. https://doi.org/10.1016/j.jaad.2008.09.008.
9. Stagg P, Prideaux D, Greenhill J, Sweet L. Are medical students influenced by preceptors in making career choices, and if so how? A systematic review. Rural Remote Health. 2012;12:1832.

10. Hudson JN, Weston KM, Farmer EA. Engaging rural preceptors in new longitudinal community clerkships during workforce shortage: a qualitative study. BMC Fam Pract. 2011;12:103. Published 2011 Sep 27. https://doi.org/10.1186/1471-2296-12-103.

11. Jackson-Williams L. IMPACT the RACE represents boon for medical education at UMMC. The Journey. Available at: https://www.umc.edu/som/Departments%20and%20Offices/SOM%20Administrative20Offices/files/Med_Journey_NL_Oct_20.pdf.

12. Barrett FA, Lipsky MS, Lutfiyya MN. The impact of rural training experiences on medical students: a critical review. Acad Med. 2011;86(2):259–63. https://doi.org/10.1097/ACM.0b013e3182046387.

13. Glasser M, Hunsaker M, Sweet K, MacDowell M, Meurer M. A comprehensive medical education program response to rural primary care needs. Acad Med. 2008;83(10):952–61. https://doi.org/10.1097/ACM.0b013e3181850a02.

14. Hunsaker ML, Glasser ML, Nielsen KM, Lipsky MS. Medical students' assessments of skill development in rural primary care clinics. Rural Remote Health. 2006;6(4):616.

15. Bennard B, Wilson JL, Ferguson KP, Sliger C. A student-run outreach clinic for rural communities in Appalachia. Acad Med. 2004;79(7):666–71. https://doi.org/10.1097/00001888-200407000-00010.

16. Denz-Penhey H, Shannon S, Murdoch CJ, Newbury JW. Do benefits accrue from longer rotations for students in Rural Clinical Schools? Rural Remote Health. 2005;5(2):414.

17. Brown B, Bushardt R, Harmon K, Nguyen SA. Dermatology diagnoses among rural and urban physician assistants. JAAPA. 2009;22(12):32–9. https://doi.org/10.1097/01720610-200912000-00008.

18. Levy BT, Merchant ML. Factors associated with higher clinical skills experience of medical students on a family medicine preceptorship. Fam Med. 2005;37(5):332–40.

19. Jones GI, DeWitt DE, Elliott SL. Medical students' reported barriers to training at a Rural Clinical School. Aust J Rural Health. 2005;13(5):271–5. https://doi.org/10.1111/j.1440-1584.2005.00716.x.

20. Thach SB, Hodge B, Cox M, Parlier-Ahmad AB, Galvin SL. Cultivating country doctors: preparing learners for rural life and community leadership. Fam Med. 2018;50(9):685–90. https://doi.org/10.22454/FamMed.2018.972692.

21. Casaletto JJ, Wadman MC, Ankel FK, et al. Emergency medicine rural rotations: a program director's guide. Annals of Emergency Medicine. 2013;61(5):578–83.

22. Azer SA, Simmons D, Elliott SL. Rural training and the state of rural health services: effect of rural background on the perception and attitude of first-year medical students at the university of melbourne. Aust J Rural Health. 2001;9(4):178–85. https://doi.org/10.1046/j.1038-5282.2001.00359.x.

23. MacQueen IT, Maggard-Gibbons M, Capra G, et al. Recruiting rural healthcare providers today: a systematic review of training program success and determinants of geographic choices. J Gen Intern Med. 2018;33(2):191–9. https://doi.org/10.1007/s11606-017-4210-z.

24. Playford D, Ngo H, Gupta S, Puddey IB. Opting for rural practice: the influence of medical student origin, intention and immersion experience. Med J Aust. 2017;207(4):154–8. https://doi.org/10.5694/mja16.01322.

25. Chan BT, Degani N, Crichton T, et al. Factors influencing family physicians to enter rural practice: does rural or urban background make a difference? Can Fam Physician. 2005;51(9):1246–7.

26. Hanson AH, Krause LK, Simmons RN, et al. Dermatology education and the Internet: traditional and cutting-edge resources. J Am Acad Dermatol. 2011;65(4):836–42. https://doi.org/10.1016/j.jaad.2010.05.049.

27. Cipriano SD, Dybbro E, Boscardin CK, Shinkai K, Berger TG. Online learning in a dermatology clerkship: piloting the new American Academy of Dermatology Medical Student Core Curriculum. J Am Acad Dermatol. 2013;69(2):267–72. https://doi.org/10.1016/j.jaad.2013.04.025.

Political Action in Rural Dermatology

8

Elizabeth Kiracofe, Claire Petitt, Erica Rusie, Tim Maglione, and Neha Udayakumar

If they don't give you a seat at the table, bring a folding chair.

Shirley Chisholm

Introduction

As debates continue throughout the United States (US) regarding a variety of approaches to the delivery of quality healthcare, it is critical to consider the needs of 46 million Americans—15% of the US population—who live in rural areas [1]. Residents of these communities face unique challenges when seeking healthcare services. They are often underinsured and must travel greater distances to visit physicians. Geographic barriers may further increase travel time to the doctor. There

E. Kiracofe (✉)
Airia Comprehensive Dermatology, PLLC,
Chicago, IL, USA
e-mail: dermMD@drkiracofe.com

C. Petitt · N. Udayakumar
University of Alabama School of Medicine,
Birmingham, AL, USA
e-mail: cpetitt@uab.edu; neahu@uab.edu

E. Rusie
Scientific Affairs, TALEM Health, Trumbull, CT,
USA
e-mail: erusie@talemhealth.com

T. Maglione
Maglione Advisors Group, LLC, Columbus, OH,
USA
e-mail: maglione.tim@gmail.com

may be limited public transportation options. Small rural hospitals may be unable to provide the latest technological and care-delivery innovations or may be closing. Finally, healthcare workers of all kinds are usually in short supply. (See Chap. 1) These factors contribute to financial insecurity and impaired health outcomes for rural Americans.

According to a study from the Centers for Disease Control and Prevention (CDC), rural Americans are at a higher risk of mortality from the top five leading causes of death—heart disease, cancer, unintentional injury, chronic lower respiratory disease, and stroke—compared to their urban counterparts [1]. Not surprisingly, with dermatologists in short supply, residents of rural areas often have an especially difficult time accessing dermatologic care. (See Chap. 1) After reviewing issues surrounding access to care in dermatology, it is the purpose of this chapter to explore the potential for political action to impact this societal problem. A series of vignettes is presented to demonstrate the effectiveness of this approach.

Rural Access to Care in Dermatology

In the United States, the maldistribution of dermatologists across the country has led to increased wait times for in-person office visits and delays in obtaining a consultation from a specialist [2, 3]. For patients living in rural areas, these problems are even more pronounced since specialists, including dermatologists, most often choose to practice in urban metropolitan areas [4, 5]. Such disparities can negatively affect the outcomes of patients residing in underserved areas who lack access to dermatologic care [4]. Patients are sometimes seen by primary care clinicians who may have little or no training in dermatology, often resulting in delays or misdiagnoses, use of ineffective treatments, and unnecessary visits [4].

Barriers to Access in Rural Areas

Broadband Access Challenges

Restrictions to travel initiated by the federal government during the Spring of 2019 in response to COVID-19 brought a Renaissance to telehealth initiatives. It is likely that teledermatology will be utilized to improve access to care, long after the COVID-19 pandemic has abated. However, telehealth programs require adequate broadband access, which is often limited in rural and underserved settings.

According to a report by the Federal Communications Commission, approximately 22.3% of Americans in rural areas and 27.7% of Americans in Tribal lands lack access to broadband connectivity (defined as actual download speeds of at least 25 megabits per second (Mbps) and upload speeds of at least 3 Mbps), as compared to only 1.5% of Americans in urban areas [6]. However without access to broadband connectivity, the benefits of healthcare through telemedicine and educational opportunities such as tele-education for primary care physicians through Project ECHO are not possible. In a recent study, 806 of 2119 counties in the United States did not have a practicing dermatologist. Adequate broad band internet access was not available in 28% of these counties [7].

Intervention

In September 2020, the US Department of Health and Human Services (HHS) released *The Rural Action Plan*, which detailed strategies for overcoming barriers to providing health care in rural areas [8]. This plan cites "leveraging technology" (ie, telehealth) as a key strategy to facilitate more efficient and cost-effective delivery of quality care to rural areas. The plan acknowledges that broadband internet access necessary for successful telehealth implementation is a challenge in rural areas, and investments to build the infrastructure are needed to close the gaps between rural and urban areas [8].

Distance to Clinic and Transportation Issues

Rural residents face many barriers to accessing medical care including transportation. While solutions to these barriers prove challenging due to extensive travel distances, low population density, geographic barriers, and infrastructure concerns, there are public transportation systems that exist in rural settings to help those living in rural communities access medical care [9].

Intervention: # 1 Transport patients in need to physician's offices

These transportation systems vary depending on a multitude of factors including location and organizations that serve a local region. Some types include demand-response or dial-a-ride services, fixed route services such as shuttles or vans, flex-route services, and reimbursement programs [10, 11]. The American Public Transportation Association (APTA) noted that 9% of public transportation rides during 2000–2005 were for medical reasons in rural locations [12]. Additionally, there are governmental programs such as the Rural Transit Assistance Program (RTAP) that exist to provide funding to transit operators in rural locations for easier access to needed products and services related to transportation services. [13, 14] For more information on creating sustainable rural transportation programs, see this Rural Transportation Toolkit at: https://www.ruralhealthinfo.org/toolkits/transportation [15].

Intervention # 2: Telehealth

Teledermatology has the potential to increase access to specialized care and address the health care needs of underserved populations [16]. Since it was first introduced, teledermatology has aided in diagnosing, triaging, and managing many dermatologic conditions, ranging from inflammatory conditions to neoplastic skin lesions [17–19]. It has also been shown to decrease in-person wait times, reduce unnecessary specialty visits, and improve collaborations between dermatologists and primary care providers [20, 21]. The AccessDerm™ teledermatology initiative of the American Academy of Dermatology (AAD), for example, has expanded access to dermatologic care by focusing on patients who are unable to travel for an in-person consultation. This service is provided without cost by volunteers from the AAD.

Government deregulation recently impacted telehealth programs. During the COVID-19 pandemic, elective medical procedures and in-person care were severely limited. Clinicians and regulatory agencies worked together to find alternative ways to help patients access needed health care [22, 23]. In March 2020, the Centers for Medicare & Medicaid Services (CMS) issued a waiver expanding reimbursements for telehealth services [24]. This waiver allowed services such as teledermatology to be used for the evaluation and management of most patients and reduced the obstacles that previously prevented its use (eg, licensing restrictions and HIPAA regulations) [22, 23].

Analysis by the Centers for Medicare and Medicaid Services (CMS) of the Department of Health and Human Services (HHS) showed a weekly jump in virtual visits for Medicare beneficiaries, from approximately 14,000 pre-pandemic to almost 1.7 million in the last week of April. Additionally, nearly half (43.5 percent) of Medicare fee-for-service primary care visits

were provided through telehealth in April, compared with less than one percent in February, before the pandemic. More importantly, studies demonstrated that telehealth visits continued to be quite frequent even after in-person visits resumed in May 2020. These findings suggest that the expansion of telehealth services is likely to be a more permanent feature of the healthcare delivery system. However, teledermatology alone without access to definitive treatments, including obtaining biopsy specimens, surgical procedures, and other diagnostic and therapeutic procedures, will not adequately serve the needs of rural populations. There must be dermatologists in the region to accept these patients on referral for teledermatology to have the greatest impact on health care outcomes.

Too Few Rural Dermatologists

Chapter 1 documents this problem nicely.

Intervention: Train more dermatologists

The number of federally funded residency training slots has not increased since 1997. If new funded slots were to be made available for rural residency training programs, it is likely that they would be directed to primary care specialties. Of course, residency training programs could decide on their own to prioritize training of rural dermatologists. (See *Chap. 7: Rural dermatology residency slots: priming the pump*). In 2004 the American Academy of Dermatology unveiled a plan to invest one million dollars per year for 3 years to establish an increased number of dermatology residency positions. Ultimately, this plan failed to gain traction because a practical plan for funneling new dermatology graduates to areas where they were most needed could not be established.

Additionally, J-1 visa waiver programs like the Conrad 30 are very important as they allow rural facilities in or near Health Professional Shortage Areas (HPSAs) or Medically Underserved Areas (MUAs) to recruit Foreign Medical Graduates (FMGs) to fill vacancies which have been difficult to fill. Eight hundred to 1000 FMGs are recruited each year through the Conrad 30 program alone. Many of these physicians choose to remain in their communities past the three-year requirement, according to a 2016 WWAMI Rural Health Research Center report. The J-1 visa waiver benefits underserved communities in need of physicians as well as FMGs who want to remain in the United States. Please see Appendix 1 at the end of the chapter for the states that participate in this program.

Primary Care Physicians Could be Encouraged to Shoulder More Responsibility for Dermatologic Care

Most primary care physicians have little dermatology training, and with the paucity of dermatologists, they must handle many dermatology problems. There have been a limited number of studies evaluating skin complaints in primary care, however a 2001 retrospective review of patient charts published in the *Journal of the American Academy of Dermatology* found that about 36 percent of patients presented with at least one skin problem (over a 2-year period). For 60 percent of those patients, the issue was their primary complaint [25].

Twenty percent of the United States population lives in rural areas, but only 9% of physicians practice rurally. As a result, rural areas represent one of the largest physician-underserved populations in the country, making it difficult for primary care physicians to take on more responsibilities for dermatology care. Still, among imperfect solutions, improving the skills of primary care physicians to handle the most common dermatologic complaints is a rational approach.

Intervention: Project ECHO – Telementoring
Though this is a private, University-based program, funding can be obtained from State and Federal sources. For instance, the State of Missouri receives Medicaid grants to provide Project ECHO service because they have demonstrated beneficial results in the areas of prevention and early intervention. (See Case Study below).

A Real Life Example

The following is an excerpt from a conversation with a Michigan-based dermatologist, Barbara M Mathes, MD, FACP, FAAD, to showcase the difficulty for patients in rural Michigan to access dermatologic care. It also provides some insight to the challenges rural clinicians face in finding time to advocate on behalf of their patients.

You may not know Michigan, so I will explain a little geography to you. Michigan is made up of two peninsulas, an upper and a lower peninsula, which are surrounded by four of the Great Lakes and connected by the five-mile Mackinac Bridge. Although the Upper Peninsula (UP) makes up a third of Michigan's land mass, only 3% of its population lives in the UP. The land is rugged, forested with few small cities and towns scattered through the region. Half the population, hardy and self-reliant, lives far from towns. Winters are cold, dark and long, averaging 20 or more feet a year with temperatures often dipping well below zero. People travel long distances for many services and in winter snowmobiles can be more reliable transport than cars. For decades mining and timber industries were major employers but as the mines closed and forests were depleted in the mid 20th century, large numbers of Yoopers (as people in the UP are called) were left unemployed. More recently the area's economic situation has improved with the opening of tribal casinos, microbreweries and increased tourism.

At one time there were 3 full time dermatologists practicing in the UP, but in recent years as these doctors got older and retired or died, there have been times when the UP has had no practicing Board-certified, full-time dermatologists. Patients are left to travel to the Lower Peninsula where the nearest dermatology clinic has one board-certified dermatologist (and several non-physician clinicians) to care for all the patients), or to towns in Wisconsin. Patients may wait up to 6 months for appointments and their visits can be all day or multiday affairs when travel time is included. A one-way drive may take 6-8 hours from the northernmost reaches of the UP and involve crossing the Mackinac Bridge, at times closed due to weather conditions.

If we think about advocacy from the perspective of the state of Michigan, it's northern Lower and Upper Peninsulas with large rural expanses and low population density, we find a small number of residency-trained dermatologists providing care. It's easy to understand that there are time-

limiting factors at play. Taking time to advocate for improved care in these areas, means time away from patient care. Time off to travel to one's congressman (6 hours for me) or to Washington, DC (at least 3 flights each way) is significant. It's time away from seeing patients, some who waited months for an appointment and drove hours to the dermatologist. It's critical to recognize that there are distinct challenges regarding advocacy in rural dermatology: the problems in the UP (and in other dermatology deserts) are different from those that plague metropolitan-practicing dermatologists.

Some programs that have supported rural medical care through political action are outlined in Table 8.1.

Advocacy: The Act or Process of Supporting a Cause or Proposal

Understanding the definition of advocacy is just the start. Learning *how* to advocate, however, is difficult and mismanagement provides a quick road to failure. In fact, advocacy in the area of health policy is tremendously challenging given the polycentric nature of health care and the competing interests in the health care universe.

This chapter focuses on one type of advocacy: influencing health care policy decisions in a legislative body or administrative agency. A simple advocacy strategy is recommended with specific action steps designed to provide a roadmap culminating in a successful advocacy campaign. Case studies are presented to increase your confidence that this process can lead to solutions that improve rural dermatologic care in America.

Preparing an Advocacy Campaign: A Blueprint that Works

The most effective advocacy strategies are based on storytelling [17].

- What is the *problem* or issue you are trying to fix?
- What has *caused* the problem or issue?
- What is the proposed *solution* and how does it address the problem?

Table 8.1 Political action to achieve public funding of programs to improve rural access to dermatology

Federal Programs

Rural Business Investment Program: Rural Business Investment Company (RBIC)

Federal Rural Health Care Telecommunications Program

Medicare Incentive Payments in Health Professional Shortage Areas

Rural HealthCare Connect Fund (Reimburses cost of broad band connectivity)

Federal Opportunity Zone Economic Development Program

State Programs

Missouri Dermatologists funding of "Show-Me" Project ECHO to support education of primary care physicians in the field of dermatology (See Case Study)

State laws supporting elimination of medical school debt in exchange for rural practice

    South Dakota Recruitment Assistance Program (Incentive to return for 3 years of rural practice)

    South Dakota Rural Healthcare Facility Recruitment Assistance Program (Monetary incentive when 3 years of practice completed)

    West Virginia Rural Health Service Program (Medical school tuition coverage for each year of rural practice commitment)

State Tax Cut Programs

    Maine Certification Program for Primary Care Tax Cut

    New Mexico Rural Health Practitioner Tax Credit

    Oregon Rural Practitioner Tax Credit Program

    Maryland Income Tax Credit for Preceptors in Areas with Health Care Workforce Shortages

    Colorado Rural and Frontier Health Care Preceptors Tax Credit

Oregon Rural Medical Practitioner Insurance Subsidy Program

Western Montana Area Health Education Center Student Rural Clinical Rotations Program (Travel reimbursement for medical students on rural rotations)

Mississippi laws requiring all health insurance companies in the State of Mississippi to pay for teledermatology and to pay at the rate they pay for "in person" visits

Fig. 8.1 Blueprint for an advocacy campaign

This approach is not foreign to physicians who use this same concept every day in discussions about medical care with their patients [17]. This blueprint, however, must be combined with the correct actions steps to advance policy issues in the legislative or administrative arenas. (See Fig. 8.1) These include:

1. **Identify an advocacy champion**

   Every issue from community stop signs to national public health legislation requires one person or an invested group who deeply believe in the purpose of the issues. They must prioritize its advocacy in their lives and use their passion to enlist the support of others. Key questions to ask in order to find a champion: Who does this problem affect? Who and where are people that agree with the need to address this problem? Are they patients? Are the specialty colleagues? Will addressing this problem provide an important solution for certain stakeholders?

2. **Create a bridge to the legislature**

   In order to advance the type of political action promoted in this chapter, advocacy champions will be most successful if they partner with a person or organization that has successfully advanced a similar legislative agenda. They can help guide the process of a bill becoming law or achieving a similar policy change in an administrative arena like the Center for Medicare Services (CMS) or the Food and Drug Administration (FDA). These are government relations specialists and, for the purpose of health care activities, consider starting with the lobbyist at your state, national, or specialty medical society.

3. **Engage a legislative champion or administrative champion**

   This is the key to carving a path for your policy through the processes in either the state or federal legislature. Ideally this is a legislator that is known personally by the advocacy champion and respected by the government relations team supporting the issue. Bills most likely to be passed start with bi-partisan support in the form of "co-sponsorship", however, that is not required for success. In the case of administrative action, do you have a contact on the inside of the targeted organization who is likely to adopt your viewpoint?

4. **Develop supportive talking points**

Messaging is critical to push a policy issue out from the marketplace of ideas, into the realm of meaningful change. Specific to public policy, the message needs to be concise and persuasive, but not overly technical. It is important to know your audience and recognize legislators have very limited knowledge of healthcare. As a physician, you know about genomics, cutaneous metastatic malignancies, and the use of nanotechnology in treating diseases, however few legislators have a healthcare background to follow at that level, so in order to keep the issue at the forefront of the conversation - keep educational background information on your issue basic. A few anecdotes can be very convincing.

5. **Build a coalition**

Getting all the stakeholders on an issue at the table together is critical. This is critical to designing effective legislation and to ensure that all those impacted by the policy – defining the rules, rolling out the changes, and enforcement – are all in agreement. The most effective coalitions think broadly about their issues. Know who may be resistant to the change and invite them to the table early. This includes patients, patient advocacy groups, other specialty societies, pharmacy boards, hospital associations, the local chamber of commerce, national organizations, and the community at large. Engagement is key. And each group needs to know they are being invited to the table because their perspective is crucial to crafting and passing a successful policy or rule.

6. **Address the opposition… *Shine a light and refute***

It is always best to anticipate the talking points of proponents on the other side of your issue proactively. Prepare persuasive talking points that are designed to convince a reasonable person that your side is provides the best solution! Controlling the message sets the narrative early, well in advance of the policy becoming public.

7. **Present your case**

Once a plan has been developed using steps 1–6 above, it's time to meet with the decision makers and their staff. Use your talking points and present the issues. Take this time to assess the support/ opposition by the legislative or regulatory body and importantly, seek feedback from those that can move the issue forward.

8. **Decide whether to compromise**

Assess whether there is enough support for your proposal. If not, is there something that can be bargained away while maintaining progress toward the goal? Will an incremental change lead to further gains down the road? It's been said in politics that "you can ask for everything, but 100% of nothing is still nothing." Meaning you can try to go for exactly what you want, but if you push too hard without compromise, you will likely walk away with nothing.

9. **Understand the long game**

Devise a timeline. Decide what is most important. Be aware how changes in the political make-up of the legislature would inform your decision to push or hold back. Compromise now could lead to a better deal later.

To better understand this approach, several case studies are presented to demonstrate that success is possible.

Case Studies

Political action in the rural setting of healthcare may not look the same as it does in the metropolitan areas. This case study demonstrates that political action can impact dermatologic problems after gaining an understanding of the problem, its cause, and developing a focused plan. The case study focuses on public funding of programs that improve dermatology access to care in rural areas. This is designed to motivate dermatologists to political action by giving confidence that this work can directly benefit patients.

Case #1

Access to Care in Rural Dermatology: Telementoring

- **Problem** – Poor health outcomes related to access to care issues
- **Cause** –A lack of geographical proximity to health care providers and a lack of alternative access such as telemedicine.
- **Solution** – Implement a telehealth program that allows specialists to provide guidance to their colleagues in service to the patient achieving the best possible health outcomes

Analysis of Problem

People living in rural settings or non-MSA counties (counties not part of a Metropolitan Statistical Area (MSA), [26] across the country experience limited access to health care including dermatology. While it is challenging to provide sufficient dermatologic care to urban and suburban populations in the United States, rural populations are a unique challenge [27]. How do we provide dermatologic care to rural populations in limited resource-settings without a dermatologist nearby? This is an access to care dilemma.

Solution

There are many possible solutions to the problem of accessing dermatologic care in the rural setting. Dr. Karen Edison, MD, FAAD, Philip C. Anderson Professor and Chairwoman of the Department of Dermatology, Medical Director of the Missouri Telehealth Network & Show-Me ECHO, and Director of the Center for Health Policy (CHP) at the University of Missouri, saw this need and acted upon it. Dr. Edison, and her colleagues, worked to bring the platform, Project ECHO (Project Extension for Community Healthcare Outcomes), to the state of Missouri [28]. (See Chap. 15) Project ECHO is a novel telementoring platform that allows clinicians to join together through collaborative, case-based teleconferences to learn specialized medicine. Originally designed by Dr. Sanjeev Arora, MD, MACP, FACG, a gastroenterologist at the University of New Mexico Health Sciences Center in Albuquerque, Project ECHO aims to serve patients by training primary care physicians to provide care where they live that would otherwise not be available [29]. There is data to suggest that receiving care from clinicians mentored through Project ECHO has just as good if not better outcomes than those treated at referral hospitals with specialized care [30, 31]. It is now being utilized in a number of areas of medicine, including dermatology.

Opposition

Time – It takes time for faculty and primary care attendees to prepare cases and participate in ECHO. Rural primary care providers must be convinced that they will benefit from e-learning and expert faculty must be willing and able to commit their valuable time.

Technology difficulties – Some primary care physicians in rural areas do not have high-speed internet access that supports Zoom/WebEx or other digital conferencing platforms and many patients in rural areas do not have internet access.

Reimbursement – Rural primary care physicians may be motivated to learn, however could not cover the recurring expenses related to faculty time and cost of technology. This is especially true in the start-up phase of a Project ECHO program.

Action

Context

Dr. Edison was working at Missouri-Columbia's dermatology program {in the 1990s}, took note of the many patients who drove long distances to see her. There was poor access to dermatology expertise both geographically and economically as many dermatologists did not accept Medicaid, an ongoing issue today [32, 33]. Dr. Edison took the initiative to ramp up her teledermatology practice during this time. For many years, she, along with other faculty at the University of Missouri including Kari Martin,

MD and Jon Dyer, MD, practiced teledermatology. They served the entire State of Missouri, from the border of Iowa to the border of Arkansas via the Missouri Telehealth Network [34]. In pursuit of the provision of adequate care for Missouri's rural population, Dr. Edison became a Robert Wood Johnson Health Policy fellow in the 106th Congress between the years of 1999–2000 [35–37]. She served on the US Senate Health, Education, Labor and Pensions Committee and then served as a senior policy analyst the following year working to improve access to telehealth through facilitating legislation, the Benefits Improvement and Protection Act of 2000 (BIPA 2000) [38]. This legislation allowed for Medicare to pay for telemedicine if the patient is in a rural area [38–40].

Political Action

In the fall of 2013, a retired Missouri state representative and primary care provider, Dr. Wayne Cooper, MD reached out to Dr. Edison and her colleagues about bringing Project ECHO to Missouri. They recognized the promise of the model to increase healthcare outcomes in the state, which still had poor health indicators despite strong health foundations and academic systems [41]. It was clear that this could be accomplished with political action. Dr. Edison and her team made strides to educate and garner support from members of healthcare and non-healthcare related sectors throughout the state for implementation of this novel healthcare delivery model in Missouri. Dr. Edison and her team hosted an event to discuss Show-Me ECHO and invited those from academic health centers, deans of medical schools, health agency directors, health and senior services leadership, members of the governor's office, health foundation leaders, state senate leaders, and Dr. Sanjeev Arora, MD, the founder of Project ECHO. A second event was organized soon after at the state capital for the state assembly, health committees, appropriation and budget, and leadership. They invited Dr. Arora to address state senate and state representative leaders at a dinner meeting in Jefferson City (Missouri's state Capital). She along with

her team went on to work at a state legislative session where they educated the State Assembly, the Governor's office, and state health officials on the role Show-Me ECHO could have in increasing access to specialty medical care in the state of Missouri. Soon, everyone was supporting this effort: the entire house, entire senate, and governor's office.

> *I have never seen anything happen in politics that fast. It was just a great idea to democratize healthcare knowledge. We got 1.5 million GO operating funds in first year. Started autism, and chronic pain ECHOs first. Now received $4.5 million every year and are doing nearly 30 ECHO programs.* — Dr. Karen Edison, MD, FAAD.

Today, Dr. Edison serves as the Senior Medical Director for the Missouri Telehealth Network and Show-Me ECHO, as well as the Missouri Center for Health Policy at the University of Missouri-Columbia.

<u>Outstanding issues</u>

While broadband access is a barrier for patients in rural settings in utilizing telemedicine, as of yet, it is less of a concern through Project ECHO's platform [42]. This is because Project ECHO utilizes a provider-to-provider connection, which operates effectively through any reliable internet connection. However, the onslaught of internet activity during COVID has resulted in an increased number of connection interruptions [43].

Other barriers to implementing Project ECHO remain recruiting primary care participants to make time to attend the ECHO sessions, as well as encouraging them to bring their own cases for review and discussion. In regard to Missouri, funding has not been an issue due to Dr. Edison and her team's diligence in securing state funding for the project, but it is a challenge in other states.

Importantly, the "O" in ECHO stands for "outcomes" and Project ECHO participants continually undergo extensive evaluation of all ECHO programs to demonstrate continued return of investment (ROI). Table 8.2 includes a number of possible programs that could benefit from political action to fund initiatives that improve rural access to dermatology.

Table 8.2 Political action leading to regulatory reform to benefit rural dermatology

Expanded and more reliable broadband access

Medicaid expansion in rural communities

Allowing greater collaboration opportunities between physicians and allied practitioners to facilitate broader expansion of services

Better protecting outdoor workers from UV damage and regulation of indoor tanning in rural areas

Support regulations allowing young children to use sunscreen in school, especially in rural areas where there are insufficient shade structures

Reducing administrative burdens for rural physicians who are often in solo private practice

Case #2

Protecting Minors & Educating the Public on the Dangers of UV Radiation from Tanning Beds: Ohio as a Model

- **Problem** – Increasing incidence of skin cancer, particularly in children
- **Cause** – Increasing promotion of tanning beds to young people, compounded by limited public understanding of the risks of UV exposure
- **Solution** – New law to protect children by restricting use of tanning beds by minors

Analysis of Problem

Ultraviolet (UV) radiation is a human carcinogen that increases the risk of developing skin cancer [44]. While a myriad of genetic predispositions and environmental effects are at play, epidemiologic studies unequivocally demonstrate that UV exposure is etiologically related to melanoma [45]. Rural residents may be less aware of effective sun protection strategies compared to their urban counterparts and may be at higher risk for UV exposure due to increased outdoor UV exposure [46]. Additionally, the myth that rural youth do not have "easy access" to indoor tanning beds which emit UV radiation has been dispelled in the last decade [47]. Easy access and wide distribution of this equipment is a significant public health issue.

Solution

Legislation proposed by dermatologists can lead to policy impacting public health as demonstrated in this case study involving Ohio, House Bill (HB) 329: Prohibit Sun Lamp Tanning Services for Those Under 18. This story emphasizes the long game, often required to promote the adoption of legislation. While HB 329 is still pending in 2020, the history of legislation attempting to regulate tanning operations and restrict the use of tanning by minors goes back many years. Figure 8.2 reviews the process of a bill becoming a law, and Table 8.3 provides the timeline for this specific legislative issue in the State of Ohio:

Opposition

In an advocacy campaign that attempts to regulate or tax a particular activity or industry, you can be sure that the affected activity or industry will not see this as a benevolent gesture. Not many like to be told to pay a higher fee/tax, to be subject to new rules or regulations, or to modify their business practices.

With the tanning legislation, it was anticipated that the indoor tanning salons, their trade organization, the Indoor Tanning Association (ITA), and their customers would "have concerns," and quickly develop a group of like-minded individuals in opposition to new regulations.

Opposition research in this case was as simple as querying the websites managed by the ITA, listed below [a, b, c], and their "talking points" in opposition the proposed regulations were readily available:

- *There are health benefits to indoor tanning.*
- *The research tying indoor tanning use and incidents of melanoma is inconclusive.*
- *Parents should be able to decide whether their minor children can use indoor tanning, not big government.*
- *We're a small business. We have employees, we pay taxes, and we provide a service that people want and these new regulations will threaten our existence.*

Knowing or anticipating these opposition talking points early in the process allowed the advocacy campaign to address them head on which dramatically helped control the messaging and set the narrative early in the process.

Fig. 8.2 How a bill becomes a law

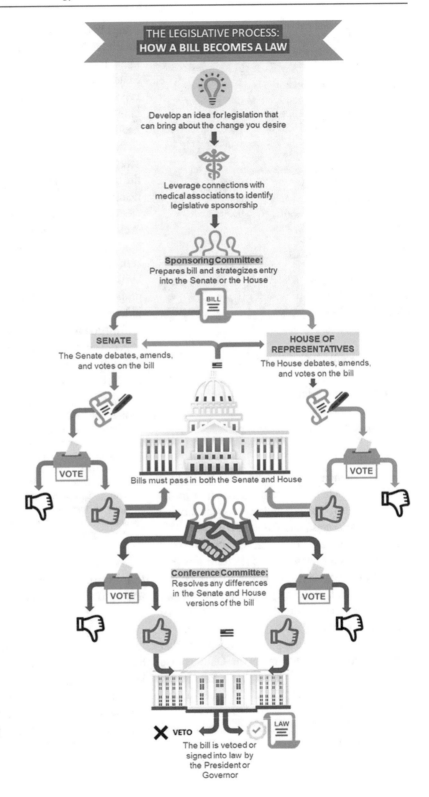

Table 8.3 Timeline for adoption of legislation in Ohio protecting minors from artificial tanning sources

2009 – HB173 introduced: required physician prescription for anyone under the age of 18.
Outcome: Failed to pass or even get a committee vote.
2011 – Same version as 2009 reintroduced in new legislative session as HB 119.
Outcome: Failed to pass or even get a committee vote.
2013 – Same version as the 2011 reintroduced in new legislative session as HB 131.
Outcome: The bill gets hearings but is amended as follows:

  Anyone over 18 must sign a consent form.
  Anyone 16–17 must get in-person parental consent; consent is valid for 45 days.

Anyone under 16 must get in-person parental consent and the parent must stay at the tanning facility while the minor uses the tanning bed.
2014 – The legislation clears both the House and Senate and is signed into law by the Governor.
2015 – Law becomes effective.

Note the misleading website titles:

(a) http://tanningtruth.com
(b) https://americansuntanning.org
(c) https://smarttan.com/salons.php

Action

Policy agendas require a "champion" and Dr. Louis Barich championed this legislation in the State of Ohio. He passionately organized his patients, office staff, and fellow physicians. With the guidance of the Ohio State Medical Association, he tirelessly advocated for this legislation to protect children. Given the difficulties Ohio faced with staunch opposition by the Chair of the House Health Committee, every ounce of Dr. Barich's strength was required, as well as numerous board-certified dermatologists that provided supportive testimony in Ohio House and Senate Committees, including doctors Robert Brodell, Brett Coldiron, Elizabeth Kiracofe, and Shannon Trotter.

Each of the dermatologists who provided testimony at the statehouse over the years had a few things in common. They were active members of their state's specialty and medical societies, the Ohio Dermatological Association and the Ohio State Medical Association. Their board-certification and affiliations with organized medicine lend credibility to the proposal for a public health initiative.

Next, it was important to reinforce the clear scientific evidence shared in testimony to bolster the measure's role in protecting the public's health. Specifically, booklets were created as "leave behinds" and shared with legislators after each meetings highlighting the association of tanning bed use (UV exposure) with increased skin cancer risk, and also debunking claims made by the Indoor Tanning Association emphasizing proposed necessity of the vitamin D derived from indoor tanning.

Testimony by Dr. Trotter detailed first-hand experience and tells a story making the message relatable and affords human context as to why this bill matters.

I have had many personal experiences with melanoma and skin cancer related to tanning bed use. I directed the melanoma clinic… many of these women reported tanning bed use starting in their teens… One of them told me, "I never would have tanned if I would have known the risk. I started at 15. My sister was only 14 and wanted to tan…" This patient passed away at age 32, three months after we met. This is just one example of how I've seen tanning beds linked to skin cancer and its devastating effects on patients.

Evidence was then provided detailing legislation measures enacted at both the state, federal and a global level including mandates already put forth by the US Surgeon General, the World Health Organization (WHO), and the US Food and Drug Administration (FDA), to support the case: The US Surgeon General noted that state laws restricting youths from tanning are an effective policy tool. The WHO supports actions to ban minor access to tanning beds. The US FDA recommends against tanning bed use by minors under 18 and black-box warnings must be placed on sunlamp products. Parallels were drawn related to other products that threatened public health in the past highlighting laws prohibiting alcohol and tobacco use in minors. Finally, the economic costs were highlighted.

Ohio Dermatological Association and the Ohio State Medical Association believe that... requiring appropriate oversight of the indoor tanning industry will have a profound effect on improving public health and reducing overall health care costs. Annually, about $3.3 billion of skin cancer treatment costs are attributable to melanoma. Of course, this figure does not begin to account for the tragic loss of life from this devastating disease.

Outstanding issues

"When a garden is planted, it needs continuous maintenance, or it soon may be overtaken with weeds." – Robert Brodell, MD.

This new law requires careful monitoring of the Board of Cosmetology which was charged with enforcement of these tanning regulations and fending off any attempts by the tanning industry to weaken the regulations. Table 8.3 provides potential targets for improving rural access to dermatology through the enactment of new laws.

Case #3
Access to Treatment: Compounding

- **Problem** – Regulations limiting the ability of physicians to compound certain medications in their offices have resulting in delays in treatment and increased costs to patients.
- **Cause** – The state board of pharmacy promulgated regulations that significantly restricted in-office compounding of medications based on updated regulations from the FDA and USP.
- **Solution** – Repeal or amend the regulations to permit in-office compounding under certain circumstances.
  Context

During the New England meningitis outbreak in September 2012, almost 100 people died due to contaminated medications from a New England compounding pharmacy. This horrific event prompted Congress to pass legislation, which directed the FDA to create more strict compounding policies and review safety measures. Both rural and non-rural clinics will be impacted by the benefits and unintended consequences of these new regulations.

Problem

Modifications of compounding policies impact all of medicine, specifically dermatology, as our daily practice involves in-office preparations of medications and treatment with compounded products.

A further complicating issue would require in-office preparations to be mixed and then immediately used for the patient. This would require dermatologists to prepare syringes one patient at a time requiring more time and effort by office staff without providing any benefits. This proposed standard came from the U.S. Pharmacopeia (USP), an independent (non-government) standard setting organization that publishes Chapter 797, which governs the use of compounded medications.

Solution

Advocacy efforts directed at education of the FDA and USP on how the unintended consequences of the new regulations will impact the timeliness and tolerability of outpatient procedures for dermatology patients, despite the clear intent of the new rules being sought for the overall protection of patients. Dermatologists educated representatives and the FDA and other stakeholders about the value of mixing lidocaine with epinephrine to enhance vasoconstriction; mixing lidocaine with sodium bicarbonate to decrease pain associated with injection of anesthetics; and, diluting triamcinolone with saline or lidocaine for intralesional steroid injections. This does not have the risks associated with compounding performed in a pharmacy for patient use at home.

In office preparations of products, such as buffered lidocaine, are a critical component of practice for a board-certified dermatologist, especially in helping reduce patient discomfort during a plethora of skin and surgical procedures. As we navigate the complexities of the US Pharmacopeia, the CDC and the FDA, it remains critical that dermatological organizations and physician leaders collaborate in as many ways as possible to ensure access to compounded medications for our specialty. – Seemal Desai, MD, FAAD.

Opposition

Federal regulators were unwilling to carve out exceptions for the in-office preparations needed to treat dermatology patients, and the state board of pharmacy took a similar interpretation of the proposed rule.

Action

In-office preparation is different from pharmacy-based compounding and has not been associated with morbidity and mortality. As such, advocacy efforts focused on the State Board of Pharmacy to modify their rule and the impact it would have on in-office preparation. Leading these efforts were the American Academy of Dermatology (AAD) and American Society of Dermatologic Surgery (ASDS) leading the charge, supported by state dermatologic societies, as well as the American Medical Association (AMA). The FDA passed guidance on in-office preparations in Fall of 2018, which stated that they would not be enforcing restrictions or audits on physicians' offices that are performing in-office preparations for patients. The FDA is continuing to work collaboratively with physicians, acknowledging the issues raised by dermatologists.

Future advocacy considerations include a 12-hour exemption. This way, practices will have an entire business day to use buffered lidocaine. Currently, advocacy efforts are focused on testing these prepared syringes now in special laboratories, and it is hopeful that this testing will show the USP, CDC, and FDA that

these prepared syringes have no microbial growth following the specified time frame and can be administered within a business day. The goal is to receive a dedicated monograph in the USP reference book.

How to Get a Seat at the Table: Get Involved

"If you don't have a seat at the table, you're probably on the menu." Elizabeth Warren

As we have demonstrated in the case studies highlighted in this chapter, in order to achieve a health policy objective, there is a need for *active, personal* participation in the advocacy process. But how to start that process can be daunting because of a lack of awareness of resources that may be available to assist in this endeavor.

Fortunately, there are a number of resources that can help you get "a seat at the table" so you are in a position to have meaningful participation and contribution in the advocacy process.

Listed below are some of those resources and tips on how to get involved:

(a) Advocacy agendas: Be aware of the local, state and federal issues that impact your profession and your patients. Check your state and national medical and specialty society websites to see if there are issues currently on the agenda that are of interest to you. https://www.aad.org/member/advocacy/priorities

(b) Participate on Committees: All of the local, state and national physician organizations have committees that specialize in advocacy or issue-specific agendas. Find a way to get involved in one of those committees. You will then have an opportunity to raise your particular advocacy issue and start the process of getting wider support and potential "official policy" of these organizations. Here are a few examples:
   (i) National medical societies like AAD, ASDS, AMA, NMA, etc.
   (ii) State medical societies
   (iii) Hospital medical staff committees

(iv) Chambers of commerce or other business organizations. Many of these groups have a health care committee.

(v) Elected official "Kitchen Cabinets" – Ask your local elected official if they have a health care advisory group. If so, ask to join. If not, ask to start one.

(vi) Health Policy Fellowships and Leadership Programs

Local, state and national medical and specialty societies are a resource for "deep dive" opportunities to get involved in politics and policy

1. American Medical Association - https://www.ampaconline.org/political-education/ampac-campaign-school/
   (a) AMPAC Campaign School offers an immersive program to learn more about the legislative process and consider running for office
2. National Academy of Medicine - https://nam.edu/programs/health-policy-educational-programs-and-fellowships/
3. Robert Wood Johnson Foundation Fellowship - https://www.healthpolicyfellows.org/
   (a) Where Dr. Karen Edison and many others learned to be effective advocates

Maxims of the Legislative Process

Things just don't happen... It is the legislature or administrative agency that makes the final decisions on policy. Kind of like finding a turtle on a fence post. Someone put it there.

Find a legislator champion... Someone that has a personal interest in an issue and someone that has the capability to engineer the proposal through the process.

Be humble and kind... You may be smarter than most legislators and you may be more righteous than most legislators. But they were elected to make decisions. Threats, or a condescending approach never works. In other words, don't curse the alligators right before you are going to cross the swamp.

Hone the message... Know your audience and don't assume legislators know your issue. You may know about genomics, cutaneous malignant melanoma and the use of nanotechnology in treating diseases. But few legislators have a health care background so keep it basic.

Build a coalition... There is strength in numbers. Find common allies and try to get them marching in the same direction.

Addressing the opposition... Shine a light and refute. You can anticipate the other side's talking points and you need to advise why their arguments against your proposal are not persuasive.

When to compromise... You can ask for everything but remember, 100% of nothing is still nothing. Most times an incremental change leads to further gains down the road.

All politics is local... Every politician will gauge how a legislative proposal will help or hurt them with the voters back home. Advocate your proposal in a way that highlights a local, hometown impact.

Conclusion

To be sure, politics and policy is a participation sport. If you want to see change, you have to get involved. But equally important, you also need a roadmap on how to achieve success. This chapter began with a blueprint to produce a successful advocacy campaign and demonstrated a roadmap for successful advancements of health policy objectives. Several suggestions and resources serve to encourage dermatologists to find their voice in political action.

Taken together – having an advocacy strategy and getting involved – physicians can have a significant opportunity to advance health policy objectives that foster the medical profession and support the health and well-being of patients.

Rural Dermatology and Telemedicine: Policy Resources

American Academy of Dermatology – Position Statement on Teledermatology	https://server.aad.org/Forms/Policies/Uploads/PS/PS-Teledermatology.pdf
American Academy of Family Physicians – Telehealth Toolkit	https://www.aafp.org/dam/AAFP/documents/practice_management/telehealth/2020-AAFP-Telehealth-Toolkit.pdf
American College of Physicians – Policy Recommendations to Guide the Use of Telemedicine in Primary Care Settings	https://www.acpjournals.org/doi/10.7326/M15-0498?articleid=2434625
American Medical Association – CMS Payment Policies & Regulatory Flexibilities During COVID-19 Emergency	https://www.ama-assn.org/practice-management/medicare/cms-payment-policies-regulatory-flexibilities-during-covid-19
American Medical Association – Ethical Practice in Telemedicine (Code of Medical Ethics Opinion 1.2.12)	https://www.ama-assn.org/delivering-care/ethics/ethical-practice-telemedicine
American Medical Association – Telemedicine Policy	file:///C:/Users/ericv/Downloads/telemed-policy.pdf
American Telemedicine Association – Action Briefs	https://info.americantelemed.org/ata-action-briefs
American Telemedicine Association – Policy Principles	https://f.hubspotusercontent30.net/hubfs/5096139/Files/Policy%20Docs_letters,%20RFI,%20etc./Policy%20Principles%202020%20FINAL.pdf
American Telemedicine Association – Telehealth Flexibilities During the COVID-19 Pandemic and the ATA's Recommendations for Permanent Policy	https://f.hubspotusercontent30.net/hubfs/5096139/Files/Policy%20Docs_letters,%20RFI,%20etc./ATA%20Permanent%20Policy%20Recommendations%20Chart_8.26.20.pdf
Center for Connected Health Policy –State Telehealth Laws & Reimbursement Policies	https://www.cchpca.org/sites/default/files/2020-05/CCHP_%2050_STATE_REPORT_SPRING_2020_FINAL.pdf
Center for Connected Health Policy – Telehealth Coverage Policies in the Time of COVID-19 to Date	https://www.cchpca.org/sites/default/files/2020-03/CORONAVIRUS%20TELEHEALTH%20POLICY%20FACT%20SHEET%20MAR%2016%202020%203%20PM%20FINAL.pdf
Medicaid – Telemedicine	https://www.medicaid.gov/medicaid/benefits/telemedicine/index.html
National Rural Health Association – Telehealth In Rural America	https://www.ruralhealthweb.org/NRHA/media/Emerge_NRHA/Advocacy/Policy%20documents/2019-NRHA-Policy-Document-Telehealth-In-Rural-America.pdf
National Rural Health Association – Telemedicine Reimbursement	https://www.ruralhealthweb.org/NRHA/media/Emerge_NRHA/Advocacy/Policy%20documents/2019-NRHA-Policy-Document-Telemedicine-Reimbursement.pdf
Taskforce on Telehealth Policy (TTP) – Findings and Recommendations	https://www.ncqa.org/wp-content/uploads/2020/09/20200914_Taskforce_on_Telehealth_Policy_Final_Report.pdf
US Department of Health and Human Services – Rural Action Plan	https://www.hhs.gov/sites/default/files/hhs-rural-action-plan.pdf
US Department of Health and Human Services, Office of Inspector General (OIG) – Waiving Telehealth Cost-Sharing During COVID-19 Outbreak	https://oig.hhs.gov/fraud/docs/alertsandbulletins/2020/policy-telehealth-2020.pdf
US Department of Health and Human Services – FAQs on OIG Telehealth Policy During COVID-19	https://oig.hhs.gov/fraud/docs/alertsandbulletins/2020/telehealth-waiver-faq-2020.pdf

Acknowledgements A special thank you Medical Doctor candidate, Neha Udayakumar, for your talent and excellent work designing and creating the graphics in this chapter. Sincere appreciation to Dr. Barbara Mathes for your time and willingness to provide insight into rural dermatology and advocacy, and Dr. Karen Edison for your incredible work advocating on behalf of our rural patient population in dermatology, so inspiring to all of us! Also, many thanks to Dr. Edison, Dr. Seemal Desai, and Rachel Mutrux for your time spent reviewing and editing the cases presented.

Appendix: States with J-1 Visa Waiver Programs to Improve Rural Access to Physicians

J-1 Visa Waiver Programs – https://www.ruralhealthinfo.org/topics/j-1-visa-waiver/funding
Virginia Conrad
California
Alabama
Alaska Conrad
Appalachian Regional Commission (ARC)
Arkansas Visa Waiver Program
Colorado
Connecticut
Delaware Conrad State
Delta Doctors
Georgia
Hawaii State Conrad State 30
Idaho Conrad
Illinois
Kansas
Louisiana Conrad 30
Mississippi Conrad State 30
Montana
Nebraska State 30
Nevada Conrad State 30
New Hampshire
New Jersey
New Mexico Conrad
North Carolina
North Dakota Conrad
Oklahoma
Oregon
Pennsylvania
South Caroline
South Dakota
Tennessee State Conrad
Utah Conrad 30
Vermont
Washington
West Virginia Conrad 30
Wisconsin Conrad 30
Wyoming J-1 Conrad 30

References

1. Potentially Excess Deaths from the Five Leading Causes of Death in Metropolitan and Nonmetropolitan Counties — United States, 2010–2017 I MMWR. https://www.cdc.gov/mmwr/volumes/68/ss/ss6810a1.htm. Accessed 24 Jan 2021.
2. JMIR Dermatology - Impact of an Intrainstitutional Teledermatology Service: Mixed-Methods Case Study. https://derma.jmir.org/2018/2/e11923/. Accessed 24 Jan 2021.
3. 2017 Survey of Patient Appointment Wait Times. https://www.merritthawkins.com/news-and-insights/thought-leadership/survey/survey-of-physician-appointment-wait-times/. Accessed 24 Jan 2021.
4. Feng H, Berk-Krauss J, Feng PW, Stein JA. Comparison of dermatologist density between urban and rural counties in the United States. JAMA Dermatol. 2018;154:1265–71.
5. Glazer AM, Rigel DS. Analysis of trends in geographic distribution of US dermatology workforce density. JAMA Dermatol. 2017;153:472–3.
6. New FCC Report Shows Digital Divide Continuing to Close I Federal Communications Commission. https://www.fcc.gov/document/new-fcc-report-shows-digital-divide-continuing-close-0. Accessed 24 Jan 2021.
7. Solomon ZJ, Ramachandran V, Kohn TP, Nichols PE, Haney NM, Patel HD, Johnson MH, Koshelev MV, Dao H. The association of broadband internet access with dermatology practitioners: an ecologic study. J Am Acad Dermatol. 2020;83:1767–70.
8. HHS Releases Healthcare Rural Action Plan. https://www.natlawreview.com/article/hhs-releases-rural-action-plan. Accessed 25 Jan 2021.
9. Bayne A, Siegfried A, Stauffer P, Knudson A (2018) Promising Practices for Increasing Access to Transportation in Rural Communities. Rural Evaluation Brief.
10. Mattson J 2017 Rural Transit Fact Book. 60.
11. Funding Rural Public Transit. https://www.aarp.org/livable-communities/act/walkable-livable-communities/info-12-2012/funding-rural-public-transit.html. Accessed 24 Jan 2021.
12. Rural Public Transportation Systems I US Department of Transportation. https://www.transportation.gov/mission/health/Rural-Public-Transportation-Systems. Accessed 24 Jan 2021.
13. About The National Rural Transit Assistance Program. https://www.nationalrtap.org/About/History-and-Mission. Accessed 24 Jan 2021.
14. (2015) Rural Transportation Assistance Program – 5311(b)(3). In: Federal Transit Administration. https://cms7.fta.dot.gov/funding/grants/rural-transportation-assistance-program-5311b3. Accessed 24 Jan 2021.
15. Rural Transportation Toolkit – Rural Health Information Hub. https://www.ruralhealthinfo.org/toolkits/transportation. Accessed 24 Jan 2021.
16. Uscher-Pines L, Malsberger R, Burgette L, Mulcahy A, Mehrotra A. Effect of teledermatology on access to

dermatology care among medicaid enrollees. JAMA Dermatol. 2016;152:905.

17. Dhaduk K, Miller D, Schliftman A, Athar A, Al Aseri ZA, Echevarria A, Hale B, Scurlock C, Becker C, editors. Implementing and optimizing inpatient access to dermatology consultations via telemedicine: an experiential study. Telemed e-Health. 2020;27:68–73.

18. Holmes AN, Chansky PB, Simpson CL, editors. Teledermatology consultation can optimize treatment of cutaneous disease by nondermatologists in under-resourced clinics. Telemed e-Health. 2019;26:1284–90.

19. Chuchu N, Dinnes J, Takwoingi Y, et al., editors. Teledermatology for diagnosing skin cancer in adults. Cochrane Database Syst Rev. 2018;12:CD013193.

20. Rajda J, Seraly MP, Fernandes J, Niejadlik K, Wei H, Fox K, Steinberg G, Paz HL. Impact of direct to consumer store-and-forward teledermatology on access to care, satisfaction, utilization, and costs in a commercial health plan population. Telemed e-Health. 2017;24:166–9.

21. Maloney M. Leveraging teledermatology for patient triage. J Dermatol Nurses Assoc. 2019;11:265–8.

22. Wang RH, Barbieri JS, Nguyen HP, Stavert R, Forman HP, Bolognia JL, Kovarik CL. Clinical effectiveness and cost-effectiveness of teledermatology: where are we now, and what are the barriers to adoption? J Am Acad Dermatol. 2020;83:299–307.

23. Gupta R, Ibraheim MK, Doan HQ. Teledermatology in the wake of COVID-19: advantages and challenges to continued care in a time of disarray. J Am Acad Dermatol. 2020;83:168–9.

24. MEDICARE TELEMEDICINE HEALTH CARE PROVIDER FACT SHEET | CMS. https://www.cms.gov/newsroom/fact-sheets/medicare-telemedicine-health-care-provider-fact-sheet. Accessed 24 Jan 2021.

25. Lowell BA, Froelich CW, Federman DG, Kirsner RS. Dermatology in primary care: prevalence and patient disposition. J Am Acad Dermatol. 2001;45:250–5.

26. Bureau UC About. In: The United States Census Bureau. https://www.census.gov/programs-surveys/metro-micro/about.html. Accessed 24 Jan 2021.

27. Vaidya T, Zubritsky L, Alikhan A, Housholder A. Socioeconomic and geographic barriers to dermatology care in urban and rural US populations. J Am Acad Dermatol. 2018;78:406–8.

28. Lewis H, Becevic M, Myers D, Helming D, Mutrux R, Fleming D, Edison K. Dermatology ECHO – an innovative solution to address limited access to dermatology expertise. Rural Remote Health. 2018;18:4415.

29. Arora S, Geppert CMA, Kalishman S, et al (2007) Academic Health Center Management of Chronic Diseases through Knowledge Networks: Project ECHO. Acad Med. https://doi.org/10.1097/ACM.0b013e31802d8f68.

30. About Project Echo. https://www.aap.org/en-us/professional-resources/practice-transformation/echo/Pages/About-Project-Echo.aspx. Accessed 25 Jan 2021.

31. Arora S, Thornton K, Murata G, et al. Outcomes of hepatitis C treatment by primary care providers. N Engl J Med. 2011; https://doi.org/10.1056/NEJMoa1009370.

32. Resneck J, Pletcher MJ, Lozano N. Medicare, Medicaid, and access to dermatologists: the effect of patient insurance on appointment access and wait times. J Am Acad Dermatol. 2004;50:85–92.

33. Alghothani L, Jacks SK, Vander Horst A, Zirwas MJ. Disparities in access to dermatologic care according to insurance type. Arch Dermatol. 2012;148:956.

34. Becevic M, Sheets LR, Wallach E, McEowen A, Bass A, Mutrux ER, Edison KE. Telehealth and telemedicine in Missouri. Mo Med. 2020;117:228–34.

35. Medicine (US) I of (2003) ROBERT WOOD JOHNSON HEALTH POLICY FELLOWSHIPS PROGRAM. National Academies Press (US).

36. Medicine (US) I of (2009) Producing Tomorrow's Health Leaders: Fellowships at the Institute of Medicine. National Academies Press (US).

37. National Health Policy Forum | Karen E. Edison, MD. http://www.nhpf.org/speakerbio_karenedison. Accessed 25 Jan 2021.

38. Benefits Improvement and Protection Act of 2000 (BIPA). NCHA.

39. Thomas WM (2000) H.R.5661 - 106th Congress (1999–2000): Medicare, Medicaid, and SCHIP Benefits Improvement and Protection Act of 2000. https://www.congress.gov/bill/106th-congress/house-bill/5661. Accessed 25 Jan 2021.

40. The Medicare, Medicaid, and SCHIP Benefits Improvement and Protection Act of 2000 (BIPA) | National Rural Health Resource Center. https://www.ruralcenter.org/resource-library/the-medicare-medicaid-and-schip-benefits-improvement-and-protection-act-of-2000. Accessed 25 Jan 2021.

41. Missouri Health Assessment | Health & Senior Services. https://health.mo.gov/data/mohealthassess/index.php. Accessed 25 Jan 2021.

42. Rural Project Summary: Project ECHO® – Extension for Community Healthcare Outcomes - Rural Health Information Hub. https://www.ruralhealthinfo.org/project-examples/733. Accessed 25 Jan 2021.

43. Knowles C Internet outages drastically increased during COVID-19 lockdowns, report finds. https://securitybrief.eu/story/internet-outages-drastically-increased-during-covid-19-lockdowns-report-finds. Accessed 25 Jan 2021.

44. Mancebo SE, Wang SQ. Skin cancer: role of ultraviolet radiation in carcinogenesis. Rev Environ Health. 2014;29:265–73.

45. Landi MT, Bishop DT, MacGregor S, et al. Genome-wide association meta-analyses combining multiple risk phenotypes provide insights into the genetic architecture of cutaneous melanoma susceptibility. Nat Genet. 2020;52:494–504.

46. Luong J, Davis RE, Chandra A, et al. A cross-sectional survey of prevalence and correlates of sunscreen use among a rural Tri-State Appalachian population – PubMed. Arch Dermatol Res. 2020;

47. Olson AL, Carlos HA, Sarnoff RA. Community variation in adolescent access to indoor tanning facilities. J Community Health. 2013;38:221–4.

Academic Rural Dermatology Offices

9

Hannah R. Badon, Stephen E. Helms, and Amy E. Flischel

With ever-evolving advances and discoveries in the fields of science and medicine, physicians are often referred to as lifelong learners. When these learners become involved with academic medicine, they also become lifelong teachers. While teaching generally conjures images of formal lectures, blackboards and desks in a school room, the final preparation of the clinical physician requires that modern medical education move outside of the didactic teaching of the classroom. In fact, physician-teachers are frequently tasked with educating future healers during clinic hours and hospital rounds, a practice largely attributed to the foresight of Sir William Osler. Moreover, including rural physicians in this corps of physician-educators leads to some notable advantages.

The Historical Basis for Modern Clinical Medical Education

Sir William Osler is credited with a number of medical achievements with his influence clearly present in the halls of today's medical schools.

"We are all teachers and we are all learners."—Drew Gilpin Faust [1]

H. R. Badon · S. E. Helms (✉)
Department of Dermatology, University of Mississippi Medical Center, Jackson, MS, USA
e-mail: hbadon@umc.edu; shelms@umc.edu

A. E. Flischel
Department of Dermatology and Pathology, University of Mississippi Medical Center, Jackson, MS, USA

Many physicians know the clinical significance of "Osler's nodes," but few have been taught about Osler's significant influence on modern medical curricula in internal medicine and medical school. William Osler lived from 1849 to 1918 and received his medical degree from McGill University in Montreal, Canada [2]. He continued his postgraduate studies in Europe under John Burdon Sanderson, a well-known physiologist during the nineteenth century [3]. His illustrious teaching career started at McGill University and was followed by professorships at the University of Pennsylvania, Johns Hopkins University, and finally Oxford University. He is perhaps most famously credited with the introduction of the concept of internal medicine to the United States, and he is remembered for the development of residency training and his passion for bedside medical student teaching [2]. Teaching at the bedside had long been an established educational methodology in Europe. Introducing medical students to hospital wards, however, was a novel idea in the United States [4]. Osler was known for conducting ward rounds with upwards of thirty students, and encouraged strong history taking and a patient-centered approach [4]. Medical educators continue to follow Osler's approaches to educating medical students and young physicians. It is becoming increasingly clear that medical students and residents can benefit when their training, at least in part, is enriched by experiencing this type of exposure and the expertise of rural academic or private practice physicians.

Recognizing Health Disparities and Working to Minimize Them

"Listen to your patient, he is telling you the diagnosis."—Sir William Osler [4]

The authors recognize that the term "rural" can be defined in many ways as it is applied to individual states in the United States or the world. The consensus utilitarian definition that best fits the purposes of the medical educator with a focus on underserved communities is simply "non-urban or nonmetropolitan" [5]. While healthcare disparities can be related to a variety of metrics in dermatology, access to dermatological care in rural parts of the United States is particularly well-documented. (see Chap. 1) Similar disparities exist for primary care physicians. (See Chap. 1) Many physicians simply prefer to reside in urban areas. Aside from big city amenities, these areas offer job opportunities at academic medical centers and large private practices favored by many young physicians. These options do not exist in many rural areas. According to the 2009 American Academy of Dermatology (AAD) practice profile survey, 38% of dermatologists reported a shortage of health care providers in their area with this shortage being greatest in rural areas [6].

Can anything be done to improve this situation? It has been found that a rural clinical rotation during medical school is the strongest predictor for a decision to practice in a rural area in the future [7]. Additionally, students who grow up in non-urban areas are more likely to return to a similar area to practice medicine [8]. Besides encouraging students to seek a career in rural medicine, education within a rural setting can provide several other distinct advantages. First, patients tend to present later in the disease state and provide a greater variety of physical presentations [9]. Secondly, rural practices provide an additional patient population for new medical students to observe and study. As medical schools grow, they may need these new sources of patients to provide clinical experiences for their students. Thirdly, the opportunity to learn procedures and perform bedside tests is increased in these prac-

tices where referral for subspecialty care is often limited. In fact, many patients in rural areas are unwilling and or unable to travel to a distant specialist for surgery or testing [10]. The rural dermatologist often becomes a "jack of all trades" and is able to share this experience with physicians in training.

Medical Student and Resident Opportunities for Rural Training

Unfortunately, there are relatively few opportunities for medical students to gain rural education exposure. Based on their experience training medical students at rural offices, the University of Mississippi Medical Center (UMMC) received a grant from the Health Resources and Services Administration in 2020 to develop a broad rural curriculum for medical students. UMMC will receive 1.9 million dollars each year for four years to fund IMPACT the RACE: Improved Primary Care for the Rural Community through Medical Education. This project's aim is to enhance the education of medical students and encourage them to choose residencies and careers in rural Mississippi. One rural faculty member working with these medical students is a dermatologist.

Rural opportunities for dermatology residents are especially rare. An assessment of 147 dermatology residency program websites revealed that only 15 (10.2%) were based in communities of 50,000 or fewer without a nearby associated metro area. Additionally, only two of the larger metropolitan based programs (<1%) offered a curriculum with a rural track. Another fifty-two programs (35%) reported optional rural rotations and/or clinical electives that could be utilized for a rural experience [11]. At UMMC, the department of family medicine has implemented a curriculum for rural medical education. During an eight-week clerkship, third-year medical students spend one month with a family physician at UMMC, and one month with a family physician in a rural area. These physicians must complete an application for review and approval to become involved in the program. Students are encouraged to return to an area of the state where they have family or friends to provide housing during this rural educational experience. Similarly, the University of Kansas School of Medicine offered a rural family medicine clerkship and conducted a study in 1999 [12]. It was of importance to note that students in the rural clerkship were not at an academic disadvantage and often had a higher clerkship grade [12]. Additionally, it was determined that urban-based students have experienced a positive change in opinion regarding rural practice following immersion in a rural education program [13]. These family medicine clerkship may serve as a model for other rural education opportunities, including dermatology rotations.

Academic Options for Rural Dermatologists

The rural dermatologist can become involved in a larger academic center's practice in a variety of ways. Serving as a bedside teacher to medical students and residents is a valuable addition to all educational curricula. This engagement with an academic center could lead to further involvement in research projects, textbook chapters, and case reports by both medical students and residents. (See Chaps. 16 and 17) Additionally, outpatient clinics at academic centers are often in need of volunteers to staff resident-managed clinics. These outside providers can offer a unique perspective and alternative modalities of diagnosis and treatment. As rural dermatologists become more connected with academic centers, these opportunities may expand and evolve to meet the needs of all involved parties.

Rationale for Medical Students and Dermatology Residents to Choose a Rural Clinical Experience

Dermatologists in rural settings can serve as valued members of a medical school's educational team providing distinctive benefits for the learner. (See Table 9.1) Rural electives provide an oppor-

Table 9.1 Advantages of Learning in a Rural Dermatology Office for the Medical Student/Resident (learner)

1. Exposure to new patient populations
2. Large numbers of patients with varied complaints provide an excellent setting for problem based learning [15, 18]
3. Greater opportunities for learning bedside tests and procedures
4. Development of a newfound awareness of the health needs of rural communities
5. Introduction of rural healthcare career opportunities
6. One-on-one learning experience that may be diluted by other medical students and residents at the medical center

tunity for students and residents to gain exposure to a patient population that is different than their home institution. Specialists practicing in remote areas are not often afforded the luxury of easily referring a patient to another provider. As such, they are often proficient in diagnosing and treating a wide variety of dermatologic diseases and syndromes without the aid of as many supportive resources. In addition, medical students and residents have the opportunity to learn alternative strategies for both diagnosis and treatment when compared to their faculty in the Ivory Tower at the medical center. Rural electives also provide an opportunity for medical students and residents who grew up in a rural area to spend a month learning while living at home. This may help them to assess the benefits of returning home to practice when their training is complete.

Rationale for Busy Rural Physicians to Teach Medical Student and Residents?

There are distinct advantages for rural clinicians to involve themselves in teaching. (See Table 9.2). Teaching students and residents can provide increased opportunities for learning for the attending physician as teaching may require a more complete understanding of the clinical problem at hand. As noted by the French author, Joseph Joubert, "To teach is to learn twice." Additionally, teaching can moti-

Table 9.2 Advantages of Teaching Dermatology for the Rural Practitioner (Teacher/Professor)

1. Collaboration with an academic center allowing: (A) participation in clinical research projects; (B) an opportunity to attend grand rounds; and, (C) the potential for developing a more personal/collegial relationship with sub-specialists
2. Opportunity to utilize the large and varied patient base of a rural practice (an ideal setting for problem based teaching) [15, 18]
3. CME credits [14]
4. Better understanding of material when preparing to teach
5. Stimulation/motivation from medical students and residents to keep up with latest literature
6. Opportunity to attend faculty development meetings at the medical school
7. Opportunity for learning from both students and residents
8. Potential extra income as a "part-time" teacher
9. Opportunity to engage in a labor of love by investing one's passion into teaching

vate the clinician in their efforts at self-directed lifelong learning to stay abreast of new dermatologic developments and treatments. Rural dermatologists engaging in educational processes, may also have special opportunities to collaborate with the sponsoring academic center through Grand Rounds events, teledermatology, and virtual or live continuing medical education (CME). Furthermore, rural dermatologists involved in medical school or resident teaching will find it easier to collaborate with faculty at an affiliated university. This may lead to the pride that comes from seeing one's ideas in print as peer reviewed journal articles and book chapters. Project ECHO (Extension for Community Healthcare Outcomes), presents remote, case-based education via video or teleconference [14]. It was initially created to cover the treatment gap surrounding chronic hepatitis C in New Mexico [14]. Project ECHO features a "hub-and-spoke" system with experts located at the hub. Not only does participation lead to continued collaboration between rural physicians and their urban counterparts, but it is also rewarded with no-cost

CME credits [14]. Finally, many physicians find that office-based teaching is just plain enjoyable and makes them feel they are "giving back."

Developing a Cadre of Rural Dermatology Clinicians

While the art of teaching is often taught and honed in medical school and residency, dermatologists outside of an academic center may feel that they have lost that skill after years of private practice. However, the word doctor is derived from the Latin term *docere*, meaning "to teach" [15]. Physicians who teach every one of their patients at every clinical encounter, are often better teachers of medical students and residents than they might realize. Involving rural clinicians may require some encouragement from their academic colleagues along with "nuts and bolts" faculty development.

A guide for developing emergency medicine residency rotations has been created for program directors to encourage rural clinical experiences in their field [16]. Many of the key features of this guide can be readily applied to rural dermatology. They suggest that guaranteeing adequate supervision is the most important consideration when developing a rural rotation [16]. In addition to having appropriate skills, education and training, a system for faculty development must be developed to guarantee that faculty supervisors are well-versed in program requirements. Furthermore, the rural faculty must be committed to provide the resident's program director with regular assessments and any urgent updates regarding performance of residents [16].

What better way can we express the benefits of students or residents retaining what they have learned during their training than what has been stated by Benjamin Franklin when he wrote, "Tell me and I forget. Teach me and I remember. Involve me and I learn." To be immersed in one-on-one experiences in a busy private practice has many advantages for both the learner and the teacher.

Improving access to care to the community, region, or state is a common component of medical school mission statements [17]. We believe that working to involve rural clinicians from private practices or clinical academic outposts will serve to benefit medical students, residents, and rural clinicians. That has certainly been the case in the State of Mississippi—each cohort has found that rural academic involvement has enriched their lives.

Conflicts of Interest There are no relevant conflicts of interest.

References

1. Faust DG. We are all teachers; We are all learners. Almanac. 1996;43(9). Accessed https://www.almanac.upenn.edu
2. Sir William Osler and Internal Medicine. ACP website. (n.d.) Accessed 22 Apr 2020. https://www.acponline.org/about-acp/about-internal-medicine/sir-william-osler-and-internal-medicine.
3. Medical Education and Early Career, McGill University, 1870–1884. NLM website (n.d.). Accessed 10 June 2020. https://profiles.nlm.nih.gov/spotlight/gf/feature/early.
4. Leach H, Coleman JJ. Osler centenary papers: William Osler in medical education. Postgrad Med J. 2019;95(1130):642–6. https://doi.org/10.1136/postgradmedj-2018-135890.
5. Talley RC. Graduate medical education and rural health care. Graduate Medical Education.
6. Buster KJ, Stevens EI, Elmets CA. Dermatologic health disparities. Dermatol Clin. 2012;30(1):53–viii. https://doi.org/10.1016/j.det.2011.08.002.
7. Rabinowitz HK, Paynter NP. The rural vs urban practice decision. JAMA. 2002;287(1):113. https://doi.org/10.1001/jama.287.1.113-JMS0102-7-1.
8. Jaret P. Attracting the next generation of physicians to rural medicine. AAMC website. February 3, 2020. Accessed 9 May 2020. https://www.aamc.org/news-insights/attracting-next-generation-physicians-rural-medicine.
9. Pitre LD, Linford G, Pond GR, et al. Is access to care associated with stage at presentation and survival for melanoma patients? J Cutan Med Surg. 2019;23(6):586–94. https://doi.org/10.1177/1203475419870177.
10. Verdon S, Wilson L, Smith-Tamaray M, et al. An investigation of equity of rural speech-language pathology services for children: a geographic perspective. Int J Speech Lang Pathol. 2011;13(3):239–50. https://doi.org/10.3109/17549507.2011.573865.
11. Streifel A, Wessman LL, Farah RS, et al. Rural residency curricula: a potential target for improved access to care? Cutis. Under review 6/28/20.
12. Maxfield H, Kennedy M, Delzell JE Jr, et al. Performance of third-year medical students on a rural family medicine clerkship. Fam Med. 2014;46(7):536–538.
13. Crump AM, Jeter K, Mullins S, et al. Rural medicine realities: the impact of immersion on urban-based medical students. J Rural Health. 2019;35(1):42–8. https://doi.org/10.1111/jrh.12244.
14. Lewiecki EM, Rochelle R. Project ECHO: telehealth to expand capacity to deliver best practice medical care. Rheum Dis Clin North Am. 2019;45(2):303–14. https://doi.org/10.1016/j.rdc.2019.01.003.
15. Brodell RT. The role of the part-time physician-teacher in dermatology. Arch Dermatol. 1996;132(7):758–60. https://doi.org/10.1001/archderm.1996.03890310040005.
16. Casaletto JJ, Wadman MC, Ankel FK, et al. Emergency medicine rural rotations: A program director's guide. Ann Emerg Med. 2013;61(5):578–83. https://doi.org/10.1016/j.annemergmed.2012.09.012.
17. Valsangkar B, Chen C, Wohltjen H, et al. Do medical school mission statements align with the nation's health care needs? Acad Med. 2014;89(6):892–5. https://doi.org/10.1097/ACM.0000000000000241.
18. Berkson L. Problem-based learning: have the expectations been met? Acad Med. 1993;68(10 Suppl):S79–88. https://doi.org/10.1097/00001888-199310000-00053.

Private Practice Rural Dermatology Offices

10

Kever A. Lewis and Ira D. Harber

Rural America comprises 25% of the United States population, yet only 10% of physicians practice in these areas [1] (See Fig. 10.1). Naturally, there are many physicians who would not dream of moving outside city beltways to the remote cultural wastelands beyond. Indeed, there are many thriving rural dermatologists, who with equal enthusiasm, are unable to fathom relocating to the heart of the big city and relinquishing their freedoms, community and fresh air. The art of medicine remains the same no matter where they practice. In fact, stepping through the doors of a private practice dermatology clinic in rural America, one would likely be unable to immediately surmise any differences from such clinics' urban counterparts. As the layers of these practices are slowly peeled away, however, it becomes readily apparent that these offices are molded by their locations among the farmlands and sprawls of rural America. Practicing dermatology in rural clinics necessitates consideration of additional variables than are typically weighed.

"Isn't it a bit unnerving that doctors call what they do "practice"?" — George Carlin

K. A. Lewis
School of Medicine, University of Mississippi
Medical Center, Jackson, MS, USA
e-mail: klewis7@umc.edu

I. D. Harber (✉)
Department of Dermatology, University of
Mississippi Medical Center, Jackson, MS, USA
e-mail: iharber@umc.edu

The Nature of Rural Private Practice Offices

Due to the low number of physicians in these rural areas, opening a large multi-partner practice is uncommon, often due simply to lack of potential partners. Thus, many young private practice entrepreneurs act as sole proprietors of their new practices [2, 3]. Although desirable to those who may wish to shape their practice around their private life, this practice style leaves the administrative duties on the shoulders of the physician whose extensive training most often lacks management, business, or accounting courses. Rural dermatologists are likely one of the few physicians for miles around, and they are often tasked with expanding their scope of practice beyond the structure, function, and diseases of the skin. Private local dermatology clinics often, by necessity, function as the patient's doorway into the healthcare system. In fact, to the local newspaper, community members, or local government they become a medical spokesperson for all things health and disease [4].

The nature of rural practice forces the physician to navigate without support of large hospital systems or nearby subspecialists to which their urban counterparts have grown accustomed. On the other hand, the minimization of the bureaucracy common to large health care entities may be attractive to some dermatologists. Rural medical practitioners do, however, rely on each other

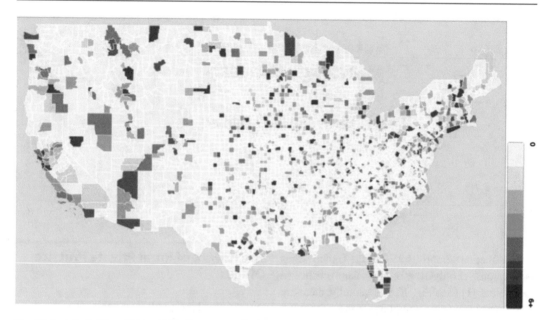

Fig. 10.1 Adapted from Vaidya 2018. County-specific dermatologist: population ratio distribution of the United States (per 100,000 persons) with darker colors representing higher population ratio of dermatologists

closely. Dermatologists, other specialists, and emergency rooms often check in with each other by phone when managing difficult patients. Such tightly knit medical communities are bolstered by the selflessness of its individual members who take such calls and offer their expertise for the benefit of their patients [5]. Young dermatologists fresh out of residency, however, may still feel isolated. Professional seclusion is a daunting challenge as physicians in surrounding areas age and grow settled into their practices. Medical hermeticism may ultimately lead to a relative lack of supervision and guidance of newer dermatologists and has been shown to affect young dermatologists adversely in that they may be less likely to initiate major management decisions [6].

There is a large burden of skin care problems in these areas. One-third of primary care visits are for a primarily dermatologic problem or the patients have an additional skin problem unrelated to the cause of the visit [7]. Since there are few specialists in rural areas, and access to dermatologic care is difficult, many patients in these areas are diagnosed late in their disease course and thus have significant delays in treatment which may lead to long-term disability [7]. This increase in severity can reasonably be attributed to both access mechanisms and cultural determinants. Having a derma-

tologist who is invested in the local community can help address such conditions, especially when many are either preventable or curable with early treatment [7]. In fact, the degree of Breslow thickness in melanoma at diagnosis was increased 10% for each 10 mile increase from diagnosing provider in rural areas [8].

As one might expect, the spread-out nature of the patient population provides a significant barrier to the provision of care. Patients often must travel long distances to see the only dermatologists for a hundred miles. This poses a significant resource strain on patients, not only financially from the standpoint of payments for services rendered, but also for wages lost traveling both to and from the appointment, arranging childcare and transportation, and charges accrued for ancillary testing or disease management. It is no secret that rural communities are often poorer. Rural physicians must be keenly aware of the effects their prescriptions and monitoring have on the patient as a whole, and it may be more successful to prescribe a less effective though less rigorously monitored regimen than the alternative. One solution adopted by many rural dermatology clinics reduces the cost burden upon patient by bringing the healthcare to them utilizing mobile health clinics [1]. While this may reduce some of the financial strain on patients, it does not overcome the fact that, historically, some rural communities have a distrust of the healthcare system. Patients from these areas are less likely to accept any type of management strategy that requires regular follow-up or higher risk of side effects [6]. However, many clinicians thrive in managing these complex patients and find the medico-socio-cultural aspects of rural healthcare both academically and professionally stimulating.

Rural communities are in great need of dermatologists [9]. Clinics are often able to operate solely on appointment-based schedules without the need to utilize walk-in appointments to supplement patient volume [10]. This consistent level of patient supply provides private practice physicians in rural areas the option to shape and mold their practice as they see fit. Due to the demand for clinical dermatology combined with a less affluent population, some practitioners find it difficult to practice cosmetic dermatology [11]. Increased

clinical practice share is of benefit to patients who enjoy shorter wait times than would be typical at clinics in an academic institution, though it requires the physician to spend more time per hour managing needs of their patients than is on average spent at an urban practice [12, 13]. This type of practice might be ideal for the dermatologist who enjoys the cognitive demands of medical dermatology. Indeed, skin conditions such as acne, chronic actinic damage, folliculitis, atopic dermatitis, warts, contact dermatitis, psoriasis and other common conditions are still the lion's share of visit types regardless of practice location. However, diagnoses not routinely seen in everyday practice such as bullous impetigo, polymorphous light eruption, erythema multiforme, and cutaneous larva migrans are more likely to be seen in these rural populations [14]. Genodermatoses such as albinism, xeroderma pigmentosum, and neurofibromatoses are also more commonly seen in rural clinics [7]. (See Table 10.1).

Table 10.1   Adapted from Tillakaratne 2016. Diagnoses made in rural versus urban clinics

Diagnosis	Rural n (%)	Urban n (%)	Total n (%)
Eczema/Dermatitis	48 (21)	47 (18)	95 (20)
Papulosquamous	19 (9)	27 (11)	46 (10)
Exanthems and Drug Eruptions	7 (3)	3 (1)	10 (2)
Urticaria, erythema, purpura, and vasculitis	5 (2)	11 (4)	16 (3)
Developmental disorders	4 (2)	0 (0)	4 (1)
Skin neoplasms	36 (16)	69 (27)	105 (22)
Infections	28 (13)	18 (7)	46 (10)
Adnexal diseases	19 (9)	15 (6)	34 (7)
Pigmentary disorders	3 (1)	3 (1)	6 (1)
Infections and bites	3 (1)	1 (0)	4 (1)
AICTD/Rheumatologic disorders	8 (4)	7 (3)	15 (3)
Disorders of hair	1 (0)	4 (2)	5 (1)
Disorders of sweat glands	1 (0)	0 (0)	1 (0)
Oral and anogenital diseases	2 (1)	4 (2)	6 (1)
Vesiculobullous disorders	6 (3)	4 (2)	10 (2)
Lympho–/myeloproliferative disorders	1 (0)	3 (1)	4 (1)
Non-infectious neutrophilic/eosinophilic	2 (1)	6 (2)	8 (2)
Disorders of Langerhans cell and macrophages	2 (1)	0 (0)	2 (0)
Disorders of dermal connective tissue	3 (1)	5 (2)	8 (2)
Disorders of subcutaneous fat	0 (0)	1 (0)	1 (0)
Vascular and lymphatic disorders	1 (0)	1 (0)	2 (0)
Genodermatoses	5 (2)	1 (0)	6 (1)
Metabolic and systemic disorders	2 (1)	0 (0)	2 (1)
Psychocutaneous disorders	1 (0)	1 (0)	2 (1)
FSE normal	16 (7)	24 (9)	34 (9)
Total	223 (100)	255 (100)	372 (100)

AICTD autoimmune connective tissue diseases, *FSE* full skin examination, *n* number of diagnoses

Private Practice Finances

Essential to the planning required before opening a private practice in a rural area is determining the financial investment required to open such a practice. Practicing in a rural location is a significant risk as the return on investment is heavily influenced by patient demand, patient volume, and competition [15]. All of these potential problems are notably less worrisome to the dermatologist who decides to practice in an underserved area. The dermatologist is choosing to invest in the community, its health, its people, and its vitality. The nature of a medical enterprise necessitates that the community be ready to reciprocate when such an investment is made. In fact, local banks, businesses, and neighbors commonly step up to the plate and support development of the infrastructure that allows for these medical services to be provided. Federal grants may be available to support the addition of a needed medical specialty to an underserved rural area and should be identified when considering practice locations (See Chap. 5). As practitioners in rural areas are often the sole proprietor of business, the burden of financial funding is placed singularly upon the dermatologist. Practice owners must be compensated relative to the level of service they provide in addition to an amount necessary to cover the overhead cost of the practice. In general, private practice dermatologists are salaried on the order of $425,000 per year with their counterparts in academia making anywhere from $291,000 - $353,000 [12]. With rural clinicians spending less of their time performing cosmetic procedures, and in view of the rural payer mix that most likely includes more self-pay and Medicaid patients, the payment for the average visit is likely to be less than in urban practices. However, many rural dermatologists see more patients in an average day than their urban counterparts and the cost of living/cost of running a practice is less in rural areas [13]. Therefore, physicians in rural private practice may not be compensated to a lesser extent than their private or academic counterparts in urban areas. The overall compensation for a rural dermatologist is, in fact, multi-factorial with the demand for services, local reimbursement levels, and other market forces all playing major parts in the risk-benefit analysis that come along with opening such a practice [16]. In addition, the commercial real estate, residential real estate, and taxes are generally cheaper in rural areas; thus, gross practice income alone may not accurately reflect the physician's income.

One of the earliest decisions that occur when opening a practice is the decision to own or rent the physical clinic space. Ownership of a practice's physical location has anecdotally been favored by those in private practice, especially those in rural locations. Of note, however, it is not uncommon for a practitioner nearing retirement age to be unable to recruit a dermatologist to take over his or her practice. Part of the reason for this is that graduating dermatology residents more frequently choose to locate in metropolitan rather than rural areas, thus opportunities for the sale of rural or remote practices are less certain [2]. On the other hand, ownership of commercial real estate is one way to build equity to be drawn upon at retirement and is a hedge against the inability to sell one's practice in the traditional sense. Present and likely future demand for dermatologists does not lend itself to the need for buying into a patient base as much as a physical practice location.

Another essential characteristic when considering opening a clinic in a rural location is the reimbursement type and styles of care. Anecdotally, patients in rural areas are less likely to have a private payer source for visits. Instead, the rural dermatologists may find themselves inundated with Medicaid red-tape or patients who lack health insurance altogether. There is a need to develop profit centers in the practice to balance the financial deficit caused by relatively large number of impoverished rural patients who are nonetheless in desperate need of dermatologic care. The notion of concierge dermatology is in our opinion less viable in rural areas than traditional insurance billing. As such, health insurance companies create a significant risk burden for newly minted practices as the push for reduction in insurance reimbursements due to the increase in healthcare spending and campaigns to reduce health-care costs move from the exception to the norm of practice [15]. This requires clinicians to develop a more frugal approach to practice management. A penny saved is a penny earned. These changes may eventually require private practice physicians to work longer hours, increase patient visit efficiency, or even delay retirement, though these issues are universal across practice locations. All private practices must adapt to the changing payer landscape that complicates an already complex system of semi-private healthcare. Future changes to evaluation and management services codes may be advantageous to high volume rural practices.

Despite the stagnant rates of insurance reimbursement and the need to pay competitive salaries for employees who become part of the "family," there are certainly financial upsides to rural practice. Rural America, while decreasing in size due to expansion of suburbia, is less susceptible to competition among businesses, especially medical specialists. Commercial real estate in small towns and boroughs is less likely to be sought after by corporations, business chains, or other physicians, and the average cost of commercial real estate is far less in rural areas. The cost of operating a practice in a rural area is subsequently less than a practice of equivalent size in a more metropolitan area. In addition, demand for expensive lasers and other cosmetic devices is often reduced, and the need to set oneself apart from competition by purchasing new devices is less of an issue in rural areas. As such, significant overhead can be allocated to additional staffing, a larger practice space, inventory, and maintenance. Together these considerations in many ways compensate for the reduced rates of reimbursement received by rural dermatology offices.

Special Characteristics of Practitioners

Choosing to leave the hustle and bustle of the big city to open a private practice in rural America requires a special type of practitioner. They are often very involved in their communities. Patients in these areas are well aware of the paucity of physicians in their communities. One unfamiliar with rural medical practice might imagine those in rural practice as having a "mom and pop" style of practice with family run clinics. While that may be true in some instances, most rural dermatology practices are notable for their efficiency, effectiveness, and pride. Dermatologists who enter academic practice often have advanced degrees or backgrounds with institutions of high levels of NIH funding. Recently graduated dermatology residents starting a rural practice most likely grew up in the area where they are settling and thus are familiar with culture, geography, and members of the community [17, 18]. This trend is especially evident in the Southern and Midwestern parts of the United States where rural areas are vast [18]. The reasons for this pilgrimage back home are multi-faceted. Having grown up in a small community lends a certain feeling of responsibility as well as desire to return home after training. Other practitioners return home because they know that they are able to provide a service to the community that otherwise would not be available. Opening and running one's own practice can be a financially, emotionally, and intellectually challenging, but, the only physician "for miles" is likely to receive a massive amount of social and professional support. This allows a practice to grow and flourish despite the challenges of practicing healthcare as a solo entity [9].

The number of trainees returning home to open practices is not as robust as might be thought. Rural dermatology is practiced primarily by older physicians. The workforce is growing older as the number of dermatologists younger than age of 55 practicing in these communities is steadily decreasing [2]. There is an increasing healthcare disparity here that is not currently being filled by the next generation of practitioners, and as the current workforce draws closer and closer to retirement age, the ability to find a dermatologist, a task already exceptionally difficult in rural communities, is becoming more difficult. Conversely, for a practitioner eager for a challenge, this very problem creates a special opportunity to thrive in the practice of dermatology.

Overcoming Disadvantages of Rural Practice

No one knows I'm here!

Along with professional isolation from the relative lack of physician mentors in rural practice, there is also an economic isolation that comes with private practice in rural settings. This can be attributed to the absence of the referral base that is typically associated with large medical centers, group practices, and urban communities [19]. This, coupled with the fraternalistic culture of small-town America, potentially spells trouble for a budding dermatology practice starting with few community connections. However, for the hometown doc now returned from residency in "the big city", it is the dermatologic golden fleece. As the referral base of these towns is based primarily on word of mouth, local familiarity of incoming physicians may reasonably be the strongest predictor for the vitality of a new practice.

For dermatologists new to an area, low budget advertising may be helpful since most primary care providers may be unaware that a specialist has located much closer than the urban referral centers and unaware of the range of services provided by the new dermatologist in the next town. A personal touch is even more helpful. Visiting one or two communities and calling on each of the primary care clinics on an "outreach day" once a month can be very helpful. Stopping by with a smile and business card leads to universally positive results for all parties. A 15-minute lecture for staff and providers on commonly encountered diseases is also very much appreciated. Soon, sufficient referrals will likely make this gesture unnecessary.

Keeping Up with the Science is difficult in the middle of nowhere!

One of the distinct challenges to practicing in a remote area is keeping up with the ever-changing science that is critical to the practice of medicine. Private practice physicians often find it difficult to ensure they are maintaining "best-practice" standards while balancing the demands of patient care in a busy practice. In general, rural dermatologists participate less frequently in continuing medical education (CME) activities when compared to their urban colleagues [3]. While the reasons for this are multi-faceted, the most commonly cited reason is the large time commitment necessary to complete such requirements [3]. Traveling to local conferences or to hear speakers at academic centers requires substantial travel and time-opportunity costs. Private practitioners in rural areas already work longer hours on average than their urban colleagues, and as such, the extra time spent away from seeing patients or the physician's family puts a larger burden upon the physician [13].

Most rural dermatologists derive the majority of their CME time for attending an annual national meeting [3]. Others attend clinical meetings held by dermatology departments at regional teaching hospitals or through presentations sponsored by national organizations on the Internet. In fact, with ever-advancing technology, the possibility for accruing continuing CME credits is becoming more and more accessible for the rural physician in private practice. Recordings and live streams allow for meetings to be watched at the viewer's convenience without the cost and time loss associated with attending meetings in person. In fact, since the implementation of electronic meeting options, the rates at which continuing medical education (CME) credits have been accrued has been shown to be equivalent regardless of if the credits were earned virtually or in person [3]. One way or another, rural physicians find ways to attend seminars, lectures, and research presentations even when they do not have the academic backing of a large institution and are able to bring evidence-based healthcare to their communities.

Few Opportunities for Scholarly Research Away from the University Campus

Many rural physicians are not involved in research because they do not have the "publish or perish" pressures of university faculty. Private practice in rural areas, however, offers unique advantages to conducting scholarly work and clinical trials. As rural physicians often have high volume practices, it is easy to recruit large numbers of patients for admission into clinical trials. The relative lack of nearby competition also provides a pipeline for the rare and severe conditions that are often difficult to find outside of tertiary academic referral centers. Additionally, as members of the community, rural physicians can serve as a "cultural broker" between research teams and study participants [20]. This enables the physician who is trusted by both parties to breach the barrier of education and culture to allow for a single common language to be spoken and understood by both parties. In truth, this common ground with community members may actually increase study participation. Researchers who are in tune with individual complexities of local culture have been shown to successfully recruit patients at a higher rate than others [20]. This is vitally important in studying previously underrepresented communities within the dermatology community. For example, in atopic dermatitis, there is an interaction with genetics and environment that contributes to the pathogenesis of the disease impacting its presentation. Unfortunately, populations most impacted by this disease, including African Americans and rural children, are notoriously underrepresented in research studies [20]. Rural physicians have a social responsibility to parlay their unique situation within these communities into desperately needed research [20]. Finally, rural practices can use a central institutional review board (IRB) provided by the sponsor of the multicenter trial enabling them to initiate clinical research studies quickly, while large academic centers must navigate their own IRBs which often takes months.

Although the potential for research in rural private practice offices is robust, there are also significant difficulties in conducting research. One of the most common barriers for clinicians to overcome is the mistrust of research that exists in some rural communities. Although multi-faceted, one could conclude that there exists a relative lack of medical literacy regarding research methods, and patients may not wish to be a "guinea pig," or may fear side effects related to treatment modalities. In fact, some of these populations have, in the past, been marginalized in the name of research, and there are heinous examples including the Tuskegee Syphilis Experiment, where African-Americans as well as central and south Americans were knowingly harmed. It is important for clinicians to acknowledge and accept the fears many patients harbor about research and attempt to establish a clear and open dialogue about research methods, findings, and purposes before obtaining study consent. This mistrust is certainly not unique to the field of dermatology or to rural populations in general. Other barriers for dermatologic research in the field of dermatology include stigmatization of a skin condition in exposed areas and lack of access to research, since most dermatology research is conducted within specialty clinics or in academic settings [20].

Limited Opportunities to Teach for Rural Dermatologists:

Rural private practice offices may not seem like a haven for medical education when compared to university clinics. They are rarely near medical schools or training programs, and the supply of trainees is rather limited to those electing to rotate in rural clinics. In our opinion the potential for medical education in rural dermatology is an underutilized resource. Teaching trainees in rural offices exposes them to rural dermatology in the private practice setting. They can see and feel the allure of rural practice. It may solidify the decision to choose a non-urban setting after graduation from residency training. In fact, it has been shown that dermatologists are more likely to return to a rural area to practice if they have had a clinical experience in these areas [4]. As such, early exposure to rural private practice could serve as a conduit for students and trainees to ultimately fill the void of dermatology providers in rural areas. At the University of Mississippi Medical Center, we have taken a hybrid approach: establishing an academic office in a rural location that functions more like a private practice. The advantages of very low staff turnover, abundant patients, and a single, highly motivated provider have led to a rural practice that is both financially and educationally successful. Rotators include medical students, pre-medical students, dermatology residents, and residents from other specialties.

The most significant barrier that inhibits teaching medical students in a rural private practice is the large geographical distance that often separates students, teachers, and resources. Medical students are not able to drive large distances from their medical school to rotate through a rural dermatology clinic with any degree of regularity; however, brief rotations are manageable and beneficial for the students. Establishing a longitudinal experience, such as an elective dermatology rotation in a private clinic near the medical student's hometown, serves as a potential remedy for this dilemma. Allowing students to learn about rural dermatology in the community that raised them is sure to be a formative experience, and it also allows students and trainees to "test out" the idea of returning home to practice long term. In addition, technological advancements of late have altered the way medical education is delivered to students. Medical students and trainees are now able to obtain first-hand experience in rural clinics while maintaining access to virtual education materials, library resources, and curriculum lectures [21].

Rural clinics should be invested in teaching the next generation of physicians and fostering an environment of curiosity. We feel that the net good from such rotations outweighs the potential for harm when considering a rotation away from main campus. Some students training in rural areas may feel their ability to track their progress in comparison to their peers is reduced, leading to a relative degree of uncertainty and distress though this has not been the experience in our university rural clinic. Other considerations with regard to rural clinic-based medical education is the potential for limited attendance in didactic teaching sessions or case conferences at the main campus. In addition, trainees may be concerned about difficulty obtaining books, journals, and research databases that medical libraries provide [21]. However, in the electronic age, these issues are amenable to solving with teleconferencing platforms and online databases.

Rural Physicians have a Quality "Life"

Private life in rural practice is not unlike that of practicing dermatology elsewhere. One's quality of life is predominantly a factor of what the practitioner decides to make of it. Anecdotally, there are certain draws to rural practice that make it ideal for some practitioners. For example, practitioners who find fulfillment in seeing large numbers of patients may very well find their niche within the world of rural private practice dermatology [22]. For others though, the practice of rural dermatology provides a great sense of belonging and being needed that is felt from being able to help those who otherwise would have difficulty accessing care [5]. One of the difficulties with private life in rural areas is the job market for a dermatologists' partner, and this may be an essential consideration when choosing to practice in rural areas. More than half of married physicians have highly educated spouses, and highly-skilled work opportunities in rural areas are sparse [2]. All in all, dermatologists in rural areas, when compared to their urban colleagues, are more likely to enjoy where they work and the practices which they operate [23]. This level of satisfaction along with decreased bureaucratic oversight and ability to mold one's practice and schedule ultimately leads to lower rates of physician burnout [24]. Additionally, medical dermatology in rural areas is constantly in high demand, and, if one chooses, one has the option to fine-tune the scope of practice with greater freedom than might be afforded in areas of higher provider density. This also means that finding a job in a rural practice is much easier should one choose to re-locate or is considering making a change from urban to rural living. The outlook for rural dermatology jobs is extremely favorable as currently the number of dermatologists seeking associates is greater even than the number of trainees annually produced [2]. Rural dermatologists also commonly have patients that are highly appreciative of the care that they receive – realizing the time and expertise required to become a physician and the assumed sacrifices made to practice in a rural area. For this reason, patients are anecdotally eager to participate in their care and place trust in the hands of their local physician. However, rural dermatology is not without its undue stressors. As specialists in a rural area, the initial startup may be challenging due to difficulty accessing higher-level diagnostic modalities, treatment options, and other tertiary care resources. However, practitioners grow accustomed to the unique aspects of rural life and are generally content [24]. Many rural practitioners have important family obligations or hobbies that attracted them to rural America. Often outdoors activities such as hunting, fishing, hiking, and water sports may be within walking distance and are available without taking a vacation.

Boundaries are important and help to safeguard against ethical dilemmas, and those in rural areas are held to the same professional standards as any physician. Being an integral member of a rural community sets the stage for interesting practice relationships: the patients treated every day are often community members with whom the practitioner is familiar. This proximity to the patient's themselves deepens the patient-physician relationship to a level that can be extremely fulfilling; however, this also presents a confounding dichotomy in that patients often feel comfortable approaching outside of office work. It is not uncommon for a rural dermatologist to be consulted on a new rash while in the grocery store or updated on the condition of a patient's acne treated at a junior high basketball game. Whether it be a request for an after-hours appointment or prescription refill without being seen, the line between friend and physician is often blurred in rural practice. It is important for practitioners to set limits and stick to them to preserve professional integrity.

Summary

As with all decisions one makes in life, tradeoffs must be made. Rural practice, by definition, necessitates greater distances from national sporting events, theatre and entertainment districts, fine dining, museums, and art galleries. This is balanced by a quieter life without the hustle and bustle of the city, more favorable commutes to and from work, lower cost of living, lower overhead, and a great sense of belonging within the community. Rural dermatologists are on average more satisfied and less susceptible to burnout. We encourage resident dermatologists and seasoned veteran dermatologists alike to consider the risks and benefits of establishing a private practice in a rural area. It does require a special set of values and ambitions, but for us, it has proven to be an extremely fulfilling approach to the practice of medicine and dermatology. There is almost no avenue of practice that is unavailable to the savvy rural practitioner regardless of career interests, including teaching, mentoring, serving the underserved, research, and clinical trials. The beauty of a rural private practice is that the dermatologist is free, and perhaps emboldened, to shape a medical practice as they see fit. Young dermatologists should avoid the pitfall of feeling that one's desired practice style is limited by geography. There are untapped opportunities for those willing to rise to the challenge in rural America.

Disclosures The authors have no relevant disclosures.

References

1. Barton M. Access to dermatology Care in Rural Populations. J Nurse Pract [Internet]. 2012;8(2):160–1. Available from: https://linkinghub.elsevier.com/retrieve/pii/S155541551100571X
2. Feng H, Berk-Krauss J, Feng PW, Stein JA. Comparison of dermatologist density between urban and rural counties in the United States. JAMA Dermatol [Internet]. 2018;154(11):1265–71. Available from: https://doi.org/10.1001/jamadermatol.2018.3022
3. Kurzydlo A-M, Casson C, Shumack S. Reducing professional isolation: support scheme for rural specialists. Australas J Dermatol [Internet]. 2005;46(4):242–5. Available from: http://doi.wiley.com/10.1111/j.1440-0960.2005.00192.x
4. Vaidya T, Zubritsky L, Alikhan A, Housholder A. Socioeconomic and geographic barriers to dermatology care in urban and rural US populations. J Am Acad Dermatol [Internet]. 2018;78(2):406–8. Available from: https://linkinghub.elsevier.com/retrieve/pii/S0190962217321801
5. Thornfeldt CR. Rural dermatologist commits to helping those who have difficulty finding care. Dermatology Times [Internet]. 2011. Available from:

https://www.dermatologytimes.com/view/rural-dermatologist-commits-helping-those-who-have-difficulty-finding-care.

6. Hajjaj FM, Salek MS, Basra MKA, Finlay AY. Nonclinical influences, beyond diagnosis and severity, on clinical decision making in dermatology: understanding the gap between guidelines and practice. Br J Dermatol [Internet]. 2010;163(4):789–99. Available from: http://doi.wiley.com/10.1111/j.1365-2133.2010.09868.x

7. Naafs B, Padovese V. Rural dermatology in the tropics. Clin Dermatol [Internet]. 2009;27(3):252–70. Available from: https://linkinghub.elsevier.com/retrieve/pii/S0738081X08002174

8. Stitzenberg KB, Thomas NE, Dalton K, Brier SE, Ollila DW, Berwick M, et al. Distance to diagnosing provider as a measure of access for patients with melanoma. Arch Dermatol [Internet]. 2007;143(8):991–8. Available from: http://archderm.jamanetwork.com/article.aspx?doi=10.1001/archderm.143.8.991

9. Ehrlich A, Kostecki J, Olkaba H. Trends in dermatology practices and the implications for the workforce. J Am Acad Dermatol. 2017;77(4):746–52.

10. Krensel M, Augustin M, Rosenbach T, Reusch M. Waiting time and practice organization in dermatology. J Dtsch Dermatol Ges [Internet]. 2015;13(8):812–4. Available from: http://www.ncbi.nlm.nih.gov/pubmed/26213818

11. Uhlenhake E, Brodell R, Mostow E. The dermatology work force: a focus on urban versus rural wait times. J Am Acad Dermatol [Internet]. 2009;61(1):17–22. Available from: https://linkinghub.elsevier.com/retrieve/pii/S0190962208011468

12. Rajabi-Estarabadi A, Jones VA, Zheng C, Tsoukas MM. Dermatologist transitions: academics into private practices and vice versa. Clin Dermatol [Internet]. 2020; Available from: https://linkinghub.elsevier.com/retrieve/pii/S0738081X20301000

13. Jacobson CC, Resneck JS Jr, Kimball AB. Generational differences in practice patterns of dermatologists in the United States: implications for workforce planning. Arch Dermatol [Internet]. 2004;140(12):1477–82. Available from: https://doi.org/10.1001/archderm.140.12.1477

14. Vallejos QM, Quandt SA, Feldman SR, Fleischer AB, Brooks T, Cabral G, et al. Teledermatology consultations provide specialty care for farmworkers in rural clinics. J Rural Heal. 2009;25(2):198–202.

15. Wang JV, Saedi N. Risk management for private practice dermatology clinics: a commentary. J Clin

Aesthet Dermatol [Internet]. 2018;11(12):40–1. Available from: http://www.ncbi.nlm.nih.gov/pubmed/30666278

16. Tierney E, Kimball AB. Median dermatology base incomes in senior academia and practice are comparable, but a significant income gap exists at junior levels. J Am Acad Dermatol [Internet]. 2006;55(2):213–9. Available from: https://linkinghub.elsevier.com/retrieve/pii/S0190962206002283

17. Shi CR, Tung JK, Nambudiri VE. Demographic, academic, and publication factors associated with academic dermatology career selection. JAMA Dermatol [Internet]. 2018;154(7):844. Available from: http://archderm.jamanetwork.com/article.aspx?doi=10.1001/jamadermatol.2018.0743

18. Chen AJ, Schwartz J, Kimball AB. There's no place like home: an analysis of migration patterns of dermatology residents prior to, during, and after their training. Dermatol Online J [Internet]. 2016;22(6):13030/qt3sf6z3pn. Available from: http://www.ncbi.nlm.nih.gov/pubmed/27617617

19. Haider A, Mamdani M, Shaw JC, Alter DA, Shear NH. Socioeconomic status influences care of patients with acne in Ontario, Canada. J Am Acad Dermatol [Internet]. 2006;54(2):331–5. Available from: http://www.ncbi.nlm.nih.gov/pubmed/16443069

20. Spears CR, Nolan BV, O'Neill JL, Arcury TA, Grzywacz JG, Feldman SR. Recruiting underserved populations to dermatologic research: a systematic review. Int J Dermatol [Internet]. 2011;50(4):385–95. Available from: http://www.ncbi.nlm.nih.gov/pubmed/21413946

21. Delaney G, Lim SE, Sar L, Yang SC, Sturmberg JP, Khadra MH. Challenges to rural medical education: a student perspective. Aust J Rural Health [Internet]. 2008;10(3):168–72. Available from: http://doi.wiley.com/10.1111/j.1440-1584.2002.tb00027.x

22. Suchniak JM. Dermatologist finds practice perks, balance in rural setting. Dermatology Times [Internet]. 2016. Available from: https://www.dermatologytimes.com/view/dermatologist-finds-practice-perks-balance-rural-setting.

23. Luman K, Zweifler J, Grumbach K. Physician perceptions of practice environment and professional satisfaction in California: from urban to rural. J Rural Heal. 2007;23(3):222–8.

24. Hogue A, Huntington MK. Family physician burnout rates in rural versus metropolitan areas: a pilot study. S D Med [Internet]. 2019;72(7):306–8. Available from: http://www.ncbi.nlm.nih.gov/pubmed/31461585

Taylor Ferris, Lisa A. Haynie, and Leslie Partridge

The nurse practitioner (NP) and physician assistant (PA) occupation emerged in the 1960s as a response to the shortage of primary care physicians across the United States. Currently, there are more than 290,000 licensed NPs and more than 131,000 PAs in the US with projected steady increases in these numbers [1–4]. NPs and PAs, often referred to as advanced practice providers (APPs), play a vital role in healthcare and have been shown to provide high-quality care in rural, urban and suburban communities in a variety of settings. While there are many similarities in the practice and utilization of NPs and PAs, there are differences in their education models and training [1, 2, 5, 6].

"Advanced practice providers can slot in easily where care is needed." Debby Renner, PhD.

T. Ferris
School of Medicine, University of Mississippi Medical Center, Jackson, MS, USA
e-mail: tferris@umc.edu

L. A. Haynie
School of Nursing, University of Mississippi Medical Center, Jackson, MS, USA
e-mail: lhaynie@umc.edu

L. Partridge (✉)
Department of Dermatology, University of Mississippi Medical Center, Jackson, MS, USA
e-mail: lahoward2@umc.edu

Nurse Practitioners

The first NP program was established in 1965 by Loretta Ford, EdD, and Henry Silver, MD, at the University of Colorado as a post-baccalaureate degree with a goal to provide health care to rural, underserved populations [7]. By 2019, there were approximately 400 academic institutions with NP programs in the U.S. with more than 30,000 new NPs completing their academic programs in 2018–2019 [8, 9]. Nurse practitioners are registered nurses (RNs) with education at least at the master's degree level and must pass a national certification examination as well as obtain state licensure to practice within their state. NPs are regulated by state nursing boards. Scope of practice is regulated by individual states and determines an NP's ability to practice and prescribe medications with or without physician collaboration or supervision [10]. Nurse practitioners, nurse anesthetists, nurse midwives' overall employment is projected to grow 45 percent from 2019–2029 [3].

Nurse practitioner education follows a nursing model and certification can be attained in one of six patient population foci:

- Family/individuals across the life span
- Adult-Gerontology
- Neonatal
- Pediatrics
- Women's health/gender-relation
- Psychiatric/mental health (AANP)

NP programs vary in length but provide extensive and advanced educational and clinical training beyond the initial professional preparation as a registered nurse (RN). For practitioners seeking the highest level of nursing education or advancement of career goals, many academic institutions now offer terminal degrees of Doctor of Philosophy (PhD) or Doctorate of Nursing Practice (DNP) [10]. The American Association of Colleges of Nursing recommends that advanced nursing practice education transition to the DNP degree as the level of entry into Advanced Practice Registered Nurse (APRN) practice [11]. To date, 348 DNP programs are enrolling students at schools of nursing across all 50 states, and an additional 98 new DNP programs are in the planning stages [11].

While over half of NPs practice in primary care, many practice in specialty and sub-specialty areas, according to the 2018 AANP National Nurse Practitioner Sample Survey Results [12]. Further specialization can occur in Nephrology, Cardiology, Psychosomatic medicine, Surgery, Holistic Care, with sub-specialty certification exams offered in Dermatology, Orthopedics, Emergency Medicine, Palliative Care, and Oncology. Specialized practice requires that the NP has adequate experience and meets the examination or certifying body requirements [12]. According to the 2013–2014 National Nurse Practitioner Practice Site Census, over 3700 NPs practice in dermatology [13, 14]. The assessment of professional competencies in dermatology nurse practitioner practice is currently through a specialty certificate examination, *Dermatology Certified Nurse Practitioner* by the Dermatology Nurse Practitioner Certification Board [14].

The Dermatology Certification Exam was established in 2008 by the Dermatology Certified Nurse Practitioner Board (DCNPB) to "promote the highest standards of dermatology nursing practice and establish credentialing mechanisms for validating proficiency in dermatology nursing" [15]. Currently, the DCNB is working toward accreditation of the DCNP certification examination through the American Board for Specialty Nursing Certification [14]. Nationally certified nurse practitioners are eligible to sit for the Dermatology Certification Exam if they have a minimum of 3000 hours in dermatology NP practice, currently practice in dermatology, hold a Masters or Doctoral degree in nursing, hold current NP state licensure, and hold current national certification as a NP. Certification is valid for a period of 3 years and recertification may be attained through contact hours of continuing education credit or by examination [16]. Currently, there are approximately 337 Dermatology Certified Nurse Practitioners practicing within the US [17].

The Dermatology Nurses' Association was founded in the early 1980's to promote excellence in dermatology care through its annual convention, Journal of Dermatology Nurses' Association, quarterly newsletter, local chapters, and shared knowledge and expertise [18]. The Nurse Practitioner Society (NPS) of DNA was established in 2005 to meet the needs of DNA's growing NP membership. The NPS established the specialty's *Scope of Practice and Standards of Care for Nurse Practitioners in Dermatology* in 2006 to help identify the primary knowledge areas for dermatology NP practice [19].

Physician Assistants

The physician assistant profession was conceptualized by Eugene Stead Jr., MD, in 1965 at Duke University to improve and expand healthcare as a response to the physician shortage. The first class of PAs consisted of four Navy Hospital Corpsmen and the curriculum of the PA program was based on knowledge of the fast-track training of doctors during World War II [2]. There are now more than 250 PA programs across the country cumulatively graduating more than 9000 new PAs each year [20]. Physician assistants complete at least 2 years of college courses similar to premedical school requirements in the basic and behavioral sciences. PAs are educated at a master's degree level with programs using a medical model-based curriculum for approximately three academic years. More recently, a clinical doctoral degree for PAs has been established, the Doctor of Science Physician Assistant or DscPA. PA students will have completed at least 2000 hours of supervised clinical practice in various settings and locations by graduation. After graduation, PAs must pass the Physician Assistant National Certifying Examination and become licensed by a state in order to practice (aapa.org). PAs practice under the supervision of a licensed physician and are regulated through state medical boards (AAPA). PAs practice in every work setting including hospitals, urgent care centers, medical offices, community health centers, nursing homes, educational facilities, correctional institutions, as well as others [2]. The employment of physician assistants is projected to grow 31% from 2019 to 2029 [4].

While the majority of PAs practice in surgical subspecialties, primary care, internal medicine subspecialties, emergency medicine, and emergency medicine, nearly 29% practice in other specialty areas with approximately 2520 in dermatology [2, 21]. The Society of Dermatology Physician Assistants (SDPA) was founded in 1994 with the mission to educate PAs practicing in the field of dermatology. The Society of Dermatology Physician Assistants offers a Diplomate Fellowship, which is an online didactic training program created by dermatology professionals and psychometricians for PAs working in the field of dermatology to standardize curriculum of education for PAs working in dermatology. This has shown to add value to the practice of both new dermatology PAs and those with experience in dermatology. The Diplomate Fellowship "represents the standard in dermatology PA education, graduates will be able to demonstrate that they have the knowledge and skills necessary to provide exceptional patient care and function as a strong member of their team" [22]. This program offers curriculum spread out across 22 modules and encompasses the following topics:

- Basic Principles of Dermatology
- Diagnostic Procedures
- Dermatopathology
- Therapeutic and Surgical Procedures
- Neoplasms
- Papulosquamous Eruptions and Eczematous Dermatoses

- Acne, Rosacea, and other Adnexal Diseases
- Infections and Infestations
- Pruritus and Psychocutaneous Disorders
- Urticarias, Erythemas, and Purpura
- Pigmentary Disorders
- Hair, Nail, Mucous Membranes
- Disorders Due to Physical Agents
- Cutaneous Manifestations of Autoimmune and Connective Tissue Diseases
- Vesiculobullous Diseases
- Vascular Disorders
- Disorders of Subcutaneous Fat
- Disorders of Langerhans Cell And Macrophages
- Cutaneous Manifestations of Metabolic and Systemic Diseases
- Genodermatoses
- Aesthetic Medicine
- Dermoscopy

APPs in Dermatology

The number of APPs specializing in dermatology has been rapidly increasing over the years, with over 6000 NPs and PAs currently practicing in dermatology across the US [13, 14, 21]. A report from the 2014 American Academy of Dermatology Practice Profile Survey showed the employment of NPs and PAs by board certified dermatologists had increased to 46%, up from 28% in 2005 [23]. This report also suggested that the increased utilization of NPs and PAs corresponded to the decreased waiting times for new patients from 34.4 days in 2005 down to 29.1 days in 2014. There are several proposed factors that are likely related to the increase in APPs in dermatology: vacant dermatology positions, long wait times for patient appointments, unmet patient demand, increases in dermatologic subspecialization (surgical, cosmetic, pediatric dermatology), low dermatology resident capacity, aging and increasing US population [24].

Skin conditions are a leading cause of disease burden worldwide, however, most APPs have little exposure to dermatology during their education and require additional knowledge and skills training [6]. Most dermatology APPs attain their dermatologic didactic and skill training through select fellowship programs, continuing education, and/or "on the job" training. "On the job" training is often provided by a board-certified dermatologist, a proficient advanced practice provider or a combination of both and often consists of journal article reviews, topic presentations, patient evaluations and discussions, and Grand Rounds presentations [14, 25]. Dermatology education and competencies have been established to provide a foundation for dermatology APPs, helping to standardize education, knowledge, and skill sets [6, 14].

Despite nearly 500 dermatologists entering the workforce each year, the number of active dermatologists in 2015 was 36 per capita (1,000,000) which equates to 3.6 per 100,000 people. Even more concerning, nearly 60% of the country had less than 3 dermatologists per 100,000 people. The number needed to adequately care for a population is estimated at 4 dermatologists per 100,000 people, with current and future projections remaining below this threshold. In addition to this undersupply, there is a maldistribution of physicians which leaves many areas underserved [21, 23, 26]. The dermatology workforce is shifting as more APPs join dermatology practices which may help to correct for the dermatology provider shortage [21, 26]. Dermatology APPs have an integral role in the dermatology care team and add value to the interdisciplinary team approach for optimal patient outcomes. Including dermatology APPs in dermatology workforce calculations increases the average US dermatology provider density to more than 4 per 100,000, potentially leading to extension of care to underserved areas [21, 23, 26].

APPs in Rural Dermatology

Approximately one in four Americans live in rural areas. Rural dwellers generally are older, poorer, likely to be uninsured or underinsured, and suffer from higher rates of chronic conditions. They often must travel great distances for health care services. Although there are significant numbers of primary care providers practicing in rural areas, the access to specialty services is extremely limited [27]. This certainly impacts the provision of essential health services needed to provide quality health care to rural populations in the United States and around the world [28]. The physician shortage hits rural communities the hardest because only 11% are working in these medically underserved areas and are responsible for covering large geographic areas. This is a concerning statistic in that approximately 20% of the American population live in rural areas [29]. Geographic location should not determine the health condition of people. Healthcare administrators in rural healthcare delivery systems face many challenges and have been forced to address these challenges with limited resources for many years. Factors such as provider availability, cost, and scalability contribute to a rural healthcare crisis. Primary care clinicians in rural areas are often required to provide specialty services that may be outside their comfort zone. This may increase the likelihood of poor patient outcomes [30]. (See Fig. 11.1).

Skilled specialty physicians may come at a hefty price for rural hospital and health care systems. This may not be feasible for many rural communities [30]. A cost-effective solution to the provider shortage is the utilization of advanced practice providers (APP's). These advanced practice providers encompass nurse practitioners (NP), physician assistants (PA), and other licensed, non-physician providers, including certified nurse midwives, clinical nurse specialists, and certified registered nurse anesthetists. In the 1960's, nurse practitioners and physician assistants emerged as a response to the primary care physician shortage. Currently, there are approximately 248,000 NPs and 115,500 PAs practicing in multidisciplinary sites across the United States. The new restrictions placed on resident duty hours has resulted in a second wave of primary care physician shortages; thereby, again putting emphasis on the need for APPs in rural settings [33].

The National Organization of Nurse Practitioner Faculties (NONPF) support new strategies that will effectively address the national healthcare provider shortage, complex-

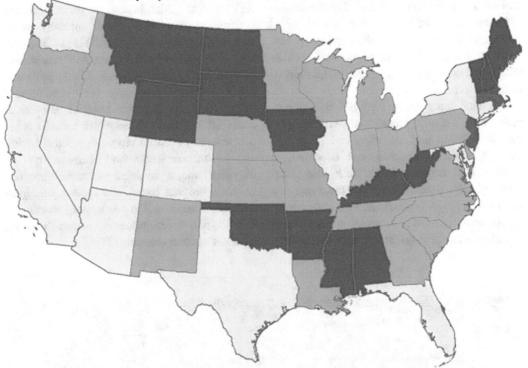

The States of Rural America
Seven out of ten states have a larger percentage of rural population than the national average

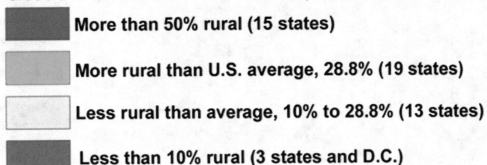

Map shows percent of residents in each state living in rural areas or "urban clusters" of between 2,500 and 50,000.

More than 50% rural (15 states)

More rural than U.S. average, 28.8% (19 states)

Less rural than average, 10% to 28.8% (13 states)

Less than 10% rural (3 states and D.C.)

Source: U.S. Census

Fig. 11.1 The States of Rural America: This map demonstrates that many Americans live in rural area where access to health care is reduced due to a paucity of health care providers [31, 32]

ity of disease, aging of our population, as well as limited access to care in rural areas. In accordance with the Institute of Medicine's (IOM) latest nursing goals, innovative solutions for providing patient care such as telehealth, and advanced specialty training of APP's offer a unique solution to these current challenges [34]. APPs are highly valued and are an integral part of a health care system and should be afforded the opportunity to practice to the full extent of their educational background and training. Research has shown that expanding the scopes of practice for APP's practicing in rural areas will improve access to specialty services, potentially decrease healthcare costs, and provide quality of care that is comparable to their physician counterparts [35].

It is vital that graduate level educational programs offer a diverse program of study and prepare students to provide safe and high-quality care [35, 36]. Interprofessional education is another key component of APP education. Interprofessional education (IPE) has become a mainstay in health professional education across the nation and assists in minimizing hierarchical models practice settings. This leads to a multidis-

ciplinary team-based approach to caring for a patient. Also, interprofessional clinical experiences utilizing telehealth technologies between providers as well as face-to-face encounters are imperative. These experiences better prepare APPs with the skills to care for patients in rural and underserved areas where specialty health care services might not be available [34, 36]. (See Fig. 11.2a and b) Furthermore, APP educational programs offer a curriculum that is typically narrowly focused, population and competency based. However, there are some commonalities across all APP curriculums which includes broad coursework: advanced physiology/pathophysiology, health assessment, and pharmacology. On the other hand, a physician's education program is broad, hospital based, and focuses on highly complex acute care. The overlapping educational skill sets of both disciplines can positively impact a patient's health outcome [37].

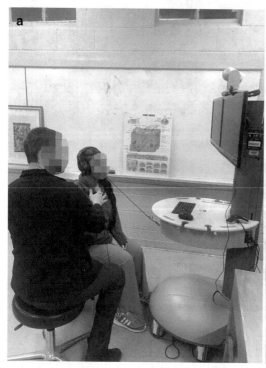
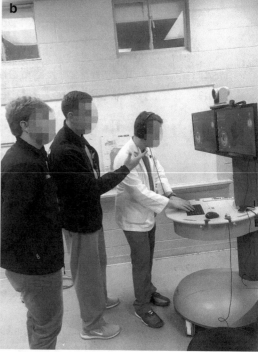

Fig. 11.2 (a and b) second year medical students and nurse practitioners at the University of Mississippi Medical Center incorporate telehealth services in their rural health clinic rotation. (1/29/2020). (Courtesy of Lisa Haynie PhD, RN, BC-FNP)

Obstacles/Challenges

Providers face a multitude of obstacles in rural dermatology on a daily basis. First, the rural workforce worldwide is limited, and recruiting and retaining qualified health professionals is an ongoing struggle [38]. Clearly many physicians are attracted to wealthier, urban areas, more than rural, poor areas [39]. Dermatologists face other obstacles in rural areas: illiteracy, poverty, prolonged sun exposure, weak health infrastructures, and poor geographical access to healthcare facilities [40]. Examining socioeconomic and geographic barriers to dermatology care, can lead to solutions to attack this problem. In a study where populations were examined in terms of race, poverty, and median income, it was determined that in rural areas, zero dermatologists practice in 88% of the areas [See Table 11.1 and Fig. 11.3] [41]. Lastly, this study illustrated that there is significantly increased income in the towns in which most dermatologists practice [41]. This data argues that there simply are not enough dermatologists to fill the void in poor rural areas, even with an increase in the number of rural physician programs. Furthermore, while telehealth in dermatology has made a small impact in reaching these rural communities, there must be a provider present to perform procedures [41]. The rural and urban dermatology access problem is also highlighted by another study that examined APP density, income, population, and number of hospitals. Like physicians, APPs gravitated to towns with higher income, larger population, and increased number of hospitals. These same towns were more likely to have at least 1 dermatologist [42]. Certainly, all communities need education and treatment of skin problems [41]. Areas with higher density of dermatologists in a geographical area is positively correlated with better outcomes and more effective diagnosing; therefore, the problem is not the quality of dermatology care, rather it is the quantity of care providers present in the country, especially rural areas such as areas in our home State of Mississippi [43]. Thus, the demand for dermatologic care is high, especially when considering the increasing rates of skin cancer and skin diseases [43].

Patients who live in rural areas also have a higher chance of developing melanoma compared to those who live in urban areas. One study examined the difference in knowledge of skin cancer, protection practices, and perceived risk by comparing rural and urban high school students in Utah [44]. Significantly, the urban county used in this study had a higher median household income than the rural counties [44]. Further, they found that rural students had lower chances of wearing ($p = 0.022$) or re-applying sunscreen ($p = 0.002$), wearing long sleeves ($p = 0.004$), and seeking shade ($p = 0.005$) compared to urban students [44]. Rural students reported spending more time outdoors and greater number or sunburns [44]. This study has shown that there are significant differences between rural and urban young adults in regards to skin cancer, and this is one of the many disparities that dermatologists face in rural communities [44].

After identifying many of the problems that dermatologists face in rural communities, it is

Table 11.1 Dermatologists distribution compared to county population, percentage of persons in poverty, median income and race

	Counties with dermatologists	Counties without dermatologists	Dermatologist: population ratio	
			≥3.5/100,000	<3.5/100,000
Population, n	287,683	22,908	409,974	59,871
Persons in poverty,%	14,9	27.2	16.1	24.7
Median income	$53,201	$44,443	$55,894	$44,848
African American majority, n (%)	13 (14)	80 (86)	6 (6,4)	87 (94)
Latino American majority, n (%)	14 (25)	56 (75)	5 (7.1)	65 (93)
Native American majority, n (%)	0 (0)	26 (100)	0 (0)	26 (100)
White majority, n (%)	930 (31)	2070 (69)	258 (8,6)	2742 (91)

With permission from Journal of the American Academy of Dermatology [41]

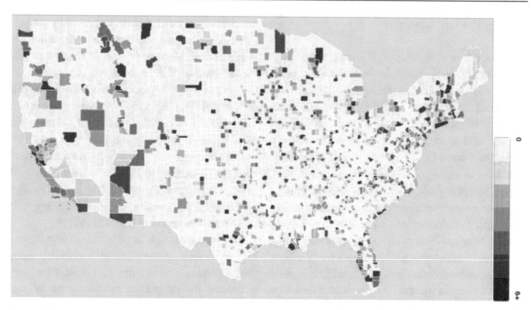

Fig. 11.3 Distribution of Dermatologists Compared to County Population, Income, and Race: County Specific Dermatology population ratio distribution of the United States (per 100,000 persons). (With permission from Journal of the American Academy of Dermatology) [41]

clear that solutions must be found. About 500 dermatologists join the work force annually, but about 325 dermatologists exit the workforce [43]. The unbalanced geographical distribution of dermatologists discussed earlier and the low number of net-new dermatologists, suggests that the problem of rural access to dermatologic care is not going to resolve without additional interventions. The large influx of advanced practice providers may be part of this solution [43]. As illustrated previously, the problem is not with the quality of care by dermatologists, but the access to care in rural America.

APPs Are Part of the Solution to Meet Rural Dermatology Needs

With proper guidance from physicians, physician assistants (PAs) and nurse practitioners (NPs) can be a great source of providers to help fill the gap in rural dermatology, providing that competent dermatologic care is achieved [44]. However, a survey examining the workforce characteristics of NPs in the dermatology setting, found that the practice patterns of the NPs were very similar to surveys of physicians. Notably, NPs most frequently reported practicing with physicians in dermatology-only groups [45]. Another survey analyzed 196 surveys from NPs where 44% reported at least starting their career in dermatology (meaning their first job was in dermatology whether or not they remained in the field of dermatology), with certifications in family practice, adult practice and multiple specialty practice [46]. While they found in this survey that 63% of the NPs reported their educational needs for keeping up with the changing field of dermatology were being satisfied, 61% of NPs desired additional training in the field of medical dermatology [46]. Further, the subjects in this study reported high job satisfaction in all perspectives of their job in dermatology [46]. The first approach to impact the rural access to care problem, therefore, may be to add APPs to existing rural dermatology practices. A study compared wait times between one clinic which employed NPs and PAs versus another that did not. The wait time of 38.9 days in a clinic without advanced practice providers was reduced to 33. 9 days [44].

Due to the fact that majority of NPs currently practicing in the field of dermatology share a similar scope of practice and working hours with dermatologists, it may be necessary to promote further advanced training including fellowship programs and residency programs to standardize training and achieve the best care possible for the patients [46]. This is especially important if APPs are working in independent practices. At the University of Cape Town, registered nurses trained and supervised by dermatologists ran a dermatology day-care center to focused on skin problems associated with HIV/AIDS to reduce the load of patients at the major hospitals of the state [47]. They developed a postgraduate dermatology qualification course to further strengthen their integrated medical nursing model [47]. Implementing this model in rural America would increase rural access to care, and reduce wait times and patient volume in the larger centers in metropolitan areas. For instance, APPs in rural communities could significantly impact screening for skin cancer. One study aimed to assess the ability of trained NPs to accurately identify possible cancerous lesions [48]. These NPs had no experience in evaluating skin cancer before they were placed in a training program and then evaluated utilizing competency assessments to differentiate between the clinical appearance of various benign and malignant lesions [48]. The results of this study concluded that NPs can accurately identify suspicious skin lesions with excellent specificity and good sensitivity with proper training [48]. Properly screening more individuals in rural communities could greatly help in identifying skin cancer much earlier and having more positive outcomes from skin cancer. This is particularly true in rural Mississippi, where our population often works outdoors.

A critical question faces APPs in their role as clinicians filling the access to care gaps in rural areas remains. Can APPs provide quality dermatology care? A randomized controlled trial was conducted that compared the quality of care as and cost-effectiveness of NPs and dermatologists [49]. In this study, an NP was substituted for a dermatologist caring for pediatric patients in an outpatient clinic [49]. There was no significant difference in the quality of care during outpatient

visits in patients between 4–16 years as judged by several parameters (See Fig. 11.4). Finally, the NP group invested an average of 100 minutes per patient versus 52 minutes pe patient in the dermatologist group [49].

When analyzing costs and cost-effectiveness between the two groups, they found that the mean annual healthcare costs were significantly higher in the dermatologist group, as well as for outpatient visits, lab testing, and medications [49]. In terms of family cost (the out of pocket expense for the family), the average annual family cost was two times higher in the dermatologist group compared to the NP group (€608 and € 302 respectively) due to higher cost per time and total healthcare costs for a patient to see a dermatologist versus a nurse practitioner, and this cost difference is significant when considering rural communities who face poverty [49]. This data affirms the ability for APPs to provide effective care and aid in filling the void in rural areas; moreover, from a health economic perspective, dealing with the issue of poverty in these underserved communities, NPs may be preferable [49]. Of course, more research of this type is required to further assess the quality and effectiveness of NP care provided both in the dermatology office and in independent practices.

Summary

In summary, implementing APPs in rural communities where there are too few dermatologists is a practical option to improve access to dermatology care in rural areas of the US. First, APPs have demonstrated their competency to care for common dermatology problems and effectively provide quality care. APPs also represent an affordable option for rural dermatology patients who often live in lower income communities. APPS can effectively work in dermatologists' offices and more states are now allowing independent practice of APPs. As more patients are seen in rural areas, patients save time and money traveling to higher density metropolitan regions, reducing the workload on urban dermatologists. Rural areas in the southern US are particularly

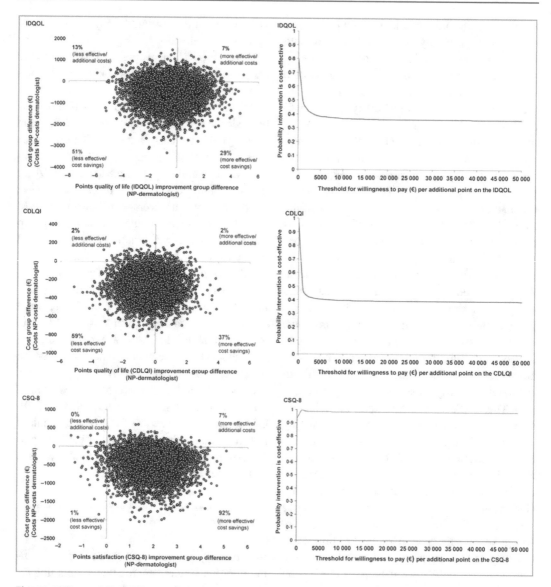

Fig. 11.4 Cost and Cost-Effectiveness Analysis per outcome parameter (left panels) and cost effectiveness acceptability curved (Nurse Practitioner vs Dermatologist Services). (With permission from British Journal of Dermatology) [49]

impacted by the perfect storm created by limited access to care and the increasing rate of skin cancer where early diagnosis has an impact on prognosis. APPs want to work together with physicians to fill access to care gaps in dermatology. The need for improved training programs as well as the standardization of training is crucial as the field of APP dermatology advances into the future [45].

Disclosures The authors have no relevant disclosures.

References

1. American Association of Nurse Practitioners. NP Fact Sheet. https://www.aanp.org/about/all-about-nps/np-fact-sheet. Published February 2020. Accessed 17 July 2020.
2. American Academy of Physician Assistants. What is a PA? https://www.aapa.org/wp-content/uploads/2020/07/WhatIsAPA-Infographic-MAY2020.pdf. Published May 2020. Accessed 17 July 2020.
3. Bureau of Labor Statistics, U.S. Department of Labor, Occupational Outlook Handbook,

Nurse Anesthetists, Nurse Midwives, and Nurse Practitioners. https://www.bls.gov/ooh/healthcare/nurse-anesthetists-nurse-midwives-and-nurse-practitioners.htm. Published September 1, 2020. Accessed 5 Sep 2020.

4. Bureau of Labor Statistics, U.S. Department of Labor, Occupational Outlook Handbook, Physician Assistants. https://www.bls.gov/ooh/healthcare/physician-assistants.htm. Published September 1, 2020. Accessed 5 Sept 2020.

5. Jones PE. Physician assistant education in the United States. Acad Med. 2007;82(9):882–7.

6. Bobonich M, Cooper KD. A core curriculum for dermatology nurse practitioners: using Delphi Technique. J Dermatol Nurses Assoc. 2012;4(2):108–20.

7. American Association of Nurse Practitioners. Historical Timeline. https://www.aanp.org/about/about-the-american-association-of-nurse-practitioners-aanp/historical-timeline. Published 2012–2020. Accessed 17 July 2020.

8. American Association of Nurse Practitioners. Planning your nurse practitioner education. https://www.aanp.org/student-resources/planning-your-np-education. Published 2012–2020. Accessed 17 July 2020.

9. American Association of Colleges of Nursing. Enrollment and graduations in baccalaureate and graduate programs in nursing. https://www.aacnnursing.org/Portals/42/News/Factsheets/Nursing-Workforce-Fact-Sheet.pdf . Published April, 2019. Accessed 17 July 2020.

10. American Association of Nurse Practitioners. What's a Nurse Practitioner (NP)? https://www.aanp.org/about/all-about-nps/whats-a-nurse-practitioner. Published 2012–2020. Accessed 17 July 2020.

11. American Association of Colleges of Nursing. DNP Fact Sheet. https://www.aacnnursing.org/DNP/Fact-Sheet. Published March 2019. Accessed 5 Sept 2020.

12. American Association of Nurse Practitioners. 2018 AANP national nurse practitioner sample survey results. file:///C:/Users/lesli/AppData/Local/Temp/2018.pdf. Published January, 2019. Accessed 20 July 2020. https://www.aanp.org/practice/practice-related-research/research-reports.

13. American Association of Nurse Practitioners. 2013-2014 National nurse practitioner practice Site Census. https://www.aanp.org/images/documents/research/2013-14nationalnpcensusreport.pdf.

14. Bobonich M, Nolen M. Competencies for dermatology nurse practitioners. J Am Assoc Nurse Pract. 2018;30(11):606–13.

15. Dermatology Nurse Practitioner Certification Board. Our Mission. https://www.dnpcb.org/. Accessed 20 July 2020.

16. Dermatology Nurse Practitioner Certification Board. Eligibility. https://www.dnpcb.org/Eligibility. Accessed 20 July 2020.

17. Dermatology Nurse Practitioner Certification Board. Alphabetical Listing of DCNP. https://www.dnpcb.

org/Find-a-DCNP. Published 20 August 2020. Accessed 5 Sept 2020.

18. Dermatology Nurses' Association. About DNA. https://www.dnanurse.org/aboutdna/. Accessed 5 Sept 2020.

19. Dermatology Nurses' Association. NP Society. https://www.dnanurse.org/aboutdna/np-society/. Accessed 5 Sept 2020.

20. Accreditation Review Commission on Education for the Physician Assistant. Program Accreditation Status. http://www.arc-pa.org/accreditation/accredited-programs/. Accessed 20 July 2020.

21. Glazer AM, Holyoak K, Cheever E, Rigel D. Analysis of US dermatology physician assistant density. J Am Acad Dermatol. 2017;76(6):1200–2.

22. Society of Dermatology Physician Assistants. SDPA Diplomate Fellowship. https://www.dermpa.org/page/DiplomateFellowship. Accessed 5 Sept 2020.

23. Ehrlich A, Kostecki J, Olkaba H. Trends in dermatology practices and the implications for the workforce. J Am Acad Dermatol. 2017;77(4):746–52.

24. Kimball AB, Resneck JS. The US dermatology workforce: a specialty remains in shortage. J Am Acad Dermatol. 2008;59(5):741–5.

25. Arnold T. Physician assistants in dermatology. J Clin Aesthetic Dermatol. 2008;1(2):28–31.

26. Feng J, Berk-Krauss J, Feng PW, Stein JA. Comparison of dermatologist density between urban and rural counties in the United States. J Am Med Assoc. 2018;154(11):1265–71.

27. Gorski, M.S. Fact Sheet. Advancing health in rural America: maximizing nursing's impact. https://assets.aarp.org/rgcenter/ppi/health-care/fs227-nursing.pdf. Accessed 12 Sept 2020.

28. World Health Organization. (2010). Increasing access to health workers in remote and rural areas through improved retention. https://www.who.int/hrh/retention/home/en/. Accessed 12 Sept 2020.

29. Ewing, J., Hinkley, K.N. (2013). Meeting the primary care needs of rural America: examining the role of the non-physician providers. The Rural Connection. National Conference of State Legislatures. https://www.ncsl.org/research/health/meeting-the-primary-care-needs-of-rural-america.aspx. Published April 2013. Accessed 12 Sept 2020.

30. Henderson K, Davis-Carlilse T, Smith M, King M. Nurse practitioners in telehealth: bridging the gaps in healthcare delivery. J Nurse Pract. 2014:845–50.

31. United States Census Bureau. The States of Rural America. https://gis-portal.data.census.gov/arcgis/apps/MapSeries/index.html?appid=7a41374f6b03456e9d138cb014711e01. Accessed 12 Sept 2020.

32. Coats, D. (2016). Can rural America rise again? Published July, 7, 2016. https://business.fullerton.edu/news/2016/07/07/the-american-economy-can-rural-america-rise-again/. Accessed 5 Sept 2020.

33. Sarzynski E, Barry H. Current evidence and controversies: advanced practice providers in healthcare. Am J Manag Care. 2019;25(8):366–8.

34. Rutledge, C., Pitts, C., Poston, R., Schweickert, P. NONPF supports telehealth in nurse practitioner education. The National Organization of Nurse Practitioner Faculties, 2018.

35. Swan M, Ferguson S, Chang A, Larson E, Smaldone A. Quality of primary care by advanced practice nurses: a systematic review. Int J Qual Health Care. 2015;27(5):396–404.

36. Institute of Medicine. (2010). The future of nursing: Leading change, advancing health. Retrieved from http://books.nap.edu/openbook.php?record_id=12956&page=R1. Published 2010. Accessed 12 Sept 2020.

37. Stanik-Hutt J, Newhouse RP, White KM, Johantgen M, Bass EB, Zangaro G, Weiner JP. The quality and effectiveness of care provided by nurse practitioners. J Nurse Pract. 2011;9(8):492–500.

38. Glazebrook R, Harrison S. Obstacles and solutions to maintenance of advanced procedural skills for rural and remote medical practitioners in Australia.

39. Thiers BH. Issues facing dermatology in the United States. An Bras Dermatol. 2006;81(6):585–9.

40. Al Kamel MA. A perspective on the obstacles to successful dermatological treatment in Yemen. J Dermatol Cosmetol. 2017;1(1):00016.

41. Vaidya T. Socioeconomic and geographic barriers to dermatology care in urban and rural US populations. J Am Acad Dermatol. 2018;78(2):406–8.

42. Feng H, Berk-Krauss J, Feng PW, Stein JA. Comparison of dermatologist density between urban and rural counties in the United States. JAMA Dermatol. 2018;154(11):1265–71.

43. Glazer AM, Farberg AS, Winkelmann RR, Rigel DS. Analysis of trends in geographic distribution and density of US dermatologists. JAMA Dermatol. 2017;153(4):322–5. https://doi.org/10.1001/jamadermatol.2016.5411.

44. Slade K, Lazenby M, Grant-Kels JM. Ethics of utilizing nurse practitioners and physician's assistants in the dermatology setting. Clin Dermatol. 2012;30(5):516–21.

45. Van Cott A, Kimball AB. Workforce characteristics of dermatology nurse practitioners. J Am Acad Dermatol. 2009;61(5):904–5. https://doi.org/10.1016/j.jaad.2009.02.026.

46. Cheng CE, Kimball AB, Cott V, Alicia A. Survey of dermatology nurse practitioners: work setting, training, and job satisfaction. J Dermatol Nurses Assoc. 2010;2(1):19–23. https://doi.org/10.1097/JDN.0b013e3181cd1de1.

47. Kelly PA. Community dermatology nursing in resource-poor communities across Africa. J Dermatol Nurses Assoc. 2012;4(4):244–8. https://doi.org/10.1097/JDN.0b013e3182617a63.

48. Oliveria SA, Nehal KS, Christos PJ, Sharma N, Tromberg JS, Halpern AC. Using nurse practitioners for skin cancer screening: a pilot study. Am J Prev Med. 2001;21(3):214–7. https://doi.org/10.1016/S0749-3797(01)00354-3.

49. Schuttelaar M, Vermeulen K, Coenraads P. Costs and cost-effectiveness analysis of treatment in children with eczema by nurse practitioner vs. dermatologist: results of a randomized, controlled trial and a review of international costs. Br J Dermatol. 2011;165:600–11. https://doi.org/10.1111/j.1365-2133.2011.10470.x.

Dermatology on American Indian and Alaska Native Reservations

12

Lucinda L. Kohn and Spero M. Manson

Introduction

American Indians (AI) and Alaska Natives (AN) experience some of the most severe healthcare disparities and worst health outcomes of any population in the developed world [1]. This chapter explores the history and political constructs from which disparities arose, outlines barriers to care, delineates the healthcare and dermatologic disparities faced by AI/AN, and reflects on an approach to serve the AI/AN population in an ethical, sustainable, and culturally respectful manner.

We know that ultimately this is not just a matter of [...] policy. It's a matter of whether we're going to live up to our basic values. [I]f we continue to work together, […] we will achieve a brighter future for the First Americans and for all Americans. President Barack Obama, Remarks at the White House Tribal Nations Conference, December 16, 2010

L. L. Kohn (✉)
Department of Dermatology, University of Colorado Anschutz Medical Campus, Aurora, CO, USA
e-mail: Lucinda.Kohn@cuanschutz.edu

S. M. Manson
Centers for American Indian & Alaska Native Health, Aurora, CO, USA

Colorado Trust Chair in American Indian Health, Aurora, CO, USA

Colorado School of Public Health at the University of Colorado Anschutz Medical Campus, Aurora, CO, USA
e-mail: Spero.manson@cuanschutz.edu

Historical Background

The relationship between AI/AN and the United States (US) federal government is unique for a minority group within the US. Not only are AI/AN distinct racially and ethnically, they also represent the only minority group in the US that is a defined political group. Today, two basic principles define this political grouping: tribal sovereignty and the federal trust relationship with tribal governments. Tribal nations are sovereign. Their authority to self-govern was established by treaties, executive orders, Supreme Court decisions, legislation and the US Constitution. Federal trust responsibility compels the US federal government to protect tribal self-governance, lands, assets, resources, and provide tribal members with healthcare and basic needs [2]. Application of the trust responsibility and attendant access to healthcare is contingent on membership in a federally recognized tribe. Membership is politically significant; requirements for membership are set by each tribe. Because of this, some people may have tribal heritage, but not meet requirements for tribal membership [3].

The complex political relationship that AI/AN have with the federal government informs the reality of being AI/AN in the US and has changed over time with political priorities of the federal government [2]. Prior to the 1800s, European settlers took Indian lands through the Discovery doctrine. Following the establishment of the US, the federal government worked to peacefully obtain land and resources from tribal nations on a government-to-government basis through treaties promising in return to provide health, education and public safety to tribal membership. Although these treaties eventually established the federal trust responsibility, non-Indians acquired land by forcibly relocating AI/AN to isolated and barren reservations [4].

In the late 1800s, US assimilation policies attempted to eradicate tribal heritage by implementing compulsory boarding schools that prohibited AI/AN children from speaking their native language, participating in traditional practices, and wearing Native attire and hair styles. The US implemented these assimilation policies through violence, which precipitated generations of mental and physical health issues [5]. In 1934, the Indian Reorganization Act supported tribal economic development and self-determination, but this policy shift was brief. In the 1950s, US policy returned to termination of tribal sovereignty as the US government dismantled parts of the federal-tribal trust relationship, ended federal recognition of at least 109 tribes, removed 2,500,000 acres of tribal land, and relocated AI/AN from tribal lands to urban areas. These actions further shattered identity, broke social and family networks, and increased mental health problems and distrust of the US government within AI/AN communities. Since the early 1970s, marked by the Indian Self-Determination and Education Assistance Act (public law 93-638), the US government has slowly and intermittently worked to restore tribal autonomy, and tribal governments have been in a phase of self-determination and self-governance [2].

Current State

As of 2020, there are 574 federally-recognized sovereign tribal nations within the United States and 334 federally and state recognized American Indian reservations located across 35 states [2].

Of the 308.7 million people who were counted in the 2010 US Census, 5.2 million people (1.7%

of the US population) identified as AI/AN alone or in combination with other races, with 2.9 million identifying as AI/AN alone [6]. Of note, the US Census estimates the actual number of AI/AN is higher than the number reported on the 2010 Census. Indian Country fast-growing with a population that is projected to reach 2% of the US population by 2050 [2, 6]. AI/AN are not uniformly distributed across the US: 41% reside in the West, 33% in the South, 17% in the Midwest, and 10% in the Northeast [7]. 54% of AI/AN live in seven states. From most to least, these are California, Oklahoma, Arizona, Texas, New York, New Mexico, and Washington [7]. The largest tribes in the US are the Cherokee with 819,000 members who identify with at least part Cherokee, followed by the Navajo with 332,129 members, then the Choctaw, Mexican AI, Chippewa, Sioux, Apache, and Blackfeet, each with populations over 100,000. The Yup'ik is the largest Alaska Native tribal grouping with 33,889 members. Navajo Nation is the largest AI reservation. Although a substantial portion of AI/AN still live on reservations or in the rural communities that border reservations, more than 78% of AI/AN live off-reservation [7].

Indian country is young compared to the US population: 36% of AI/AN are under 18 years of age versus 21.7% of the general population. The median age of AI/AN on reservations is 26 years old versus 37 years old in the general population [2].

General Health Disparities

AI/AN suffer some of the worst health inequities of any minority population within the US [1]. The average life expectancy of AI/AN (73.7 years) is 4.4 years less than those in the general US population (78.1 years) [8]. AI/AN die at higher rates from heart disease, diabetes, chronic liver disease, and chronic lower respiratory disease than the general population [8]. AI/AN youth have highest suicide rate – 2.5× higher than national average for young people [7]. AI/AN children have the worst access to medical care of any racial group in the US. Less than two-thirds of AI/AN children have access to all their needed healthcare, and about one-third find it difficult to access specialty care [9].

Current Healthcare landscape for American Indians and Alaska Natives

American Indians and Alaska Natives from federally recognized tribes have a legal right to healthcare provided by the Indian Health Service (IHS). The IHS was established in 1955 as part of the United States Public Health Service. Today, the IHS is a sub-agency of the US Department of Health and Human Services and oversees the delivery of health services through over 2000 facilities that provide primary care and dental services to AI/AN. Not only does the IHS run its own clinics and healthcare facilities, it oversees tribally operated healthcare facilities that are operated through public law 93-680, and 33 Urban Indian Health organizations.

The IHS serves approximately 2.2 million AI/AN from 574 federally recognized tribes. Sixty-four percent of the population served by the IHS lives on reservations or in rural communities, mostly in the western US and Alaska. Although the IHS's mission is to provide health services to all members of federally recognized tribes, the IHS is a severely underfunded federal health program. In 2017, the IHS spent $4078 per capita, compared with $13,185 by Medicare, $10,692 by the Veterans Health Administration, and $8109 by Medicaid [10]. Most strikingly, per capita spending available to IHS is less than the per capita healthcare obligations available to Federal Bureau of Prisons, which is $8602 [11]. Because of funding shortfalls, the IHS focuses its efforts on primary and emergency care services, some specialty services, and limited prescription drug coverage. Currently, the IHS faces a 25% clinician vacancy rate, which includes core clinicians such as primary care clinicians, midwives, dentists, and pharmacists [8].

Locums tenens personnel are utilized to temporarily ameliorate the paucity in core clinicians. This approach is more costly and interrupts continuity of care [12]. For services not routinely available within IHS, like dermatology, the Purchased/Referred Care program was created. However, the funds for this program are not sufficient to pay for all necessary care, and most often only medical services categorized as necessary to prevent immediate threat to life, limb or senses are approved [13].

Dermatologic Diseases in AI/AN

Not surprisingly, there is very little data on the ability of AI/AN to access board-certified dermatologists for the care of their skin, and there is limited data on the prevalence and severity of dermatologic diseases amongst AI/AN. The difficulties AI/AN must overcome to access specialty care, particularly dermatology, are considerable [14]. Extrapolating from available data, 27% of the population are uninsured or on Medicaid, yet only 5% of the population visited a dermatologist in 2002. Not all dermatologists accept Medicaid, and among those who do, Medicaid patients may be treated differently with longer wait times than privately insured patients [15]. In a study of the National Ambulatory Medical Care Survey in 1990, AI/AN patients utilize dermatologic services much less than Whites and Asians. This study also demonstrated a positive correlation between family income and the utilization of dermatologic services [16]. There is also a nationwide shortage of dermatologists, which is exaggerated in rural areas [17]. Lastly, AI may use the emergency room for dermatologic care more than any other group and access this care more frequently on the weekends. This suggests that they cannot afford to miss work during the week to access healthcare [18].

Despite the low rates of access to dermatologic care by AI/AN, they may suffer from a significant skin disease burden, and sometimes from more severe skin diseases. There is some evidence that acne is more frequent, inflammatory, and severe in AI/AN. One questionnaire-based study of 158 AI/AN young adults who self-identified as members of a federally recognized tribe and have lived on a reservation revealed

79% had a history of acne and 55.1% had acne scarring. Acne scarring was much more prevalent in AI/AN than in African American (34%), Hispanic (25%) Asian (8%) and Whites (3%). Thirty-one percent still had active acne. This incidence of active acne parallels the higher incidence seen in the African American (37%), Hispanic (32%) and Asian (30%) populations as compared to Whites (24%) [19]. Despite the high prevalence and disease severity, very few respondents sought care from a dermatologist. Half of survey respondents sought treatment from a healthcare professional, but only one saw a dermatologist.

Skin cancers are less common in minorities with darker pigmentation, but are often diagnosed later and at more advanced stages, resulting in a worse prognosis. The incidence of non-melanoma skin cancers in AI/AN is largely unknown, however from personal experience (LK) in directing the American Academy of Dermatology Association's resident rotation at Chinle Hospital, an IHS hospital on the Navajo reservation, both non-melanoma and melanoma skin cancers occur regularly [20]. A substantial portion of AI believe skin cancers do not occur in AI/AN and non-Whites, which may predispose AI/AN to not engage in skin cancer prevention practices, including sun protection and skin cancer screenings [21].

One study of invasive melanoma cases occurring between 1999 and 2006 reported that AI/AN cases comprised 0.2% of total cases and had an incident rate of 4.52 per 100,000 (versus 20.3 in Whites, 1.04 in Blacks and 4.68 in Hispanics). Overall, AI/AN were younger at the time of diagnosis with a median age of 54 years old than White (59 years old) and Black patients (63 years old). 51.5% of AI/AN with invasive melanoma were male. Predominant sites included the trunk (28%), upper limb and shoulders (23.5%) and lower limbs and hips (21.9). Twenty-five percent of melanomas diagnosed in AI/AN were of the superficial spreading subtype; 4.2% were lentigo maligna; 7.8% were nodular; 55.7% were not otherwise specified. It is possible, like in other populations with more pigmentation, acral melanomas are more frequent in AI/AN. One series of

18 melanoma cases in the Navajo described nine acral melanomas, of which 9 were subungual and three occurred on the palm or sole [22]. The incidence of melanoma within AIAN may be increasing. Between 1992–2002, the incidence was reported to be 1.6/100,000 and in 2001–2005, incidence increased to 3.1/100,000 [23]. Melanomas diagnosed in AI/AN are more likely to be distant, thicker (>4 mm) than those of Whites (5%). The 5-year melanoma specific survival for AI/AN was 84.9% compared with 90% of Whites, 87% for Hispanics, and 78.2% for Blacks [24].

Melanoma disparities have also been observed in AI/AN children. Analysis of the New Mexico Tumor Registry revealed that AI/AN children presented with significantly thicker melanomas (3.79 mm) than Hispanics (1.54 mm) or Whites (1.03 mm). This study reported an incidence of 3.2 per million for AI/AN children vs 7.5 per million for non-Hispanic White children and 2.1 per million for Hispanic children. It is difficult to draw conclusions based on these data since healthcare access is not equal for all children.

Photodermatoses, which include actinic prurigo and polymorphic light eruption, are more common in peoples indigenous to America, including AI/AN and those of Mestizo heritage. Actinic prurigo was first described under the term "solar prurigo" in Mexico by the late Dr. Escalona in 1954 in the first edition of his textbook on dermatology. Reports of actinic prurigo have been published in the Navajo population and numerous other AI tribes, including the Choctaw, Delaware, Shawnee, Creek, Cheyenne, Kickapoo, Kiowa, Chippewa, and Inuit at high prevalence rates [25–27]. In the Inuit, it was observed there was later onset with an average age of diagnosis of 29 years, more ocular involvement (62%) and a less favorable prognosis with improvement in 27% of patients. Actinic prurigo impacts both children and adults and has significant impacts on quality of life [28, 29].

Rheumatologic diseases and metabolic diseases are more prevalent among AI/AN, and as such it is not surprising that skin disorders associated with metabolic disease such as acanthosis nigricans are more prevalent among AI/

AN. Acanthosis nigricans, which typically reflects a state of hyperinsulinemia, has a high prevalence among AI children. Prevalence rates range from 10–19% in Plains Indian children living on reservations [30, 31]. This is in contrast to prevalences of 0.45%, 5.5% and 13.3% in non-Hispanic Whites, Hispanics and Black children, respectively [30]. The higher prevalence of acanthosis nigricans in AI/AN communities is most likely due to both predisposition to metabolic disease and life circumstances. Reservations and border towns are food deserts, offering sugary drinks and junk food in abundance, but with limited produce and meat available to families [32]. Supporting genetic predisposition to metabolic disease, one study that directly compared AI children with children of other races living in the same communities observed that 10% of AI children developed acanthosis nigricans versus 3.2% of children of other races [33]. In other reports that include adults, the true prevalence may be even higher. Thirty-four percent of Cherokee members aged 5–40 years old had acanthosis nigricans, which was more likely to occur with a family history of diabetes or and increasing degree of Cherokee heritage [34]. Forty percent of members of the Alabama-Coushatta tribe in eastern Texas between the ages of 10 and 49 had acanthosis nigricans [31].

A number of studies have demonstrated an increased rate and worse outcomes of connective tissue diseases, including rheumatoid arthritis (RA), systemic lupus erythematosus (SLE), inflammatory myopathies, spondyloarthropathies, and systemic sclerosis than the general population. The higher prevalence of connective tissue diseases is disease and tribe-specific. This is possibly related to higher frequency of certain class II major histocompatibility complex alleles and polymorphic T cell antigen receptors in each tribal population.

In general, AI have high rates of RA, whereas AN, specifically Eskimos and Athabaskan-Eyak-Tlingit Indians, have high rates of spondyloarthropathies (SA). This increased frequency may be genetic in nature as there are high rates of the HLA-B27 antigen in AI/AN, but particularly high in the Athabaskan-Eyak-Tlingit and Eskimo-Aleuts. Consequently, SA are found in high rates in AI/AN, but especially among Athabaskan-Eyak-Tlingit and Eskimo-Aleut groups [35].

Rheumatoid arthritis is also found in increased prevalence in the AI than in the general population. In the Chippewa and Blackfeet tribes, the prevalence of RA is 5.3% and 5%, respectively, versus 1% in the general white North American and western European populations. There is also a predisposition for RA to present at an earlier age, be more severe, with high frequency of rheumatoid factor and anti-nuclear antibody seropositivity, and have more frequent extra-articular manifestations in AI/AN. Relevant cutaneous manifestations include rheumatoid nodules, and alopecia.

Dermatologists should be mindful of the increased rate and severity of connective disease among AI/AN, as the index of suspicion should remain high to identify atypical presentations when treating AI/AN.

Other, more rare diseases have been reported in association with specific tribes. One such example is the genetic disease poikiloderma with neutropenia (Clericuzio type), which was first described in the Navajo in 1991, resembles Rothmund Thompson and is characterized by a poikilodermatous rash, noncyclical neutropenia, small stature, pachyonychia pulmonary disease, and calcinosis cutis [36]. Since its original description, it has been described in other ethnicities showing that this is not unique to the Navajo, including in a Turkish family, a Morrocan family, and a patient from India [37–39].

Another genodermatosis, xeroderma pigmentosum (XP), was not originally described in the Navajo, but has recently received press for affecting several Navajo families in an incidence far greater than the general population. Typically, the incidence of XP is 1 in 1 million, but amongst members of the Navajo Nation, the incidence is 1 in 30,000. Xeroderma pigmentosum is a photosensitive disorder where there is an inability of the body's cells to repair DNA damage, specifically the type of damage caused by UV radiation, and results in numerous, aggressive skin cancers and sometimes neurologic deterioration. Since this is not a disease that was historically seen in the Navajo, some have postulated this increase in incidence is due to the effects of uranium mining on Navajo land [40]. Between 1944 and 1986, 4 million tons of uranium was mined out of Navajo land, which the federal government used to purchase atomic weapons. As the Cold War petered out, so did the 500 uranium mines, but they did not shut down responsibly. The harmful effects of uranium mining linger on Navajo Nation today. Since 2008, the Environmental Protection Agency has worked to properly dispose of mine waste and rebuilt contaminated homes [41]. There is some scientific evidence to support the claims that XP in Navajo children is caused by the effects of uranium mining. Uranium causes zinc loss from DNA repair proteins, including the complex that is affected in XP, and people who live near uranium mining have exhibited DNA repair deficiency [42, 43].

Certain infectious diseases are also more common in AI/AN. Some studies have shown increased rates of MRSA infectious in certain groups of AI/AN, which are likely due to low socioeconomic status, including crowded living conditions, poor sanitation, lack of running water and limited access to healthcare [44, 45]. Lastly, it is well-established that AI/AN, along with Filipinos and African Americans, are more likely to develop disseminated coccidioidomycosis, typically a pulmonary infection caused by the soil saprophyte Coccidioides immitis, which is endemic to the arid US Southwest [46].

Barriers to Care

The health inequities faced by AN/AI are not only due to the quality of health care available to this population, but also stem from many practical, infrastructural and socioeconomic factors. These so-called social determinants of health include geographical hurdles; poverty; economic, employment and educational opportunities; safe communities and suitable housing; lack of basic community infrastructure; underrepresentation in the US physician workforce; and an inadequate health system to meet individual and tribal health needs [5].

By their very design, AI reservations lie in remote areas. Because of this, it can be time consuming and expensive for those who reside on reservations to access healthcare specialists in major cities [14]. These geography-related barriers, when in the context of prolonged waiting room times and multiple follow-up appointments needed for optimal medical care, may seem insurmountable to patients.

Poverty is the greatest barrier AI/AN face in accessing healthcare and engaging in practices that would prevent disease. AI/AN suffer 2.5 times the poverty rate of the general US population: 28.4% of AI/AN are living in poverty versus 11.3% of the general population. On reservations, the proportion of people living in poverty reaches 39%. The median household income of AI/AN is $43,635 versus $50,046 in the general US population.

Unemployment is strongly related to poverty among AIAN. Indian joblessness as measured by a Bureau of Indian Affairs labor force report was 49%, and the unemployment rate for AI/AN on reservations is 19%. Without employment, most AI/AN are at a loss for acquiring work-sponsored health insurance. Indeed, in 2013, 30% of non-elderly AIAN adults were uninsured (versus 17% of the non-elderly general population). Of those who had health insurance, one-third were covered by Medicaid, one-third had private insurance versus the general non-elderly population in which 21% had Medicaid and 62% had private insurance [47]. Many AI/AN would qualify for Medicaid, but there are barriers to enrollment, such as lack of knowledge about such benefits, difficulty of the enrollment process, and literacy and language barriers. In theory, the Affordable Care Act should ameliorate these disparities by its expansion of Medicaid coverage; however, many AI/AN adults live US states that did not expand Medicaid [47].

Many AI/AN reside in communities with homes that lack basic infrastructure required to prepare healthy meals, practice sanitary behaviors, and access healthcare. The 2011–2015 Census American Community Survey estimated that 23.6% of AN homes and 7.7% of AI homes lack complete plumbing. Seventeen percent of

AN and 6% of AI homes lack a complete kitchen. Just under 10% of AI and 4.1% of AN do not have a telephone. Fourteen percent of AI and 28.2% of AN reside in overcrowded homes [2].

Higher education is also limited among AI/AN, which leads to a cascade of effects that compromises a community's ability to be autonomous. Eighty-two percent of AI/AN adults over 25 years old have a high school degree or equivalent versus 86% of the US population. Eighteen percent of AI/AN have a bachelor's degree versus 29% in the general population.

Many non-AI/AN assume that because AI/AN are native to the US, they are fluent in English. However, almost every tribe has its own native language—the use of which is common amongst tribal elders— and as many as 40% of AI/AN report that they do not speak English well [48].

Even if AI/AN had adequate resources, lacked geographic barriers, and could access healthcare easily, AI/AN are likely to utilize healthcare at lower rates than the general population [49]. This may be due to distrust of the US government and the medical system, lack of facilities and physicians who accept Medicaid, lack of clinicians in IHS healthcare facilities and lack of culturally competent clinicians in their communities [20]. The IHS reports a clinician vacancy rate of 25% in AI/AN communities [8]. Only .08% of the US physician workforce and 10–15% of the IHS physician workforce is composed of AI/AN physicians. More AI/AN physicians serving AI/AN communities would highly benefit the AI/AN population, as AI/AN physicians may have more cultural sensitivity and be more readily accepted by their patients.

The paucity of AI/AN physicians is a recently recognized national crisis. In 2018, the Association of American Medical Colleges (AAMC) and Association of American Indian Physicians (AAIP) jointly released a report revealing that the number of single-race AI/AN applicants to medical school ranged from 150–200 per year in the 2010s despite steady increases in US medical school matriculants averaging about 21,000 per year. The reality is that AI/AN students are ill-prepared to face the academic and financial challenges of pursuing a medical degree. The AAMC, AAIP, American Medical Association (AMA) and medical schools have pledged to address these educational and financial disparities. Mentorship programs, scholarship options, and Science Technology Engineering Mathematics (STEM) partnership programs between universities and tribal communities aim for higher rates of medical school enrollment among AI/AN.

Interventions are required at all levels of the medical education ladder. Not only is it important to recruit more AI/AN into medical school, but it is equally important to teach them and their non-AI/AN peers the cultural nuances of caring for AI/AN patients. Some medical schools have developed partnerships with tribes, integrated AI/AN culture into medical training, and exposed their students to healers, and sweat lodges, to address AI/AN recruitment resistance. Examples of robust programs include the College of Medicine at the University of Arizona's Rural Health Professions Program, and the University of Washington Medicine's Indian Health Pathway Certificate for medical students [42]. These partnerships are important for many reasons. They inspire AI/AN students in applying to medical school, and help to develop a culturally competent AI/AN and non-AI/AN physicians who are motivated to care for AI/AN communities.

Addressing the Dermatologic Healthcare Disparities for AI/AN

As discussed above, access to dermatologic care for AI/AN is poor. In addition to the IHS's rare dermatologic services, there are a few scattered dermatology clinics and university-tribal partnerships providing care. These include the one board certified dermatologist employed by the IHS, telehealth clinics offered through a private telehealth company to tribes in the Dakotas, a dermatology clinic staffed by board certified dermatologists at the Alaska Native Health Center in Anchorage, "free" telehealth clinics put forth by volunteer dermatologists as part of the American Academy of Dermatology Association's AccessDerm platform, and consults sent through the IHS's purchased and referred care program [14, 20, 50].

Offering free healthcare services to a community by clinicians who do not live in nor are not personally invested in those communities has been called "voluntourism" and has been met with criticism and dismay within the global health realm. Although there are immediate, tangible benefits to this practice, these temporary free clinics erode local economies and weaken motivation to construct sustainable, high quality healthcare programs within the communities they aim to help. Although dermatologists may help patients with immediate questions and health concerns, these free telehealth services run by dermatologists in distant cities send the wrong message to IHS, tribal governments, and major health care funders and key players in hospital systems that dermatology services do not need to be built or funded in tribal communities. Furthermore, dermatology services offered through telehealth are generally substandard, because they are offered by a remote, sometimes volunteer physicians who are unfamiliar with the population, who does not possess the motivation nor physician presence to establish the community relationships needed with primary care clinicians and specialists, and who do not have the ability to provide the same quality of care they would to the patients in a communited-based brick and mortar clinic. Lastly, these free programs, run with good will and altruism, are usually short-lived and eventually disappear, harming the people they were meant to help.

Instead of offering direct dermatologic care given by remote dermatologists through telehealth or periodic visitorships, our goal should be towards sustainable and ethical change. This means increasing AI/AN enrollment in medical schools and nursing schools, and training more AI/AN in dermatology. Unlike direct patient-care teledermatology programs, telehealth education programs enlist specialists who partner with local primary care clinicians who do know the healthcare system and local community. We need to build partnerships between universities and tribes in which the latter may actively participate in the programs erected through collaboration and promote solutions to problems created by voluntourism.

Some residencies have also established partnerships with tribes [8]. Examples of these include Massachusetts General Hospital's partnership with the Rosebud Sioux tribe in South Dakota and University of California, San Franciso's Health, Equity, Action, Leadership Global Health Fellowship, which has a site in Chinle, Arizona on the Navajo Reservation [8]. Dermatology-specific programs include the American Academy of Dermatology Association's Native American Health Service Resident Rotation in Chinle, Arizona on the Navajo Reservation; Brigham and Women's Hospital's partnership with Shiprock, Arizona on the Navajo Reservation, and University of Utah Dermatology Department's resident continuity clinic with Navajo Nation.

Operating these programs requires dedication, effort, and resources. The ideal program is deeply rooted in trusting, equitable relationships between the tribe and university. These relationships take time to solidify, depend upon frequent communication and in-person visits to maintain personal relationships. Tribes must be seen as full partners, not mere consultants, in these partnerships. Universities must respect the requirements for trainees and supervising physicians to participate in these programs. There must also be a commitment to rigorous and high-quality education experience for the trainees, high clinical standards and competent medical care provided to clinic patients. In other words, while tribal clinics are responsible for the quality of the preceptorship offered to trainees; trainees should not practice beyond their level of training; all clinicians must abide by federal and state regulations in regards to medical licensure. These programs must also educate trainees about serving the AI/AN population in a culturally sensitive manner and issues that arise in rural health more generally. Topics such as tribal history, cultural sensitivity training, advice on working with patients specific to the community, possible conflicts between Western and traditional medicine should be covered. Universities, in turn, need to appropriately invest in these programs by supporting faculty members' outreach, administrative and supervision time, support staff to navigate the regulations and paperwork, and underwrite student expenses like housing and transportation. It is the responsibility of the university to verify that doctors on staff who want to teach are able to maintain the same standards required of faculty on campus and carry continuous quality improvement that incorporates the experiences of all stakeholders.

Telemedicine clinics that prioritize equity, and partnership with the local tribe, and mutual collaboration do exist. One successful model is the telepsychiatry clinics offered by the University of Colorado Health Sciences Center's American Indian and Alaska Native Programs (AIANP) to the Rosebud Sioux tribe reservation through the Department of Veterans Affairs (VA) [51]. This program was built with a stepwise approach over several years. In this development model, the steps to building a successful, sustainable specialty clinic include an initial needs identification, infrastructure survey, partnership organization, and structure configuration, pilot implementation, and finally, solidification. The approach has spawned 12 other telepsychiatry clinics across tribal communities that reflect these key elements.

Assessing local capacity is equally important to developing a telehealth clinic. This requires more than simply technological resources such as network and software needs. One also needs to identify the services and resources available to patients in the community. In the dermatologist's case, these include the medications available at the local pharmacy, the soaps and moisturizers in stock at local stores, the surgeons who may be willing to perform skin cancer excisions, primary care physicians and rheumatologists who are knowledgeable in managing biologic treatment regimens, and local clinicians able to perform skin biopsies as warranted.

Once deemed feasible, all key players in the AIANP telepsychiatry clinics— namely, tribe, university, and the VA— jointly discussed clinic design and implementation. With plan in hand, there was a pilot period in which patient and staff were asked for feedback to ensure quality care. Lastly, having modified the pilot program to accommodate patient and staff preferences, the program emerged as a fully functional, long-term successful and sustainable clinic [51]. This development model represents a roadmap that dermatologists may follow to build a successful, sustainable specialty clinic that prioritizes equity, partnership with the local tribe, and mutual collaboration.

Conclusion

Health inequities exist for AI/AN, and they are even more pronounced for dermatologic care. To bridge this gap, reform must be made at all levels. Our immediate goals should be to train more AI/AN dermatologists, increase dermatologists' competence and comfort at treating AI/AN skin diseases and develop an understanding of cultural and societal predispositions and practices. Over time, we should work towards building culturally respectful and sustainable dermatologic programs for AI/AN that partner with tribal nations.

Disclosures The authors have no relevant disclosures.

References

1. Healthy People 2030 [Internet]. [cited 2021 Jan 12];Available from: https://health.gov/healthypeople.
2. National Congress of American Indians. Tribal nations and the United States: an introduction. [Internet]. [cited 2020 Dec 30];Available from: https://www.ncai.org/tribalnations/introduction/Tribal_Nations_and_the_United_States_An_Introduction-web-.pdf.
3. Hd. J, Sandefur GD, Rindfuss RR, Cohen B, Lee RD. Changing numbers, changing needs. Population (French Edition); 1997.
4. Seminole Nation v. United States; 1942.
5. Fortney JC, Kaufman CE, Pollio DE, Beals J, Edlund C, Novins D, AI-SUPERPFP Team. Geographical access and the substitution of traditional healing for biomedical services in two American Indian tribes. Med Care. 2012;50(10):877–84.
6. Census Bureau releases estimates of undercount and overcount in the 2010 census [Internet]. [cited 2021 Jan 12]. Available from: https://www.census.gov/newsroom/releases/archives/2010_census/cb12-95.html.
7. Norris T, Vines PL, Hoeffel EM. The American Indian and Alaska Native Population, 2010 Census Briefs; 2012.
8. Indian Health Service. Quick look. Fact sheets.

9. Flores G, Tomany-Korman SC. Racial and ethnic disparities in medical and dental health, access to care, and use of services in US children. Pediatrics [Internet]. 2008;121(2):e286–98. Available from: http://pediatrics.aappublications.org/cgi/doi/10.1542/peds.2007-1243.

10. United States Government Accountability Office. Indian Health Service: spending levels and characteristics of IHS and three other federal health care programs. 2018. Available from: https://www.gao.gov/assets/700/695871.pdf.

11. United States Government Accountability Office. Bureau of Prisons: better planning and evaluation needed to understand and control rising inmate health care costs report to congressional requesters United States Government Accountability Office. 2017;(June). Available from: https://www.gao.gov/assets/690/685544.pdf.

12. United States Government Accountability Office. Indian Health Service: agency faces ongoing challenges filling provider vacancies. 2018;(August). Available from: https://www.gao.gov/assets/700/693940.pdf.

13. Indian Health Service. Purchased/Referred Care (PRC) [Internet]. Available from: https://www.ihs.gov/newsroom/factsheets/purchasedreferredcare/.

14. Morenz AM, Wescott S, Mostaghimi A, Sequist TD, Tobey M. Evaluation of barriers to telehealth programs and dermatological care for American Indian individuals in rural communities. JAMA Dermatol [Internet]. 2019;02114. Available from: http://archderm.jamanetwork.com/article.aspx?doi=10.1001/jamadermatol.2019.0872.

15. Resneck JS, Isenstein A, Kimball AB. Few Medicaid and uninsured patients are accessing dermatologists. J Am Acad Dermatol. 2006;55(6):1084–8.

16. Fleischer AB, Feldman SR, Bradham DD. Office-based physician services provided by dermatologists in the united states in 1990. J Invest Dermatol [Internet]. 1994;102(1):93–7. Available from: https://doi.org/10.1111/1523-1747.ep12371739.

17. Feng H, Berk-Krauss J, Feng PW, Stein JA. Comparison of dermatologist density between urban and rural counties in the United States. JAMA Dermatol. 2018;154(11):1265–71.

18. Chen CL, Fitzpatrick L, Kamel H. Who uses the emergency department for dermatologic care? A statewide analysis. Journal of the American Academy of Dermatology [Internet]. 2014;71(2):308–13. Available from: https://linkinghub.elsevier.com/retrieve/pii/S0190962214012407.

19. Perkins AC, Cheng CE, Hillebrand GG, Miyamoto K, Kimball AB. Comparison of the epidemiology of acne vulgaris among Caucasian, Asian, Continental Indian and African American women. J Eur Acad Dermatol Venereol. 2011;25:1054–60.

20. Kohn LL, Introcaso CE. A Cultural Context for Providing Dermatologic Care to American Indian and Alaskan Native Communities Through Telehealth. JAMA Dermatol. 2019;155(8):884–86. https://doi.org/10.1001/jamadermatol.2019.0860. PMID: 31215957.

21. Maarouf M, Zullo SW, DeCapite T, Shi VY. Skin cancer epidemiology and sun protection behaviors among native Americans. J Drugs Dermatol JDD. 2019;18:420–3.

22. Black WC, Msph CW. Melanoma among southwestern American Indians. Cancer. 1985;55:2899–902.

23. Shoo BA, Kashani-Sabet M. Melanoma arising in African-, Asian-, Latino- and native-American populations. Semin Cutan Med Surg. 2009;28:96–102.

24. Wu X-C, Eide MJ, King J, et al. Racial and ethnic variations in incidence and survival of cutaneous melanoma in the United States, 1999–2006. J Am Acad Dermatol [Internet]. 2011;65(5):S26.e1–S26.e13. Available from: https://linkinghub.elsevier.com/retrieve/pii/S0190962211006098.

25. Magana M, Lane PR, Martel MJ, Reeder B. Actinic or solar prurigo [8]. J Am Acad Dermatol. 1997;36(3):504–5.

26. Lain ES. Skin diseases among full-blood Indians of Oklahoma. JAMA J Am Med Assoc [Internet]. 1913;61(3):168. Available from: http://jama.jamanetwork.com/article.aspx?doi=10.1001/jama.1913.04350030008003.

27. Kryatova MS, Okoye GA, Ginette Okoye CA. Dermatology in the North American Indian/Alaska Native population.

28. Jong CT, Finlay AY, Pearse AD, et al. The quality of life of 790 patients with photodermatoses. Br J Dermatol. 2008;159:192–7.

29. Rizwan M, Reddick CL, Bundy C, Unsworth R, Richards HL, Rhodes LE. Photodermatoses: environmentally induced conditions with high psychological impact. Photochem Photobiol Sci. 2013;12:182–9.

30. Stuart CA, Pate CJ, Peters EJ. Prevalence of acanthosis Nigricans in an unselected population. Am J Med. 1989;87:269–72.

31. Stuart CA, Smith MM, Gilkison CR, Shaheb S, Stahn RM. Acanthosis Nigricans among native Americans: an indicator of high diabetes risk. Am J Public Health. 1994;84:1839–42.

32. O'Connell M, Buchwald DS, Duncan GE. Food access and cost in American Indian communities in Washington State. J Am Diet Assoc [Internet]. 2011;111(9):1375–9. Available from: https://www.ncbi.nlm.nih.gov/pmc/articles/PMC3624763/pdf/nihms412728.pdf.

33. Brown B, Noonan C, Bentley B, et al. Acanthosis Nigricans among Northern Plains American Indian children. J Sch Nurs. 2010;26:450–60.

34. Stoddart ML, Blevins KS, Lee ET, Wang W, Blackett PR. Association of acanthosis Nigricans with hyperinsulinemia compared with other selected risk factors for type 2 diabetes in Cherokee Indians: the Cherokee diabetes study. Diabetes Care. 2002;25:1009–14.

35. Peschken CA, Esdaile JM. Rheumatic diseases in North America's indigenous peoples. Semin Arthritis Rheum. 1999;28:368–91.

36. Clericuzio C, Hoyme HE, Aase JM. Immune deficient poikiloderma: a new genodermatosis. Am J Hum Genet. 1991;49(Suppl):A661.

37. Mostefai R, Morice-Picard F, Boralevi F, et al. Poikiloderma with neutropenia, Clericuzio type, in a family from Morocco. Am J Med Genet A. 2008;146:2762–9.

38. Bishnoi A, Jamwal M, Das R, et al. Clericuzio-type poikiloderma with neutropenia in a patient from India. Am J Med Genet A. 2021;185:278–81.

39. Patiroglu T, Akar HH. Clericuzio-type poikiloderma with neutropenia syndrome in a Turkish family: a three report of siblings with mutation in the C16orf57 gene. Iran J Allergy Asthma Immunol. 2015;14:331–7.

40. Bender A. Rare disease suddenly arises on Navajo Reservation[Internet]. People's World.[cited 2021 Jan 10]. Available from: https://www.peoplesworld.org/article/rare-disease-suddenly-arises-on-navajo-reservation/.

41. Morales L. For the Navajo Nation, Uranium mining's deadly legacy lingers [Internet]. NPR. [cited 2021 Jan 10]. Available from: https://www.npr.org/sections/health-shots/2016/04/10/473547227/for-the-navajo-nation-uranium-minings-deadly-legacy-lingers.

42. Au WW, Salama SA. Cytogenetic challenge assays for assessment of DNA repair capacities [Internet]. In: DNA repair protocols. Totowa: Humana Press. p. 25–42. Available from: http://link.springer.com/10.1385/1-59259-973-7:025.

43. Cooper KL, Dashner EJ, Tsosie R, Cho YM, Lewis J, Hudson LG. Inhibition of poly(ADP-ribose)polymerase-1 and DNA repair by uranium. Toxicol Appl Pharmacol [Internet]. 2016;291:13–20. Available from: https://doi.org/10.1016/j.taap.2015.11.017.

44. Elston DM. Community-acquired methicillin-resistant Staphylococcus aureus. J Am Acad Dermatol. 2007;56(1):1–16.

45. Groom AV, et al. Community-acquired methicillin-resistant Staphylococcus aureus in a rural American Indian community. JAMA [Internet]. 2001;286(10):1201. Available from: http://jama.jamanetwork.com/article.aspx?doi=10.1001/jama.286.10.1201.

46. McCotter O, Kennedy J, McCollum J, et al. Coccidioidomycosis among American Indians and Alaska Natives, 2001–2014. Open Forum Infect Dis. 2019;6(3):1–8.

47. Artiga S, Arguello R, Duckett P. Health coverage and care for American Indians and Alaska Natives [Internet]. The Kaiser Commission of Medicaid and the Uninsured. [cited 2021 Jan 6]. Available from: https://www.kff.org/racial-equity-and-health-policy/issue-brief/health-coverage-and-care-for-american-indians-and-alaska-natives/.

48. Laveist TA. Minority populations and health: an introduction to health disparities in the United States. San Francisco: Jossey-Bass; 2005.

49. Wong ST, Kao C, Crouch JA, Korenbrot CC. Rural American Indian Medicaid health care services use and health care costs in California. Am J Public Health. 2006;96(2):363–70.

50. Zullo SW, Maarouf M, Shi VY. Acne disparities in Native Americans. J Am Acad Dermatol [Internet]. 2018. Available from: https://doi.org/10.1016/j.jaad.2018.09.037.

51. Shore JH, Manson SM. A developmental model for rural telepsychiatry. Psychiatr Serv (Washington, DC) [Internet]. 2005;56(8):976–80. Available from: http://www.ncbi.nlm.nih.gov/pubmed/16088015.

Project ECHO: Improving Rural Dermatology Through Digital Primary Care Education

Catherine Clare Gloss, Eric Grisham,
Adam Rosenfeld, and Kari Lyn Martin

ECHO Basics

Teledermatology has evolved into a powerful tool used to address decreased access to specialty services in rural communities [1]. ECHO is a collaborative, interdisciplinary educational tool that trains community-based and rural primary care providers to manage and diagnose complicated diseases in a community setting [2, 3]. Since its inception in 2003, this innovative program has become a global enterprise, addressing over 70 health conditions and public health concerns in 241 academic, government, and other partners in 34 countries [4].

Complex specialty medical care in rural, minority, and underserved communities is often severely limited [5]. Over one-third of primary care patients seek care for at least one skin problem [1]. The maldistribution of dermatologists in

---

"Often, we are too slow to recognize how much and in what ways we can assist each other through sharing expertise and knowledge."—Owen Arthur

---

C. C. Gloss · E. Grisham · A. Rosenfeld
University of Missouri School of Medicine,
Columbia, MO, USA
e-mail: ccg5z6@health.missouri.edu;
eaggzd@health.missouri.edu;
ahrpp9@health.missouri.edu

K. L. Martin (✉)
Department of Dermatology, University of Missouri
School of Medicine, Columbia, MO, USA
e-mail: martinkar@health.missouri.edu

the United States of America, combined with the challenges associated with obtaining specialty care in rural areas, can result in delays in treatment and significant morbidity and mortality for these patients [1]. Improving access to specialty care has become a vital component to addressing disparities in healthcare [5]. Rural and underserved patients often have manageable skin disorders. However, the scarcity of specialists in rural areas results in an inability to provide rapid and reliable treatment and follow up care. Some patients do not receive any treatment at all. ECHO has addressed one of the American Academy of Dermatology's (AAD) Special Positioning Workgroup's core areas of impact by treating common, severe skin conditions with innovative, team-based teledermatology [6]. While policymakers and healthcare leaders recognize the value of applications such as teledermatology, and more work is needed to remove barriers to utilization that still exist (e.g., reimbursement, legal and regulatory challenges), ECHO has already helped to educate community-based providers in the provision of healthcare to patients who would otherwise receive lower quality care [6, 7].

ECHO Beginnings

ECHO was developed in 2003 at the University of New Mexico Health Sciences Center to address the inadequate treatment of hepatitis C in rural and underserved patients [8, 9]. ECHO connected local providers with specialists at large treatment centers up-to-date in current evidence-based medical practices. During weekly virtual sessions, specialists taught local practitioners how to screen, prevent, and treat hepatitis C through didactic lectures and discussion of de-identified patient cases [4]. Participants addressed systemic barriers to care by developing holistic treatments tailored to their patients' unique social, cultural, and economic circumstances [4]. The initial results were promising. ECHO resulted in a decrease in adverse events in patients treated at ECHO sites as compared to patients treated at academic medical centers (6.9% and 13.7%, respectively) [9]. Remote learning through ECHO provided up-to-date, evidence-based medical practices to primary care providers with pre-existing provider-patient relationships and knowledge of local culture and available patient resources, resulting in improved outcomes [9]. Since its inception, ECHO has expanded to include the treatment of HIV/AIDS, chronic pain, heart failure, substance use disorder, endocrinology, autism, multiple sclerosis, psychiatric problems, dermatology, and the list continues to grow [8, 9].

In 2015, the University of Missouri (MU) implemented Dermatology ECHO, the first non-military Dermatology ECHO in the United States of America, into the Missouri Telehealth Network (MTN) as part of a more extensive state-wide ECHO implementation, Show-Me ECHO. Dermatology ECHO attempted to approach care gaps in rural and underserved populations similar to those of the original model. Before its inception, patients in Missouri not in one of the four major urban areas – St. Louis, Kansas City, Columbia, and Springfield – faced significant barriers to seeing a dermatologist, including but not limited to financial limitations, unreliable means of transportation, long wait-times between appointments, and work constraints [1]. Those unable to travel for care often received suboptimal treatment and testing, or no treatment at all, resulting in high morbidity and increased healthcare costs [1]. Dermatology ECHO has used videoconferencing technology to

overcome geographical limitations and create a "community of practice" in Missouri, providing specialists with the opportunity to mentor local providers, facilitate continued medical education, suggest diagnoses, and recommend evidence-based treatments for dermatological diseases [1].

Applying ECHO to Dermatology

The Dermatology ECHO hub team at MU, in its inception, consisted of seven providers: two general dermatologists, two pediatric dermatologists, a dermatopathologist, a clinical psychologist, and an advanced practice nurse [1]. They met with PCPs for 1 hour weekly [1]. Over time, the team has slightly changed, and the frequency decreased to biweekly to best fit the participants' needs. PCPs present patient cases, and subsequent diagnosis and management discussions occur. In addition to these cases, each ECHO session consists of a brief CME-approved didactic session as part of a comprehensive dermatology curriculum geared towards topics most needed by primary care providers [1].

Before Dermatology ECHO began, many rural and underserved patients had limited access to dermatologists [1]. Referrals to dermatology are frequent when the service is available [1]. Combined with the maldistribution and shortage of dermatologists, this results in long delays in referral access [1]. Dermatology World's August 2016 issue reported that 32% of physicians report having difficulty referring patients to dermatologists, and 84% said their patients needed access to dermatologists [1].

Facilitating a dermatology ECHO session is relatively straightforward. It involves providers participating in dermatology ECHO sessions use Zoom video or other web-based conferencing platforms on their desktop or mobile devices to share de-identified patient history and clinical images of dermatologic disease with the other providers and dermatologists on the ECHO hub team [10]. The providers and hub team experts discuss the case presentations, diagnosing skin diseases, and discussing the hub team's recommendations for continued management.

Dermatology ECHO at MU has provided accessible dermatological expertise to primary care providers to diagnose challenging cases appropriately and make treatment recommendations. However, it has also fostered a team-based approach where primary care providers can discuss cases with the hub dermatology team. These patients often present with rare conditions that a PCP may not be trained to diagnose. Other times, cases may be discussed that have an unusual presentation of a common disease. Both scenarios allow for fruitful discussions laced with many learning points to enhance the primary care provider's dermatologic knowledge.

Before ECHO, many patients had been seen by their PCP numerous times before the correct diagnosis was obtained – undergoing unnecessary testing and utilizing more healthcare resources than needed [1]. A retrospective cross-sectional study examining 137 adult patient cases presented at Dermatology ECHO, revealed 43.8% were incorrectly diagnosed by their PCP. The dermatology hub team's discussion of morphologic clues in the skin, differential diagnoses, appropriate and cost-effective work-ups, clinicopathologic correlation of any cases with biopsy and/or results, and treatment recommendations benefited 83.6% of adult cases and 72.5% of pediatric cases [2].

An anecdotal case serves as an excellent illustration of the value of ECHO. A 69-year-old Caucasian woman presented with a non-healing wound on her right lower extremity [1]. This case was described in "Dermatology ECHO – an innovative solution to address limited access to dermatology expertise," published in the *Journal of Rural and Remote Health*. The patient, spurred by a rooster, initially presented to a rural medical clinic with an erythematous plaque proximal to her medial malleolus. She was prescribed cephalexin for presumed cellulitis. Twelve days later, the patient was found to have an eschar over her right posterior leg but noted improvement since the initial injury. Tetanus, diphtheria, and pertussis toxoid vaccine were given, but the patient was informed to remain off antibiotics. One month later, she visited her PCP with a 2.5–3 cm erythematous plaque with associated fluctuance and tenderness. The abscess was drained, and the patient was started on levofloxacin. She was seen twice more for presumed recurrent cellulitis. She had now visited her PCP five times in two months but was still having an inadequate response to treatment. This case was discussed at MU Dermatology ECHO, with digital images of the affected area. The ECHO hub team diagnosed this as an atypical Mycobacterium skin infection with a later culture positive for *M. chelonae*. During this session, the dermatologists recommended azithromycin 250 mg daily and applying a heating pad to the infection site 2–3 hours per day. Within one month, the patient had dramatic improvement using heat and azithromycin as recommended.

The didactic curriculum is an essential component of the dermatology ECHO as well. Compared to PCPs, dermatologists have much greater accuracy in diagnosing skin lesions and are superior at detecting melanoma [1]. Since a patient is more likely to visit their primary care doctor before ever seeing a dermatologist, improving PCP diagnostic ability is likely to improve patient mortality [1]. A feasibility study on melanoma screening in the primary care setting recruited ten primary care providers to participate in melanoma-specific ECHO sessions. In a year, PCPs diagnosed 36 biopsy-proven melanomas [11]. The study observed that primary care providers often felt a lack of confidence in performing a full-body skin exam that improved with frequent discussions and education with dermatologists through Dermatology ECHO [11]. The AAD's Special Positioning Work Group found that non-dermatologists view dermatologists as "smart, thorough, and patient-friendly." It is believed that these perceptions support the success of MU Dermatology ECHO and will continue to facilitate recruitment efforts [7].

Benefits

Project ECHO benefits both patients and providers with the opportunity to increase the quality of care through increasing knowledge, competence, and patient health. Through ECHO, dermatologists can teach primary care providers to manage and treat many patients in their region. ECHO educates providers in rural areas and areas with poor or limited access to dermatology at low cost to the providers and the patients. According to a Hepatitis C ECHO program, the average savings per person were $1352 for a total of $352,872 for 261 patients [12]. Providers that may participate in and benefit from ECHO are not only limited to primary care providers. A provider is anyone involved in treating a subset of patients, such as nurse practitioners, physician assistants, lab technicians, school nurses, health-care field trainees or anyone in the health care field. Participation in ECHO requires only a willingness to improve their knowledge and care of patients. The Bone Health TeleECHO initially intended to involve PCPs who were professionally isolated but resulted in the involvement of providers who had a desire to expand their knowledge about bone health, including specialists in endocrinology, rheumatology, and orthopedics [12]. Providers' knowledge about topics presented through ECHO increased according to both surveys and pre- and post-test comparisons. Bias may have been present due to the provider's higher self-assessment of their knowledge than their actual knowledge in both their pre-analysis and post-analysis for how they view their knowledge. However, the surveys and pre- and post-test comparisons both indicated that knowledge strengthened. Surveys also indicated that the participants felt an increase in confidence in treating their patients properly [13].

Patient health has improved with Project ECHO, as demonstrated with diseases that have been covered, such as Hepatitis C, dementia, and diabetes. It was found that Hepatitis C positive patients treated by primary care providers who participated in ECHO for Hepatitis C had a sustained viral response rate similar to those treated by a specialist [13]. Other studies have also been published with similar results [13]. Another study reviewed patients with dementia after dementia and behavioral issues were taught to providers through ECHO, and identified improvement in the care of 74% of 44 patients. The other patients with dementia who did not improve either did not

follow the recommendations because of distrust of the provider's judgment or due to resistance of the patient's family [13]. Diabetes was also better managed with the assistance of ECHO [13]. Following ECHO training on diabetes care, the A1C levels of patients with diabetes significantly decreased. These results were compared with data from two nearby clinics which did not have providers participate in the ECHO training, and both had a rise in the A1C levels of their patients [13].

Other benefits include providing American Medical Association (AMA) category 1 continuing medical education (CME) credits to providers, preventing professional isolation, and sharing information. CME is a benefit and an incentive in involving health providers in Project ECHO [13]. Dr. Karen Edison, MD, a dermatologist at the University of Missouri, has dedicated much of her time to Project ECHO and understands the applications and benefits Project ECHO can provide to providers and patients [14]. While brief pithy lectures at the start of each ECHO session are beneficial in improving learning and patient care, a survey of ECHO participants revealed that both the presenting provider and the participants learn best through case presentations [14]. When someone is actively presenting on a topic, genuinely listening for input that will directly benefit their patients, they will solidify their learning and memory stronger than if they were listening to a lecture that required less involvement [14].

Barriers

Barriers to ECHO include limited time, costs, and provider's participation [13]. The main limiting factor for providers is finding time to add in meetings and more teaching to their already busy clinical day [13]. Dermatology faculty participation in ECHO takes time away from their management of their own clinic patients [13]. While ECHO training will provide a long-term and future benefit to both the providers and their patients, it may be difficult for participants to weigh the future improvement in population health with the immediate loss of time in the care of dermatology clinic patients [13].

Another limitation is the cost inherent in the operation of ECHO [13]. While the overall cost of Project ECHO is relatively low, a dependable source of funding is important [13]. Since technology plays a large role, support staff must be engaged and available to answer questions and give assistance to those who have difficulties or complications with the program [13]. ECHO is also restricted to teaching and active involvement of primary care participants [13]. Efforts to recruit willing and able participants is essential [13]. Fortunately, the need for ECHO drives the ability for ECHO to run efficiently. Providing CME credit is one way to incentivize participation [14].

Best Practices

Best practices in ECHO telementoring represent an emerging topic of discussion with no agreed-upon consensus. It is critical to develop a hub team of skilled academic dermatologists with diverse interests who are respected by the medical community and enjoy teaching. As mentioned above, the dermatology ECHO hub team at MU originally consisted of seven providers: two general dermatologists, two pediatric dermatologists, a dermatopathologist, a clinical psychologist, and an advanced practice nurse [1]. However, hub teams may vary by location and represent the expertise needed by its participants.

Various theories of learning contribute to the organization and structure of the ECHO learning model. The principles of ECHO development are based on social cognitive, situated learning, and community of practice theories [3]. First and foremost, ECHO participants must believe that the time spent in ECHO sessions will benefit their patient populations. ECHO sessions must enable participants to diagnose and manage skin conditions in their practice. This learning must occur in a specific environment (e.g., ECHO session) - also known as situated learning - and must be tailored to the participants' communities' needs. ECHO learners must have the opportunity to develop their current knowledge and understand the standard of care. Sessions must also be engaging and motivating for the learners.

However, Kara Braudis, MD, current lead of the MU Dermatology ECHO, states "we want (local providers] to be experts in basic dermatology" and that the purpose of ECHO is not to train dermatologists. She states, "if (a patient) has basic plaque psoriasis, acne, or bacterial folliculitis, these are things that primary care doctors need to recognize." Most importantly, ECHO must promote a collaborative community of practice with providers and specialists promoting open sharing of information and cases. ECHO combines didactics (e.g., lectures, slides, and handouts) with case presentations from community providers and discussions led by ECHO hub team experts. These give-and-take ECHO sessions take advantage of these teacher-learner relationships and apply an "all-teach, all-learn" approach to ensure that participants are actively engaged in learning how to diagnose and manage skin diseases [1].

ECHO facilitators should evaluate each ECHO session's effectiveness by surveying participants and measuring indicators for success regarding three key themes: relevance of the curriculum to providers, the quality of relationships between ECHO providers and educators, and the format of the ECHO [14]. Curriculum relevance is defined as the relevance of the content in the ECHO sessions (e.g., didactics and their relationship to problems in the participants' communities) [14]. It was typically regarded as the primary benchmark for evaluation of the quality of a program. The higher the relevance to local practice, the higher the participant engagement in the ECHO program [14]. The ability to apply new skills and knowledge to local practice was highly dependent on curriculum relevance. Measuring participant engagement is a vital metric for determining the quality and effectiveness of an ECHO program. Before participants can use the information learned in an ECHO in their community, there needs to be a firm understanding of that community's educational needs, an emphasis on registration and communication with participants, accountability and institutional support for attendance, and strategies to maintain engagement of participants [15].

The educational material must be relevant to the ECHO participants. ECHO sessions will not succeed if ECHO facilitators fail to emphasize effective learning strategies. They must also help participants network with other providers, stimulate additional learning opportunities, and promote existing intra-ECHO relationships between providers and other hub members [14]. When these conditions are met, ECHO programs typically have a higher level of community involvement, registration, attendance, and engagement [14]. When novel tools for learning (e.g., Zoom video communications) are used alongside engaging discussions, small ECHO sizes (1–2 faces per tile), and participants are kept involved and engaged between sessions, participant engagement in the ECHO program increases [14]. Engagement is a pre-requisite for assimilating information from ECHO for the benefit of the participants' patients.

The ECHO participants' various relationships - particularly among participants, between participants and ECHO educators, and their work colleagues and managers - influenced participant likelihood in participating in the ECHO [14]. Many of the participants surveyed indicated that their primary reason for registering for the ECHO was to "grow individual professional networks" [14]. Participants who had more workplace support for continuing medical education attended more ECHO sessions and reported that a supportive agency was an important factor in attending sessions [14]. Lastly, the better the quality of the relationship between ECHO participants and educators, the more comfortable many participants felt during sessions, and the more likely they were to attend sessions [14]. This may also induce appropriate referrals to the dermatology clinic.

The ECHO format and delivery played a crucial role in whether or not participants attended, engaged in, and used the ECHO session in their communities [14]. The format is described as curriculum design, participants' ability to engage in the session, and tools used for learning (e.g., Zoom video services for remote learning) [14]. Participants lauded Zoom and other video services as increasing the ease of use of the ECHO platform [14]. Scheduling was typically a challenge for most participants, but having agency support and dedicated continuing medical education time typically increased the chances of high attendance and engagement [14]. Short and focused case presentations also aided in participants' ability to learn the material and engage in the sessions [14]. Lastly, other aspects that impacted performance included the number of participants in a session, the number of participants on their screen, communications with new registrants, information technology support for equipment support, and communication and support for participants between sessions [15].

The Future of Dermatology ECHO

ECHO is a new idea within telemedicine, having only begun in 2003. Even more recent, is the application of ECHO for dermatologic education. Dermatology ECHOs are growing in popularity, but still quite scarce around the country. Dr. Braudis states "I think that a grand vision… would be to be able to facilitate dermatologic care in rural areas where we (MU) would not be the only hub, but multiple physician-led hubs that are purposefully and intentionally connected with primary care providers in that area which could cover the whole state. Autism ECHO has been intentional about that, working on getting an 'Autism expert' in every county in Missouri. We have had a grassroots approach instead of specific outreach, county by county. I want to do something similar" [16].

According to Dr. Braudis "There are huge gaps in care, which is why we are working with [nurse practicioners] and primary care doctors" [15]. There are many strategies in which dermatology as a specialty can use to move forward on a national level, addressing barriers in access to care as well as gaps in the quality of that care. ECHO is a powerful tool that should be incorporated into these strategies moving forward.

Bibliography

1. Lewis H, Becevic M, Myers D, Helming D, Mutrux R, Fleming D, et al. Dermatology ECHO - an innovative solution to address limited access to dermatology expertise. Rural Remote Health. 2018;18(1):4415.
2. Arora S, Kalishman S, Thornton K, Dion D, Murata G, Deming P, et al. Expanding access to hepatitis C virus treatment--Extension for Community Healthcare Outcomes (ECHO) project: disruptive innovation in specialty care. Hepatology. 2010;52(3):1124–33.
3. Socolovsky C, Masi C, Hamlish T, Aduana G, Arora S, Bakris G, et al. Evaluating the role of key learning theories in ECHO: a telehealth educational program for primary care providers. Prog Community Health Partnersh. 2013;7(4):361–8.
4. Arora S. Project ECHO: democratising knowledge for the elimination of viral hepatitis. Lancet Gastroenterol Hepatol. 2019;4(2):91–3.
5. Sequist TD. Ensuring equal access to specialty care. N Engl J Med. 2011;364(23):2258–9.
6. Torres A. Academy to firmly establish breadth and depth of specialty with robust campaign. Am Acad Dermatol. 2016. Available at: https://www.aad.org/members/publications/member-to-member/2016/july-29-2016/academy-to-firmly-establish-breadth-anddepthof-specialty-with-robust-campaign.
7. Telehealth policy trends and considerations. National Conference on State Legislatures; 2015.
8. Arora S, Thornton K, Komaromy M, Kalishman S, Katzman J, Duhigg D. Demonopolizing medical knowledge. Acad Med. 2014;89(1):30–2.
9. Arora S, Thornton K, Murata G, Deming P, Kalishman S, Dion D, et al. Outcomes of treatment for hepatitis C virus infection by primary care providers. N Engl J Med. 2011;364(23):2199–207.
10. Zoom. Available from: https://zoom.us/.
11. Becevic M, Smith Hoffman E, Bysani V, Rosenfeld AH, Hoffman E, Edison K. Melanoma ECHO: a feasibility study of perceived facilitators and barriers of skin cancer screening implementation in primary care settings.
12. Lewiecki EM, Rochelle R. Project ECHO: telehealth to expand capacity to deliver best practice medical care. Rheum Dis Clin North Am. 2019;45(2): 303–14.
13. Zhou C, Crawford A, Serhal E, Kurdyak P, Sockalingam S. The impact of project ECHO on participant and patient outcomes: a systematic review. Acad Med. 2016;91(10):1439–61.
14. Edison K. Benefits of dermatology ECHO. In: Gloss C, editor; 2020.
15. Shimasaki S, Bishop E, Guthrie M, Thomas JFF. Strengthening the health workforce through the ECHO stages of participation: participants' perspectives on key facilitators and barriers. J Med Educ Curric Dev. 2019;6:2382120518820922.
16. Braudis K. The future of dermatology ECHO. In: Grisham E, editor; 2020.

Delivering "Store and Forward" Teledermatology to Rural Primary Care Practices: An Efficient Approach to Provision of Rural Skin Care

14

Anastasia Mosby, Ruth McTighe, Ira D. Harber, and Chelsea S. Mockbee

Introduction to Teledermatology

Advances in electronic medical records and near universal access to telecommunications in the United States have ushered in the age of electronic evaluation and management of patients. Teledermatology platforms expand access to dermatologic care when in-person evaluation is not readily available or otherwise not preferable. As technology has sharpened video and voice during the digital age, so too will telehealth platforms continue to improve. Virtual evaluation is particularly well suited for dermatology since our patients wear the organ of interest on the outside

of their body! Of course, the visual nature of the specialty also leads to technological challenges.

Teledermatology encounters can be conducted via real-time (RT), store-and-forward (SAF), or hybrid modalities. RT teledermatology allows the dermatologic provider to interact synchronously with the patient or the primary care provider and patient through audio-visual telecommunication [1]. SAF teledermatology is an asynchronous modality in which information about the patient is paired with an image and sent to the dermatologist for evaluation. The hybrid modality combines aspects of both real-time and SAF teledermatology. For example, a dermatologist might speak to the patient or referring provider via audio or audio-visual communication and then have them send high-quality digital images of the affected skin. This allows the dermatologist to efficiently gather history and pose follow-up questions to the patient, while also utilizing the higher quality images available via photography versus a video feed [2].

Models of Teledermatology Practice

Any of the above technological modalities may be utilized in a variety of practice models: triage, consultative, and direct care. The triage teledermatology model allows for the assessment of a patient's need for in-person dermatologic evaluation, a con-

"The next generation of healthcare will be decentralized, mobilized, and personalized."—Anita Goel, MD PhD

A. Mosby
School of Medicine, University of Mississippi Medical Center, Jackson, MS, USA
e-mail: amosby2@umc.edu

R. McTighe
Department of Internal Medicine and Dermatology, University of Mississippi Medical Center, Jackson, MS, USA
e-mail: rmctighe@umc.edu

I. D. Harber · C. S. Mockbee (✉)
Department of Dermatology, University of Mississippi Medical Center, Jackson, MS, USA
e-mail: iharber@umc.edu; csmockbee@umc.edu

sideration of further evaluation utilizing a consultative form of teledermatology, and an assessment of the urgency of appointment scheduling. In view of the shortage of dermatology providers in rural areas and the long waiting time for dermatology appointments, implementation of this model helps to reduce unnecessary in-person dermatology visits and shortens the "wait time" for patients with more urgent dermatologic needs.

The consultative model in teledermatology mirrors the format of a traditional "in-person" specialty consult requested by a primary care provider. A referring physician poses a specific question to the dermatologist or needs assistance with diagnosis and/or management of a dermatologic complaint. The dermatologist receives information about the patient via real-time, SAF, or hybrid modality and provides recommendations back to the referring provider, who is responsible for implementing the dermatologist's recommendations [2]. The direct care model, on the other hand, allows patients to initiate care directly with a dermatologist, and the dermatologist assumes responsibility for obtaining laboratory tests, prescribing medications, and making plans to see the patient at a follow-up visit. This method may be especially useful for follow-up of patients already under the care of the dermatologist [3].

The rural dermatologist is not unique from his or her more urban counterparts in providing care to patients from a vast geographic area and from varied socioeconomic backgrounds. Some primary care providers in every community will embrace non-traditional forms of consultation, including teledermatology. It is in the dermatologist's best interest to be prepared to meet the demand for specialty care utilizing teledermatology. This is especially true for rural dermatologists who can extend their reach across broad areas providing access to care for patients and serving the needs of referring providers.

Store-and-Forward Teledermatology

SAF is an asynchronous virtual evaluation and management system that is based on a limited amount of information about a patient [4]. These patient encounters may be between providers or direct from the patient to the dermatologist, so-called "direct-to-patient" (DTP) teledermatology. In addition, the follow-up SAF may be requested by the dermatologist to provide continuing care related to a condition identified previously without requiring the patient to return to the office. The provider to specialist model begins with a primary care provider (or other provider who has a relationship with the dermatologist being consulted) seeing the patient in their office and obtaining history of a dermatologic complaint as well as a photo or photos of the involved area. The information is used to populate a SAF consultation sheet by the provider or medical staff which is then sent electronically to the dermatologist. Sometimes the patient is recruited to provide some of this information to save time for the referring provider and their staff. The dermatologist utilizes a pre-formatted template to record history, examination findings, the differential diagnosis and recommendations. Suggestions to the primary care physician may include management tips or a suggestion for an expedited "in-office" visit with the dermatologist. In our experience, it is important to dedicate time to provide software and hardware training to the primary provider and staff to avoid frustration that is common with automated systems.

DTP SAF teledermatology is either initiated by the dermatologist in order to follow-up with an established patient after an "in-office" visit or initiated by an established patient through various cell phone apps or electronic medical record interfaces. The AAD has published recommendations regarding establishment of a physician-patient relationship as it relates to DTP SAF teledermatology.

For direct-to-patient teledermatology, the Academy believes that the consulting dermatologist must either: i. Have an existing physician-patient relationship (having previously seen the patient in-person), or ii. Create a physician-patient relationship through the use of a live-interactive face-to-face consultation before the use of store-and-forward technology, or iii. Be a part of an integrated health delivery system where the patient already receives care, in which the consulting dermatologist has access to the patient's existing medical record and can coordinate follow-up care [4].

Advantages

There are several advantages to using a SAF approach in detecting and managing dermatological conditions. These will be considered individually.

Reduced Time to Initial Evaluation

One of these advantages is a reduced time to initial evaluation, especially for rural patients. Hsaio et al. found remote patients were evaluated more quickly when referred by a teledermatology program, in contrast to patients who used conventional clinic-based referrals [5]. In addition, this study found initial biopsy, diagnosis, and initiation of treatment occurred earlier with SAF teledermatology [5]. A similar study published in 2002 evaluated the time to initial definitive diagnosis when using stored images versus traditional, "in-clinic" visits. For patients accessing care through the SAF system the median time to initial definitive diagnosis was 41 days. Patients who accessed care by way of a traditional clinic visit had a median time to initial definitive diagnosis of 127 days. Additionally, need for clinic-based appointments was reduced by 18.5% [6]. SAF technology can lead to rapid interventions for ambulatory patients, with some studies demonstrating completed consult time from initial referral in an average of 16 hours. The majority of the consults were completed within 24 and 48 hours [7]. Of course, the traditional clinic-based system of referrals also routinely has a "rapid track." A phone call between a concerned provider and the dermatologist leads to an "add on" visit in the dermatology clinic. These interactions are crucial and superior to SAF visits for patients with severe desquamating rashes, blistering diseases, and frank melanomas since procedures can be performed during the "in-person" visit. It is the undiagnosed and unrecognized urgent and semi-urgent condition for which SAF teledermatology can expedite care for patients that otherwise would wait a prolonged period to obtain an "in-office" appointment.

Cost Effectiveness

A distinct advantage in using SAF teledermatology includes the cost effectiveness of this practice. Yang et al. found that SAF encounters cost approximately 10–80 dollars less per visit than an in-person encounter/visit [8]. The cost-effectiveness of SAF for patients living great distances from a dermatology office is apparent considering: (1) the cost of transportation to and from the office; and, (2) the lost productivity tied to the patient, and possibly a spouse or companion, missing work [9]. There is, however, a more subtle financial benefit. Using SAF to triage and manage patients naturally results in a portion of patients who do not require an "in-person" visit. This results in a greater number of higher acuity, medically complex, or procedure-heavy patient encounters. Thus, the net effect is not only a cost savings to the health system as a whole, but also an increase in billable services provided per patient that are seen "in-person."

This theoretical benefit was or has been confirmed in a large study that revealed SAF teledermatology reduced the need for in-person dermatology visits, as well as emergency department visits by almost 30% and decreased the need for referrals by 27%. This results in an overall cost reduction of $7000 to $37,000. Just as importantly, access to care for patients in rural areas is increased and appointment wait-times are reduced [8].

Triage of Complex Patients to Rapid Access "In-Person" Clinic Visits

There is value in SAF teledermatology being used to evaluate complex patients who might otherwise need to wait for weeks or months to see a skin specialist.

Increased Patient Satisfaction

Patient satisfaction is increased in association with time and cost savings related to decreased travel to "in-person" appointments. Many patients feel very comfortable asking questions about their condition when utilizing this modality. In fact, a majority of patients feel as though they are receiving the same quality of care as in the dermatology clinic [10]. SAF teledermatology is convenient for patients who have demanding schedules, as they can receive care in thirty minutes to an hour, instead of the two to three hours it could take to get ready for an appointment, drive, and sit in a waiting room to be seen. Of course, SAF teledermatology is only appropriate for patients who/that do not have a strong preference for in-person visits.

Maintenance of High Quality Care with SAF (Asynchronous) Teledermatology

Perhaps the most important question related to SAF services is whether or not dermatologists are able to diagnose complaints through these asynchronous encounters as well as they can during "in-person" visits. In fact, the diagnostic accuracy using SAF teledermatology is quite high. The first study to assess diagnostic concordance of SAF and face-to-face visits, performed in the low-megapixel era of the late 1990s and early 2000s, found concordance rates of 81–89% among 106 conditions observed in 92 patients [11]. It seems that the ability of dermatologists to recognize digital impressions of in-vivo diseases was honed by projector lamp and kodachrome slides well before the high-resolution digital age. In a study focused on cutaneous neoplasms utilizating a smart phone app for SAF teledermatology, the diagnostic sensitivity of SAF teledermatology (80%) was similar to traditional clinic visits (81%). The specificity between the two approaches was also identical [12]. Agreement between dermatologists evaluating the same SAF consultations has been found to be in excess of 81% with only a 4–8% disagreement in terms of clinically relevant decisions between three clinicians [11]. A study has also found similar clinical outcomes between both SAF teledermatology and traditional in-person dermatology visits [13].

Maintenance of High Quality Care with DTP Teledermatology

DTP store-and-forward teledermatology has the capacity to be personalized by the provider and tailored for specific complaints. For example, the subjective information required to assess an acne patient is quite different from the information required from a patient with a "mole." For established patients with a known diagnosis, short, disease specific questionaires can be formulated to provide information to accompany a photograph. This approach can be utilized for patients requiring wound checks after Mohs surgery or for stable acne patients. Some health systems choose to make this type of encounter a low-cost, cash-only service, avoiding insurance billing. Many patients are happy to pay a small fee for the convenience of avoiding an in-patient visit for a stable issue. The payment is accomplished with a credit card paywall. It was not difficult to integrate this with the EHR patient portal allowing prescriptions to be sent electronically and notes saved so that they are readily accessible for future encounters.

A number of studies demonstrate that the diagnostic accuracy of DTP store-and-forward encounters is similar to consultative or referral SAF encounters. One such study found dermatologists were able to make a diagnosis from patient-initiated encounters using a digital photograph and a standard form about 95% of the time with a confidence level of 79% [14]. Integration of these consultations into the electronic medical record was valued because it permitted more comprehensive evaluation and management. Furthermore, other providers are readily able to access previous medical records [14].

Other Advantages of SAF Teledermatology

Tertiary benefits of conducting virtual visits in general include the conservation of supplies and personal protective equipment, the low cost associated with performing a visit (low overhead), and the rapid speed with which an encounter can be completed. Specific to asynchronous encounters, the dermatologist can complete the visit at the pace and location of their choosing. In addition, virtual visits can be completed during downtime in clinic or even remotely, perhaps after young children are put to bed! Thus, the freedom afforded by digital access is enjoyed not only by patients and referring physicians but also by enterprising dermatologists.

SAF teledermatology and synchronous teledermatology can be used to provide healthcare when other approaches are not applicable. For instance, teledermatology was used during the COVID-19 pandemic when all but emergency in-person clinic visits were cancelled [15]. In fact, during the initial phase of the pandemic teledermatology became the primary system to both diagnose and treat dermatology patients. Taking advantage of teledermatology services also helped to preserve personal protective equipment during the COVID-19 pandemic [16].

Disadvantages

There are several barriers preventing widespread adoption of SAF teledermatology. Collectively, these can be classified as follows:

Equipment Acquisition and Cost

First, the cost of technology and equipment can be significant. The ongoing cost of high bandwidth connections are more important for RT encounters, as opposed to SAF teledermatology, but speed is valued in any digital system. Additionally, there is a commitment of time and money to train referring physicians to efficiently take clinical images and complete and transmit SAF patient intake forms. Another disadvantage in using teledermatology is the need for a large secure browser to keep patient files safe, while permitting providers to communicate effectively with patients. This is associated with an increased risk of inadvertently disseminating private patient information, especially when telehealth is using smartphone platforms [9].

Image Quality

An important disadvantage of SAF teledermatology platforms is the widely variable quality of submitted images. While there are no universal standards for imaging technology, the quality issues are primarily with the skill of the photographer in the submitting physician's office. A three-year SAF study demonstrated that poor image quality frequently hindered dermatologists in making a diagnosis. Specifically, 66% of undiagnosed cases were attributed to this problem, and 13% of the consultations included a low image quality. Image quality can also influence a dermatologist's confidence in their diagnosis [17]. Having standardized instructions and detailed instruction sheets can help to limit the more common errors primary care providers make when obtaining clinical images.

Appropriate Use of SAF Teledermatology

To be most effective, referring physician must recognize appropriate clinical scenarios for SAF teledermatology. This platform is most appropriate for straightforward complaints and for established patients. Most often the patient or referring physician knows the specific area of concern they need to photograph. One of the shortcomings of SAF teledermatology is when a more comprehensive history or examination is needed. In these cases, the limited history received from a universal questionnaire might not suffice, and the format of SAF visits makes it difficult to obtain timely follow-up information. The nature of the complaint itself may increase the likelihood of needing an "in-office: follow-up visit. Patients with inflammatory disorders and rashes are less likely to require follow-up in the dermatologist's office, while patients with skin lesions are more likely to need in-person follow-up [18].

Similarly, there are dermatologic conditions for which a patient or primary care physician may only notice a singular lesion. Photographing this area of obvious concern may miss other findings that would provide significant clues to the nature of the disease process that would only be identified in an "in-person" examination of the entire patient [19]. Deacon et al. reported a case in which a patient sent in pictures of cutaneous herpes simplex virus but did not include an image of a melanoma on another part of the body. Fortunately, this patient scheduled a follow-up appointment for a full-skin exam, at which time the lesion was identified [20].

Physician Reimbursement Issues

Each state in America has its own laws that impact teledermatology through regulation of insurance companies. In Mississippi, telehealth services are required by law to be reimbursed at the same rate as in-person visits by all insurance companies. Still, there is debate on what constitutes a telehealth visit. Some third-party payers cover only synchronous visits as equivalent to an in-office evaluation and not SAF encounters. Without adequate reimbursement providers are deterred from participation in telehealth. In addition, Medicare and Medicaid coverage varies from state to state, resulting in differing levels of reimbursement for telehealth services [9]. Membership in the American Medical Association, American Academy of Dermatology, and one's state dermatologic associations will help to improve legislation that impacts patients and physicians.

Physician Disatisfyers

SAF teledermatology may result in a decrease in satisfaction among clinicians. This is attributed to a lack of follow-up, and doubts about the helpfulness of telehealth services. Some providers derive less satisfaction in telehealth encounters versus in-person visits [21]. In our opinion, this is primarily the result of altering the physician-patient relationship in SAF teledermatology encounters. Integrating SAF teledermatology with standard in-office clinical encounters is a good way to reap the benefits of both and provide a balanced and fulfilling practice for the dermatologist.

Dermatologists are also leary of the widespread adoption of SAF telehealth apps by non-dermatologists. In addition, the image of dermatology within the house of medicine can be impacted by disreputable practitioners. An adherence to the AAD's position statement on teledermatology, is a starting point for maintaining the high-quality care that is expected by our patients and referring providers.

The UMMC Experience

The SAF dermatology program at the University of Mississippi Medical Center (UMMC) was born out of the early successes of telehealth in other fields of medicine. Due to the large rural areas in Mississippi that do not have a trauma center (the only Level 1 trauma center in the state is UMMC), it became clear that a significant improvement in patient outcomes could be gained through evaluation of trauma and emergency cases which presented to remote county hospitals. University physicians, through live interactive telehealth, supervised in-person providers. Telehealth was also used to triage patients to appropriate hospitals which offered varying levels of care. It is easy to appreciate that in emergency situations the nearest hospital may not be the best location for transfer if the patient appears to be having a stroke and the nearby rural emergency room does not offer interventional stroke care. This program was such a success that the University expanded telehealth into many ambulatory departments. Additionally, the state legislature passed laws requiring insurance companies to reimburse telehealth encounters at the same rate as in-person visits. This provided the boost to confidently expand telehealth operations. Prior to the Covid-19 suspension of HIPPA regulations, SAF encounters were felt to be less vulnerable to the breach of patient privacy than live interactive audio-visual encounters.

The current method of SAF encounters is through the consultative model, using a standard form for providers to send for consultation with our dermatologists. This process is being streamlined to improve efficiencies for both the referring physicians and the consulting dermatologist. One initiative being piloted enlists patients to fill out the subjective portion of the telehealth form in the primary care physician's office in an effort to shorten the time and burden on the referring provider's staff.

The University of Mississippi also has a telehealth department which aids greatly with the workflow of intake and integration of the referral forms into our EHR. Schedulers devoted to telehealth are able to field consultations and manage these virtual scheduling templates so that the dermatology front office staff are not overburdened by telehealth encounters in addition to their usual responsibilities.

UMMC is also expanding into the DTP model for established acne patients to use SAF every other visit. There are plans to add routine Mohs surgery follow-up care in the next few months. Diagnosis-specific forms targeted to the specific disease state are designed to simulate the general questions often asked in "live" visits. Two or three patient-supplied photographs are then submitted from a smartphone. This model was implemented through the patient portal interface app, with the completed form being forwarded to the physician's inbox in the EHR. Prescriptions, referrals, and detailed instructions are delivered electronically through the typical EHR encounter, and the patient has access to telehealth instructions through the patient portal. The app also has electronic payment capability for collection of copays at the time of the encounter.

Tips and Tricks

4th year medical students in their required rural clerkship are used to train primary care providers in remote areas of the state to use UView (our current portal for referrals), as well as provide training for taking optimal digital photographs. Standardized patient intake forms are used by referring providers. Efforts were initially concentrated in the most underserved areas of the state, in our case the Mississippi River delta region,

and expanded gradually to other rural areas. Placing the forms in the primary care physician's offices and providing equipment and training up front gave the project an initial boost. An office manager or telehealth coordinator is available to take troubleshooting calls providing immediate support for busy primary care offices.

Store-and-Forward in the Future

Continued use of store-and-forward teledermatology will allow for increased access to dermatologic care for more patients, especially those distant from the specialist's office. This communication also provides education for primary care clinicians. As teledermatology grows, more uniform standards of care will help facilitate high quality SAF encounters. Rural patients, and others without access to dermatologist's services, will benefit from primary care physicians that are better equipped to expedite teledermatology encounters. This patient population may also benefit from advances in DTP store-and-forward mobile applications. SAF technology has already affected the way UMMC dermatologists evaluate and diagnose patients from an ambulatory setting, but its use could be expanded in a variety of settings, including the emergency department. A study was conducted in a pediatric Emergency Department in Wisconsin to determine whether providing photographs along with a brief patient history to a pediatric dermatologist could produce as accurate diagnoses. This study showed a 82% concordance rate between in-office dermatologist evaluation and diagnostic interpretation with shared photographs. There was also a 70% inter-rater agreement between an in-person diagnosis versus the store-and-forward [22]. Many emergency departments encounter both pediatric and adult dermatologic complaints, so expansion of SAF consultations to this setting might provide a useful diagnostic tool, as well as cost savings by potentially decreasing unnecessary admissions.

Many dermatologic conditions can greatly impact the quality of life. Long wait times due to limited access to a dermatologist are expected to lead to increased growth of tele-dermatology [23].

Digital information capabilities have expanded more quickly than ever before during the Covid-19 pandemic, leading to market pressure for quick expert diagnostic care in the virtual format. From a SAF teledermatology standpoint, it remains to be seen whether this practice will become more common or will be replaced by increasing real-time telehealth visits that were popular at the beginning of the COVID-19 pandemic; Market forces will continue to play a role. Decreases in reimbursement for telehealth services as the pandemic winds down could limit these visits in the future to a fee-for-service cash payment system. Moving forward, legislation in all fifty states requiring universal coverage of teledermatology would be very helpful.

Conclusion

Currently, a large number of people, especially those in rural locations, have limited access to dermatologic care, as evidenced by the long wait times and large distances some people have to travel to see a dermatologist. We are having success overcoming these barriers at UMMC by using SAF teledermatology. Teledermatology has proven to be useful in reducing wait times and time to diagnosis, reducing costs for patients, and increasing access to care for dermatology patients. The less apparent benefits of integrating virtual care into one's practice are realized through clinic schedules filled with more procedures and higher acuity visits. There are potential benefits of this technology to all involved: the patients receive more timely and appropriate care; the referring physicians have another tool in their arsenal to address dermatologic complaints; and the dermatologist is able to easily incorporate quick SAF visits into a busy clinic schedule making it efficient, fulfilling, and profitable.

Disclosures The authors have no relevant disclosures.

References

1. Roman M, Jacob SE. Teledermatology: virtual access to quality dermatology care and beyond. J Dermatol Nurses Assoc. 2014;6(6):285–7. https://doi.org/10.1097/JDN.0000000000000086.
2. Coates SJ, Kvedar J, Granstein RD. Teledermatology: from historical perspective to emerging techniques of the modern era: part I: history, rationale, and current practice. J Am Acad Dermatol. 2015;72(4):563–74. https://doi.org/10.1016/j.jaad.2014.07.061.
3. Pathipati AS, Lee L, Armstrong AW. Health-care delivery methods in teledermatology: consultative, triage and direct-care models. J Telemed Telecare. 2011;17(4):214–6. https://doi.org/10.1258/jtt.2010.010002.
4. American Academy of Dermatology. 2016. https://server.aad.org/Forms/Policies/Uploads/PS/PS-Teledermatology.pdf. Accessed 5 Oct 2020.
5. Hsiao JL, Oh DH. The impact of store-and-forward teledermatology on skin cancer diagnosis and treatment. J Am Acad Dermatol. 2008;59(2):260–7. https://doi.org/10.1016/j.jaad.2008.04.011.
6. Whited JD, Hall RP, Foy ME, et al. Teledermatology's impact on time to intervention among referrals to a dermatology consult service. Telemed J. 2002;8:313–21.
7. Rajda J, Seraly MP, Fernandes J, Niejadlik K, Wei H, Fox K, Steinberg G, Paz HL. Impact of direct to consumer store-and-forward teledermatology on access to care, satisfaction, utilization, and costs in a commercial health plan population. Telemed J E Health. 2018;24(2):166–9. https://doi.org/10.1089/tmj.2017.0078.
8. Yang X, Barbieri JS, Kovarik CL. Cost analysis of a store-and-forward teledermatology consult system in Philadelphia. J Am Acad Dermatol. 2019;81(3):758–64. https://doi.org/10.1016/j.jaad.2018.09.036.
9. Wang RH, Barbieri JS, Nguyen HP, et al. Clinical effectiveness and cost-effectiveness of teledermatology: where are we now, and what are the barriers to adoption? J Am Acad Dermatol. 2020;83(1):299–07. https://doi.org/10.1016/j.jaad.2020.01.065.
10. Williams T, May C, Esmail A, Ellis N, Griffiths C, Stewart E, Fitzgerald D, Morgan M, Mould M, Pickup L, Kelly S. Patient satisfaction with store-and-forward teledermatology. J Telemed Telecare. 2001;7(1):45–6. https://doi.org/10.1177/1357633X010070S118.
11. High WA, Houston MS, Calobrisi SD, Drage LA, McEvoy MT. Assessment of the accuracy of low-cost store-and-forward teledermatology consultation. J Am Acad Dermatol. 2000;42:776–83. https://doi.org/10.1067/mjd.2000.104519.
12. Silveira CG, Carcano C, Mauad EC, Faleiros H, Longatto-Filho A. Cell phone usefulness to improve the skin cancer screening: preliminary results and critical analysis of mobile app development. Rural Remote Health. 2019;19:4895. https://doi.org/10.22605/RRH4895.

13. Pak H, Triplett CA, Lindquist JH, Grambow SC, Whited JD. Store-and-forward teledermatology results in similar clinical outcomes to conventional clinic-based care. J Telemed Telecare. 2007;13(1):26–30. https://doi.org/10.1258/135763307779701185.

14. Pathipati AS, Ko JM. Implementation and evaluation of Stanford Health Care direct-care teledermatology program. SAGE Open Med. 2016;4:2050312116659089. https://doi.org/10.1177/2050312116659089.

15. Gisondi P, Piaserico S, Conti A, Naldi L. Dermatologists and SARS-CoV-2: the impact of the pandemic on daily practice. JEAVD. 2020;34:1106–201. https://doi.org/10.1111/jdv.16515.

16. Cartron AM, Rismiller K, Trinidad JCL. Store-and-forward teledermatology in the era of COVID-19: a pilot study. Dermatol Ther. 2020;33:e13689. https://doi.org/10.1111/dth.13689.

17. Lasierra N, Alesanco A, Gilaberte Y, Magallón R, García J. Lessons learned after a three-year store and forward teledermatology experience using internet: strengths and limitations. Int J Med Inform. 2012;81(5):332–43. https://doi.org/10.1016/j.ijmedinf.2012.02.008.

18. Dobry A, Begaj T, Mengistu K, Sinha S, Droms R, Dunlap R, Wu D, Adhami K, Stavert R. Implementation and impact of a store-and-forward teledermatology platform in an urban academic safety-bet health care system. Telemed J E Health. 2020. https://doi.org/10.1089/tmj.2020.0069.

19. Grenier N, Bercovitch L, Long TP. Cyberdermatoethics II: a case-based approach to teledermatology ethics. Clin Dermatol. 2009;27(4):367–71. https://doi.org/10.1016/j.clindermatol.2009.02.009.

20. Deacon DC, Madigan LM. Inpatient teledermatology in the era of covid-19and the importance of the complete skin examination. J Am Acad Dermatol. 2020;6(10):977–8. https://doi.org/10.1016/j.jdcr.2020.07.050.

21. Marchell R, Locatis C, Burgess G, Maisiak R, Liu WL, Ackerman M. Patient and provider satisfaction with teledermatology. Telemed J E Health. 2017;23(8):684–90. https://doi.org/10.1089/tmj.2016.0192.

22. Heffner VA, Lyon VB, Brousseau DC, Holland KE, Yen K. Store-and-forward teledermatology versus in-person visits: a comparison in pediatric teledermatology clinic. J Am Acad Dermatol. 2009;60(6):956–61. https://doi.org/10.1016/j.jaad.2008.11.026.

23. Whited JD. Quality of life: a research gap in teledermatology. Int J Dermatol. 2015;54(10):1124–8. https://doi.org/10.1111/ijd.12909.

Overcoming Barriers to Implementation of Teledermatology in Rural America

15

Caroline Doo, Curtis Petruzzelli, Karen Dowling, Amanda S. Brown, and Robert T. Brodell

Obstacles are those frightful things you see when you take your eyes off your goals
—Henry Ford

Introduction

The promise of teledermatology is clear. Technology has the capacity to support audio-visual communication with patients in distant areas and permits the delivery of dermatologic care to individuals in the comfort of their own homes. Teledermatology saves time, the cost of travel, and permits gathering of essential information that often eludes physicians including the name of the medication sitting in the patient's medicine cabinet, and information on their insurance card they may forget to bring with them to their in-office visit. The value of teledermatology is magnified in underserved rural areas.

At first blush, one might assume that dermatology is the best field for telehealth applications: patients wear their skin on the outside of the body where it is readily visualized! The COVID-19 pandemic taught us that this is not always the case since teledermatology may not provide the high-resolution, focused, steady images of the patient's skin that are required to make a definitive diagnosis. The application of effective and efficient technology is an important consideration, but as COVID-19 fades, we are more concerned that a raft of regulatory barriers that stymied telehealth in the past will snap back into place. The benefits of telehealth must be preserved, at least for individuals who are unable to travel from nursing homes, hospitals, prisons, and especially distant rural areas where dermatologists rarely practice. It is the purpose of this chapter to consider these barriers to the provision of real-time, audio-visual telehealth platforms and store-and-forward systems. Of course, solutions to overcome these barriers and permit effective dermatologic care to rural areas will also be proposed.

C. Doo (✉) · A. S. Brown
Department of Dermatology, University of Mississippi Medical Center, Jackson, MS, USA
e-mail: asbrown3@umc.edu

C. Petruzzelli
Florida State University College of Medicine, Tallahassee, FL, USA
e-mail: cjp12f@med.fsu.edu

K. Dowling
Children's of Mississippi, University of Mississippi Medical Center, Jackson, MS, USA
e-mail: kdowling@umc.edu

R. T. Brodell
Department of Dermatology and Pathology, University of Mississippi Medical Center, Jackson, MS, USA
e-mail: rbrodell@umc.edu

R. T. Brodell et al. (eds.), *Dermatology in Rural Settings*, Sustainable Development Goals Series,
https://doi.org/10.1007/978-3-030-75984-1_15

Barriers Related to Specific Teledermatology Platforms

Live Interactive (Synchronous) Teledermatology

Real-time two-way digital interaction between the health care provider and patient through a variety of video chat platforms allow visualization of the skin and the opportunity to discuss the diagnosis and treatment of skin problems. Real-time, audiovisual teledermatology was the mainstay of telehealth during the COVID-19 pandemic, either as a stand-alone platform, or in hybrid systems supported by still digital pictures [1]. Technical issues associated with live video conferencing platforms quickly became apparent, and efforts were made to overcome these problems since in-office visits were reserved for only the most urgent situations in patients with skin disease. These problems included lack of clarity of the video, inability to stabilize cameras, and difficulties examining areas that the patient was unable to easily show on camera.

These problems can be partially overcome by supplementation of live audio-video images with still digital images taken by patients or their families. However, some patients do not possess the time or technical expertise to capture and store images. The physician or staff must also be proficient in downloading these images into the electronic health record (EHR). In addition, audio and video were often interrupted due to connectivity issues, and echoes sometimes occurred requiring speaking in short phrases. Most importantly, "teeing up" the patients sometimes proved more difficult than actually "seeing" the patient. Patients sometimes had to download additional software to their computer or smartphone, borrow a smartphone, gain access to reliable internet connection, and make a series of clicks to establish audio-video connection. Gupta et al. described similar barriers to live video teleconferencing. They acknowledged that patients require specialized guidance prior to visits, may struggle to display lesions properly, and visits may lag due to technology issues [2]. All of this requires more staff at a time when clinical practices are trying to cut overhead.

In addition to these technical issues, it is impossible to perform procedures (KOH preparation, scabies preparation, cryotherapy, biopsies, and excisions) which represent a significant portion of overall workload and income in a dermatology office. According to the National Ambulatory Medical Care Survey conducted between 1993–2010, excisions or destruction of lesions or tissues of skin and subcutaneous tissue (ICD-9 code V86.30) were performed in nearly 21% of visits. Skin biopsies (ICD-9 code V86.11) were performed in 6.5% percent of visits [3]. At our institution, telehealth visits generated on average only 40% of work relative value units (wRVUs) when compared with an in-office visit, the difference primarily being due to the lack of procedures in telehealth visits since the evaluation and management (E&M) codes for telehealth and in-office visits are the same. In addition, in our clinical practic at the University of Mississippi Medical Center, cosmetic procedures and the majority of Mohs surgery cases were temporarily postponed. This constituted substantial loss of revenue for the department of dermatology and our university.

Productivity was reduced by at least 50% at our institution over the 2 month period when live synchronous teledermatology was our main form of visits. Similarly, Yale Department of Dermatology was seeing 41% of their pre-COVID-19 volume with the ramp-up of teledermatology 6 weeks after clinics were largely closed [1]. Since most dermatology offices had trouble keeping up with patient demand even before COVID-19, this rapid shift to predominantly teledermatology temporarily markedly reduced patient access.

The cost of these systems must also be considered. When comparing live interactive synchronous teledermatology to store-and-forward teledermatology, the cost disparity is significant. A study in the United Kingdom analyzed these costs by characterizing both "variable" and "fixed" costs [4]. Variable costs describe the product of the estimated hourly wage of both the primary care provider and dermatologists and the time it takes for each of these physicians to participate in these modalities. It also incorporates the time and travel costs of patients to participate in these consultations. The fixed costs describe purchasing the equipment and depreciation costs. The researchers also included savings into their calculations, notably that live interactive teledermatology consultations resulted in fewer face-to-face referrals than store-and-forward consultations. However, when both costs and savings were taken into account, it was shown that the store-and-forward approach was less expensive. This is largely due to the fact that store-and-forward consultations are much quicker and consume less physician time. Additionally, the cost of equipment required to perform live synchronous teledermatology is more expensive.

Privacy also remains a concern with teledermatology. With the relaxation of HIPAA regulations during the COVID-19 pandemic, some of our providers resorted to using video teleconferencing applications such as FaceTime™ since this application is already built into most iPhones™, iPads™, and Mac™ computers. In our experience, most of our providers were comfortable with using their personal phone numbers or emails to make such calls. However, we acknowledge this is not the case with all providers and that this may be a reason for excluding some of these applications in their workflow. Similar concerns have been reported with platforms such as Whatsapp [5]. On the contrary, this has given opportunity for technology growth. Many HIPAA-compliant teleconferencing platforms have been used during the COVID-19 pandemic. However, while some of these offer free versions (i.e. doxy.me™, Doximity Dialer™), it is unclear if these services will remain free of charge. If companies start charging for these services, that will be an additional cost that providers will incur.

Store-and-Forward Teledermatology (Asynchronous)

Asynchronous store-and-forward (SAF) teledermatology offers a patient-friendly solution to busy practices who do not have time to perform scheduled "live" audio-visual telehealth services.

The patient or their primary care provider submits an image or pre-recorded video for the dermatologist to review. The dermatologist then provides the patient or primary care provider with information on the suspected diagnosis, work-up, and treatment. Dermatologists can complete stored consultations by accessing them at their convenience.

Though this form of teledermatology offers convenience for both dermatologists and patients, there are significant drawbacks hindering its utility and implementation. When images are taken to be used in the store-and-forward method, it is the primary care provider or the patient who determines which lesion to photograph. In a study conducted in Australia, participants were recruited to measure the efficacy of using mobile dermatoscopy store-and-forward self-examinations compared to traditional face-to-face visits with a dermatologist. Among the 49 participants who participated in both store-and-forward self-examinations and then completed face-to-face interactions, a total of 40 lesions of concern on 25 people were discovered in face-to-face visits that were not included in photographs taken in self-examination store-and-forward images [6]. Fortunately, none of these lesions turned out to be melanoma. However, another quality improvement study conducted on store-and-forward consultations within the Veterans Affairs (VA) system revealed a significant number of unimaged melanomas. In these consultations, it was primary care providers taking images of skin lesions. The data revealed the frequency of unimaged melanomas to be 10.1 per 10,000 consultations [7]. In summary, the absence of a live interactive feature produces greater room for error by omission with regard to concerning lesions.

Image quality is another barrier that must be overcome with store-and-forward teledermatology technology. Accurate diagnoses in dermatology require high quality images. A study of image quality of both macro and dermatoscopic images sent from primary care providers to dermatologists revealed significant variation in the quality of images. Thirteen primary care providers sent 108 store-and-forward teledermatology consultations that were evaluated by 4 dermatologists. The dermatologists rated 36% of images as bad, 28% of images as reasonable, and 36% of images as good [8]. They concluded that the substandard images taken by primary care providers were due to lack of time to focus on taking a quality image, lack of training or atrophy of imaging skills over time, and problems with equipment, such as failing batteries in a camera that may be rarely used.

It has also been demonstrated that the accuracy of a store-and-forward diagnosis is significantly enhanced in cases where a mobile dermatoscope is used [9]. Despite the added diagnostic value of a mobile dermatoscope, the price of this technology is a significant barrier to primary care providers and patients. Additionally, there is a learning curve required to effectively use this device.

Barriers Related to Approaches to Rural Teledermatology

There are three teledermatology approaches that have been shown to improve access to care in rural areas: (1) triage/consultative store-and-forward teledermatology; (2) direct-to-consumer (DTC) store-and-forward teledermatology, and (3) live synchronous teledermatology.

Triage/Consultative Store-and-Forward Teledermatology

This form of teledermatology could be used to *support* access to care for both inpatients and outpatients in rural areas. However, there are some specific barriers to implementing this platform with primary care providers. This includes lack of awareness of this technology by primary care providers. This is somewhat dependent on how connected rural practices are to dermatologists/academic medical centers. Another key problem is finding a teledermatology program that accepts a particular patient's payer source. This is especially challenging for improving access for patients in rural primary

care practices or designated rural health centers (RHCs) as many of these patients either have no insurance coverage or Medicaid insurance [10–12]. Poor understanding of reimbursement and medical malpractice issues may also play a role [12–15]. There may also be a lack of adequate training for primary care providers to effectively use teledermatology in their daily practice [15].

Technological issues also present barriers to implementing telehealth in rural practices. Rural practices may be challenged by unreliable high-speed internet impacting their willingness to adopt telehealth [16]. Incompatible EHR between the referring provider and specialist can create technical challenges and process inefficiencies [15, 16]. Additionally, negative attitude regarding electronic health records (EHRs) and the perceived difficulty of use impacts adoption because of the practice of submitting eConsults through an EHR [17].

It may also be difficult for primary care providers to incorporate this teledermatology method into their practice model. Time demands of communicating treatment plans and prescribing recommended medication or follow-up is a potential barrier of consultative store-and-forward teledermatology compared to traditional referral pathways. Oftentimes, this usage of teledermatology is conducted on existing informal channels such as texting and phone calls, though this technology is not HIPAA compliant [18]. Furthermore, research has shown poor follow-up from the primary care provider following teledermatology consults [19]. Lastly, primary care providers also have no financial incentive to support this added workload.

The use of store-and-forward teledermatology as a triage service may also be impeded by attitudes and beliefs commonly held by primary care providers. Many rural primary care providers are self-reliant and have a high level of comfort treating skin disorders on their own through a trial and error approach [18]. They may prefer to have the ability to see, touch, and personally engage the patient in person [18, 20]. Some may even see telehealth as a competitive threat to their practice [18].

Patient Initiated Direct-to-Consumer (DTC) Store-and-Forward Teledermatology

In this form of teledermatology, a request for service is made from the patient followed by an electronic communication which may be initiated by either the provider or the patient. This specific type of teledermatology presents its own additional barriers. (Table 15.1).

Resnick et al. investigated the quality of DTC websites in California in 2016. The researchers created simulated patient cases with photographs and a scripted history and sent them to various DTC websites. These cases described various neoplastic, inflammatory and infectious conditions. In total, 62 simulated patient cases were reviewed by clinicians. Their findings revealed shortcomings with these DTC services [21].

They discovered that patients often lack the ability to choose their clinician, with only 32% of simulated cases being offered any choice. Additionally, many of the clinicians on these sites were not board-certified dermatologists. Simulated patient cases were reviewed by dermatologists, internists, emergency medicine physicians, family medicine physicians, OB/GYNs, a cardiologist, a pain management physician, and a physical medicine and rehabilitation physician. These sites also featured international clinicians such as primary care providers in India and dermatologists in Sweden. Non-physician providers such as family nurse practitioners and physician assistants also provided care. Of physicians who were identified to be US-based physicians, only 26% disclosed information about licensure. The geographic location of the clinician was also not always clear, as this was disclosed in 61% of encounters [21].

There were also problems with the history taking process and counseling. Each patient was asked various questions regarding their medical history; however, only in 34% of cases did the clinician perform a review of systems. Most, though not all, did inquire about current medications and allergies. In only 52% of cases did the clinician ask, when appropriate, about the pregnancy or lactation status of a female patient. There were

Table 15.1 Inadequacies of direct-to-consumer (DTC) teledermatology

Inadequacies of DTC Teledermatology	Summary
Patient Choice	Most patients using DTC are unable to choose their clinician.
Clinician Credentials, Licensure, and Location	There is a lack of adequate disclosure of a clinician's credentials and licensure. Patients are often uninformed of their clinician's credentials. Many patients were seen by non-dermatologist physicians, international physicians, and family nurse practitioners.
Quality of Care	There is a record of inconsistency in the history intake process. Questions about ROS, pregnancy status, medications, and allergies were not always inquired. Patients who were prescribed medications were not always informed of side effects. Clinicians commonly misdiagnosed classic presentations of various dermatological diseases.
Integration with Health Care Systems	Few clinicians inquired about the patient's primary care provider or dermatologists in the local area. Of the patients who were told to see a clinician in person, few were given suggested names.

Source: Resneck et al. [21]

four encounters where the clinician made a diagnosis only based on the submitted history. In these instances, the clinician did not ask or offer the chance to submit photographs. Diagnoses were given in 77% of cases, the rest were referred to see local physicians for a diagnosis. Among the patients who received a diagnosis and prescribed medication, only 32% of the cases received counseling on the side effects of the medications. In the cases where a female was prescribed a pregnancy class C or higher drug, pregnancy risks were disclosed 43% of the time [21].

The diagnostic accuracy and treatment plan for the simulated cases were also of significant concern. For example, in the patient cases with polycystic ovarian syndrome with inflammatory acne, each clinician diagnosed the case with acne, and none asked questions about hirsutism, androgen excess, or polycystic ovarian syndrome.

In another simulated case with a female patient presenting with life threatening eczema herpeticum, this patient received a diagnosis of ordinary eczema flare or contact allergies in 7 of 9 instances. In two of these cases, the clinician prescribed prednisone which can worsen eczema herpeticum if given without antiviral medication. There were additional instances of misdiagnosis including 7 of 8 clinicians diagnosing secondary syphilis with unusual plaques as psoriasis [21].

Lastly, there was a seeming lack of integration between DTC providers and local healthcare systems. DTC providers largely did not ask the name of an existing primary care provider or dermatologist. For patients who were told to see a local physician for a diagnosis, only 2 of 14 instances was the patient given the name of a practicing local physician. For patients who were given a diagnosis but instructed to see a local physician for treatment, only 3 of 9 were given a name of a local practicing physician. No patients were sent to a laboratory for studies. While DTC teledermatology has the potential to expand quality dermatological care, it seems that there are issues of choice, transparency, quality of care, and integration into local health systems that impede its progression [21].

Live Synchronous Teledermatology

This form of teledermatology could be performed for patients from distant rural sites and hospitals/nursing homes that do not have access to in-person consultative services, although it does have limitations. At the onset of the COVID-19 pandemic, this form of teledermatology became routine as many dermatologists were permitted to see only the most urgent patients in the office. In the Department of Dermatology at the University of Mississippi Medical Center, patients who were previously scheduled for in-person appointments, were called and offered live synchronous teledermatology visits instead. However, as the pandemic waned, our providers predominantly returned to in-person visits because: (1) many patients required procedural dermatology services which could not be duplicated through tele-

dermatology; (2) the available platforms and technological savvy of patients and physicians made it difficult to "tee up" patients efficiently. Additionally, we could only see about 50% of the number of patients we saw during pre-COVID-19 in-person scheduling.

Barriers Related to Laws and Regulations

There are a variety of legal considerations relative to the telemedicine environment with jurisdiction at both the state and federal levels. Federal laws cover patient privacy and security of data and patient records (Health Insurance Portability and Accountability Act), and prescriptions relating to controlled substances, which are enforced by the Drug Enforcement Agency. However, states have their own laws covering these areas of concern in addition to laws and regulations covering licensing requirements, establishment of the provider-patient relationship, standards of care for practicing, and requirements for prescribing medications [22].

Regarding telemedicine in general, each state has specific laws related to the services that can be provided via telemedicine, and these laws continue to evolve. Such laws focus on the provider-patient relationship, clinical diagnosis and treatment of the patient, and prescribing of medications. All states allow for an initial evaluation of a new patient to be conducted via telemedicine if the technology that is used is sufficient to diagnose and treat the patient in accordance with applicable state laws. However, relating specifically to telemedicine, the question arises as to which state laws apply when the provider is located in one state while the patient is located in a different state. When this occurs, it is the state law where the patient is located at the time of the encounter that prevails. Acceptable technology is also defined differently among states. While some states have only general definitions, other states prescribe very specific technology platforms such as real time audio or video technology only versus store-and-forward technology, and other states do not address the tech-

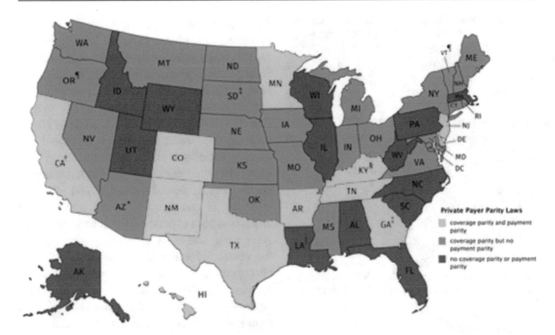

Fig. 15.1 Private payer parity laws. Map of the United States colored according to the status of private payer telemedicine parity laws as of October 15, 2019. Created with mapchart.net
* No coverage parity until January 1, 2021, but requirement for telehealth coverage in dermatology and other conditions/settings
† Effective January 1, 2021. Does not apply to Medi-Cal Managed Care.

‡ Effective January 1, 2020
§ Unless the health provider and payer contractually agree to a lower reimbursement rate
| Payment must be at least 75% of traditional services
¶ Coverage parity for live video, but not store-and-forward
Source: Chuchvara et al. [30]

nology issue at all [22]. Reference can be made to Fig. 15.1 for specifics of the state laws applicable to telemedicine. (Fig. 15.1).

Once the provider-patient relationship has been established, the requirements applicable to the diagnosis and treatment of the patient are usually the same as those that apply when the provider is actually treating the patient in a normal in-person setting. Again, however, some states do impose additional restrictions and requirements for telemedicine such as requiring additional certifications for the provider, obtaining informed consent for procedures, providing information about the provider to the patient, and maintaining medical records, etc. [22] (Table 15.2).

Prescribing medications via telemedicine may be covered under medical laws of the state or under the pharmacy laws. Some states have specific requirements for when medications can be prescribed based on a telemedicine encounter

only, and some states will not allow pharmacists to fill those prescriptions unless certain requirements are met. As some states have requirements for prescriptions in both their medical and pharmacy laws, care must be taken to ensure that all requirements are met and that there are no potential conflicting issues that result [22].

While state laws may also cover patient privacy and data security, federal laws take precedence. Telemedicine may present some unique concerns that must be addressed as they can result in legal problems for the provider as well as ramifications for the clinics and hospitals where they are employed. Prior to COVID-19, a telemedicine encounter was likely to occur with the provider in the clinical setting and utilizing only the technology provided by the hospital or clinic. However, since then, a telemedicine encounter may very well occur with the provider

Table 15.2 Covered Medicare telemedicine services (and their CPT/HCPCS codes) of relevance to teledermatology

Service	CPT/ HCPCS Code
Telehealth consultations, emergency department or initial inpatient	G0425– G0427
Follow-up inpatient telehealth consultations furnished to beneficiaries in hospitals or SNFs	G0406– G0408
Office or other outpatient visits	99201– 99215[a]
Prolonged service in the office or other Outpatient setting requiring direct patient Contact beyond the usual service	99354– 99355[a]
Prolonged service in the inpatient or observation setting requiring unit/floor time beyond the usual service	99356– 99357

CPT Current Procedural Terminology, *HCPCS* Healthcare Common Procedure Coding System, *SNF* skilled nursing facility
Source: Chuchvara et al. [30]
[a]New codes for 2019

at home or outside the clinical setting, which presents other issues that must be addressed [23].

Secure methods of communication are of particular importance. Unfortunately, the Health Insurance Portability and Accountability Act of 1996 (HIPAA) has greatly hindered the availability of teledermatology services because of arcane provisions requiring end-to-end encryption of data with audit controls requiring monitoring of sites where protected health information is created, modified, accessed, shared, or deleted [24]. These audit controls require healthcare providers to enter into a Business Associate Agreement with the third party handling protected health information. Technology companies are loath to enter into these agreements so the liability falls squarely on the healthcare provider [24]. Thus, text messaging is not HIPAA compliant and was rarely used until relief from regulations during the COVID-19 pandemic.

In the event that personal computers and other personal devices are utilized in the provider's home setting, security that is typically provided in the hospital or clinical setting may not be available. Therefore, safeguarding all communication devices and records is especially impor-

tant in the event they are lost or stolen. Important practices include the habit of logging out of devices at the end of each session, password protecting devices to prevent unlocking by unauthorized individuals, and even the possibility of remotely eliminating all medical information from the device in the event it is stolen. Patients must be made aware that there is measured risk treating patients utilizing teledermatology and using electronic documentation in that a hacker could access communications and records. However, if communications are sent through text messages and personal email accounts, there is a greater possibility of this occurring. Finally, the possibility exists that some files, including photographs, could be uploaded to social media accounts, resulting in a particularly egregious lapse of patient privacy. Care should be taken to ensure that settings on the devices do not allow for unintended automatic uploads of photographs or other release of protected patient information. In addition, all photographs should always be deleted immediately after they are uploaded to the patient's health record [23].

The aforementioned concerns raise other legal questions as well. Professional liability insurance does not always cover the telehealth environment with regard to any breach of responsibilities [23]. For telemedicine to be viable, it is crucial that all providers follow the specific policies and protocols established by their employer in order to avoid such an occurrence.

Barriers Related to Reimbursement

Each state varies in the coverage allowed for specific types of telehealth with some states only providing coverage for certain types of telemedicine. Mississippi is the only state that provides coverage for all forms of telehealth including live video, store-and-forward, and remote patient monitoring [25]. Prior to COVID-19, the location of the patient was important in determining whether Medicare or Medicaid covered telehealth services. The individual had to be in a rural setting, and they had to be located at a des-

Fig. 15.2 Medicaid live video coverage. Map of the United States colored according to status of Medicaid live video coverage as of October 15, 2019, with a focus on teledermatology. Created with mapchart.net
* Professional relationship must exist between the patient and the distant site provider
† If deemed by the commissioner as appropriate, cost effective, and likely to expand access
‡ When services are ongoing, the patient should receive a traditional clinical evaluation at least once per year; the distant site provider should coordinate with the primary care provider

§ Mental/behavioral health only
| Some restrictions apply (beneficiary cannot travel, imminent health risk)
¶ Must be at an eligible site, in a rural area with a health professional shortage (same as Medicare)
** Distant site must be of a sufficient distance from originating site; prior approval required in some cases
†† Only for initial and follow-up inpatient telehealth consultation
‡‡ Only by TN Managed Care
Source: Chuchvara et al. [30]

ignated health facility. Rural was generally confined to areas outside of a metropolitan statistical area or in a medically underserved census tract. The designated health facility had to be a doctor's office, hospital, rural health clinic, etc. In other words, the patient could not be at home or at some other non-health setting. In addition, the patient encounter had to be a real-time encounter with both the patient and physician present [26]. As a result of the COVID-19 pandemic, a Stafford Act declaration waived certain federal and state regulations that previously created a barrier in providing telehealth services and reimbursement. Limitations placed on the environment and location for the provider and patient

were temporarily lifted in an effort to promote social distancing while still allowing patients to receive health care [27]. Since that time, there has been a significant increase in teledermatology services.

Similar to these limitations, individual companies determine their reimbursement policies for telemedicine services in many states. Issues include coverage of telemedicine services, the actual reimbursement rate, and billing/coding for services. (Figs. 15.2 and 15.3) Typically, outside the state of Mississippi, reimbursement rates for telehealth services have been 40% lower than non-telehealth reimbursement rates. The state of Mississippi has been more for-

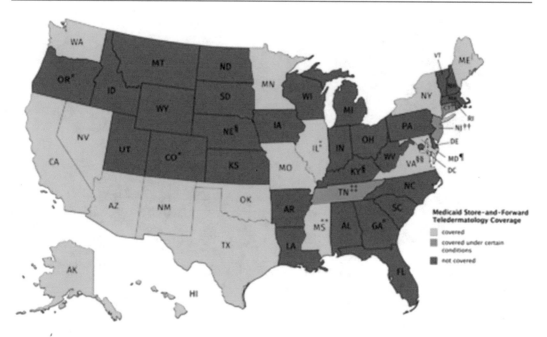

Fig. 15.3 Medicaid store-and-forward coverage. Map of the United States colored according to status of Medicaid store-and-forward coverage as of October 15, 2019, with a focus on teledermatology. Created with mapchart.net
* Teledentistry only
† Only in the case of provider-to-provider communication (electronic consults)
‡ Teledermatology only
§ Teleradiology only
| Effective January 1, 2020

¶ Teledermatology, teleophthalmology, and teleradiology only
** Patient must be notified of the right to interactive communication with the distant site provider, who must respond within 30 days of the patient's request
†† Not explicitly included, but may be covered under the definition of telemedicine
‡‡ Only by TN Managed Care
§§ Outpatient teledermatology, teleradiology, and diabetic retinopathy screening only
Source: Chuchvara et al. [30]

ward-thinking with its reimbursement laws. Mississippi also requires that that the state's Medicaid program pay the same rates for store-and-forward services as those paid for in-person services [25]. Mississippi has also passed laws requiring health insurance plans in the state to cover telemedicine services in the same manner as they cover in-person consultations and requires health insurance and employee benefit plans to reimburse for remote patient monitoring and store-and-forward telemedicine services [28, 29]. Documentation within the service note should include obtained consent to perform the visit via agreed upon service method as well as an attestation statement verifying service method and duration of the visit [26].

Solutions

There are no simple solutions to breaking down the barriers that have impeded the implementation of teledermatology in rural America. Since the COVID-19 pandemic led to a rapid shift to teledermatology practice, dermatologists across the country gained insight into methods to overcome the barriers to better facilitate care for patients. These solutions will prove to be especially beneficial in caring for patients in rural communities who may otherwise not have access to dermatology care.

First, the importance of provision of only the highest quality dermatologic care must be paramount when choosing an approach to teledermatology in rural America. If a synchronous or store-and-forward direct-to-consumer approach is taken, medical record keeping should be equivalent to in-person notes; there must be provisions for in-person referrals when required for complex problems or procedures; there must be a method for prescribing medications that can be accessed by the patient; the patient should receive careful education about their diagnosis and instructions regarding their treatment plan; and, primary care providers should receive a copy of the medical record to keep them informed of their patients status when new diagnoses are made or treatments are initiated. In order to overcome the technological barriers of live synchronous teledermatology, providers must have adequate support staff to help set up the visits. There must be clear instructions and plans in place for how to conduct the visit with the patient to minimize the amount of time needed to troubleshoot technology during the actual visit. At our institution, we implemented a triage workflow where the patient and/or guardian would be contacted by telephone prior to the appointment to confirm the type of phones available, availability of high speed internet and a computer, up-to-date insurance information, and "check in" the patient in the EHR (Figs. 15.4, 15.5, and 15.6).

This was followed by another telephone conversation the day of the appointment to "check in" the patient and obtain consent to proceed. After any initial hurdles establishing the call, subsequent visits were often easier to conduct. If a store-and-forward approach to primary care providers is taken, an efficient system should be instituted to support the needs of the referring physician in an efficient manner while respecting patient privacy. It would also be helpful to provide instructions and feedback to the provider submitting the images to ensure the image is of high quality.

Secondly, advocacy efforts should be focused on enacting state laws, like those in the State of Mississippi, which require insurance companies to universally cover teledermatology services and reimburse providers at the same rates as in-person care. Similar efforts should be invested in the federal Medicare program so that the elderly, who represent a greater proportion of rural populations, can be served by teledermatology services.

Next, as we emerge from the COVID-19 period, the easing of federal HIPAA regulations that led to a teledermatology renaissance should not be permitted to "snap back" to the previous provisions that limited adoption of this technol-

Patient Access Workflow

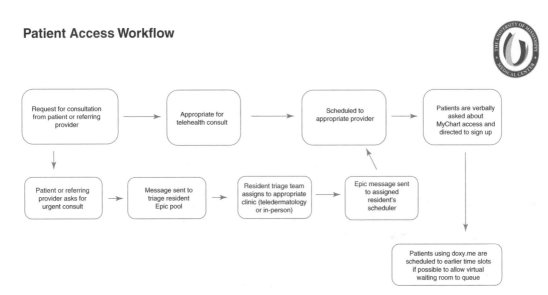

Fig. 15.4 UMMC Dermatology patient access workflow for teledermatology during COVID-19

Nursing Pre-Charting Workflow

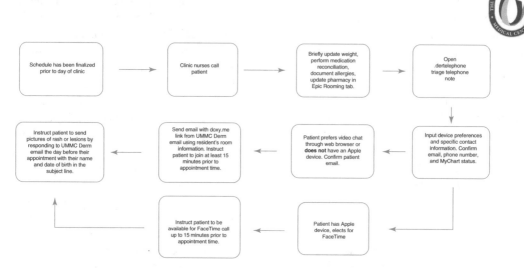

Fig. 15.5 UMMC Dermatology nursing pre-charting workflow for teledermatology during COVID-19

Patient Clinical Care Workflow

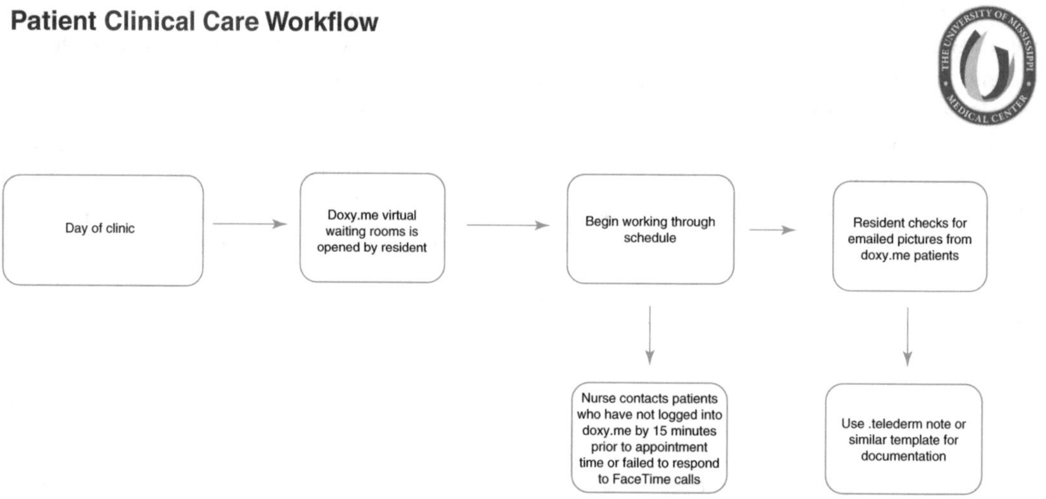

Fig. 15.6 UMMC Dermatology patient clinical care workflow for teledermatology during COVID-19

ogy. Most particularly, audit control rules that require healthcare workers to enter into a Business Associate Agreement with third parties when handling protected health information should be eased. Physicians and patients may come to terms with the technology they choose to utilize for teledermatology visits, just as they have done during the COVID-19 pandemic.

Finally, health care providers should continually adapt and upgrade their systems as new technologies emerge that permit the highest quality image transmission for both synchronous and store-and-forward technologies. This is the barrier to be broached to improve access to dermatology in rural areas through this technology.

Acknowledgements We thank Ross Pearlman, MD, for providing Figures 15.4–15.6.

References

1. Perkins S, Cohen JM, Nelson CA, Bunick CG. Teledermatology in the era of COVID-19: experience of an academic department of dermatology. J Am Acad Dermatol. 2020 Jul;83(1):e43–4. https://doi.org/10.1016/j.jaad.2020.04.048.
2. Gupta R, Ibraheim MK, Doan HQ. Teledermatology in the wake of COVID-19: advantages and challenges to continued care. J Am Acad Dermatol. 2020;83(1):168–9. https://doi.org/10.1016/j.jaad.2020.04.080.
3. Ahn CS, Allen MM, Davis SA, Huang KE, Fleischer AB Jr, Feldman SR. The National Ambulatory Medical Care Survey: a resource for understanding the outpatient dermatology treatment. J Dermatolog Treat. 2014;25(6):453–8. https://doi.org/10.3109/09546634.2014.858409.
4. Loane MA, Bloomer SE, Corbett R, Eedy DJ, Hicks N, Lottery HE, et al. A comparison of real-time and store-and-forward teledermatology: a cost-benefit study. Br J Dermatol. 2000;143(6):1241–7. https://doi.org/10.1046/j.1365-2133.2000.03895.x.
5. Black SM, Ali FR. Secure communication conduits during COVID-19 lockdown. Clin Exp Dermatol. 2020;45(6):748–9. https://doi.org/10.1111/ced.14244.
6. Manahan MN, Soyer HP, Loescher LJ, Horsham C, Vagenas D, Whiteman DC, et al. Br J Dermatol. 2015;172(4):1072–80. https://doi.org/10.1111/bjd.13550.
7. Gendreau JL, Gemelas J, Wang M, Capulong D, Lau C, Bratten DM, et al. Unimaged melanomas in store-and-forward teledermatology. Telemed J E Health. 2017;23(6):517–20. https://doi.org/10.1089/tmj.2016.0170.
8. van der Heijden JP, Thijssing L, Witkamp L, Spuls PI, de Keizer NF. Accuracy and reliability of teledermatoscopy with images taken by primary care providers during everyday practice. J Telemed Telecare. 2013;19(6):320–5. https://doi.org/10.1177/1357633X13503437.
9. Senel E, Sabancılar E, Mansuroğlu C, Demir E. A preliminary study of the contribution of telemicroscopy to the diagnosis and management of skin tumours in teledermatology. J Telemed Telecare. 2014;20(4):178–83. https://doi.org/10.1177/1357633X14533885.
10. Anderson D, Villagra VG, Coman E, Ahmed T, Porto A, Jepeal N, et al. Reduced cost of specialty care using electronic consultations for medicaid patients. Health Aff (Millwood). 2018;37(12):2031–6. https://doi.org/10.1377/hlthaff.2018.05124.
11. Caldwell JT, Ford CL, Wallace SP, Wang MC, Takahashi LM. Intersection of living in a rural versus urban area and race/ethnicity in explaining access to health care in the United States. Am J Public Health. 2016;106(8):1463–9. https://doi.org/10.2105/AJPH.2016.303212.
12. Martin AB, Probst JC, Shah K, Chen Z, Garr D. Differences in readiness between rural hospitals and primary care providers for telemedicine adoption and implementation: findings from a statewide telemedicine survey. J Rural Health. 2012;28(1):8–15. https://doi.org/10.1111/j.1748-0361.2011.00369.x.
13. Armstrong AW, Kwong MW, Chase EP, Ledo L, Nesbitt TS, Shewry SL. Why some dermatologists do not practice store-and-forward teledermatology. Arch Dermatol. 2012;148(5):649–50. https://doi.org/10.1001/archdermatol.2012.42.
14. Barbieri JS, Nelson CA, Bream KD, Kovarik CL. Primary care providers' perceptions of mobile store-and-forward teledermatology. Dermatol Online J. 2015;21(8):13030/qt2jt0h05w.
15. Moore MA, Coffman M, Jetty A, Klink K, Petterson S, Bazemore A. Family physicians report considerable interest in, but limited use of, telehealth services. J Am Board Fam Med. 2017;30(3):320–30. https://doi.org/10.3122/jabfm.2017.03.160201.
16. Lin CC, Dievler A, Robbins C, Sripipatana A, Quinn M, Nair S. Telehealth in health centers: key adoption factors, barriers, and opportunities, Health Aff (Millwood). 2018;37(12):1967–74. https://doi.org/10.1377/hlthaff.2018.05125.
17. Mansouri-Rad P, Mahmood MA, Thompson SE, Putnam K. Culture matters: factors affecting the adoption of telemedicine. HICSS '13: Proceedings of the 2013 46th Hawaii International Conference on System Sciences. 2013; 2515–24. https://doi.org/10.1109/HICSS.2013.157.
18. Dowling KH. Exploring Perceptions of Dermatology Access and the Adoption of Consultative Tederm Among Primary Care Providers in Rural Mississippi Submitted in partial fulfillment of the requirements for the degree of Doctor of Health Administration. School of Health Related Professions University of Mississippi Medical Center Jackson, Mississippi March 2020.
19. Bertrand SE, Weinstock MA, Landow SM. Teledermatology outcomes in the providence veterans health administration. Telemed J E Health. 2019;25(12):1183–8. https://doi.org/10.1089/tmj.2018.0242.
20. Marchell R, Locatis C, Burgess G, Maisiak R, Liu WL, Ackerman M. Patient and provider satisfaction with teledermatology. Telemed J E Health. 2017;23(8):684–90. https://doi.org/10.1089/tmj.2016.0192.
21. Resnick JS, Abrouk M, Steuer M, Tam A, Yen A, Lee I, Covarick CL, Edison KE. Choice, transparency, coordination, and quality among direct-to-consumer

telemedicine websites and apps treating skin disease. JAMA Dermatol. 2016;152(7):768–75. https://doi.org/10.1001/jamadermatol.2016.1774.

22. Goodspeed TA, Page RE, Koman LE, Hollenbeck AT, Gilroy AS. Legal and regulatory issues with teledermatology. Curr Dermatol Rep. 2019;8(2):46–51. https://doi.org/10.1007/s13671-019-0254-0.

23. Stevenson P, Finnane AR, Soyer HP. Teledermatology and clinical photography: safeguarding patient privacy and mitigating medico-legal risk. Med J Aust. 2016;204(5):198–200. https://doi.org/10.5694/mja15.00996.

24. Bhate C, Ho CH, Brodell RT. HIPAA and teledermatology in the age of COVID-19. J Am Acad Dermatol. Accepted 6/25/20.

25. Trout KE, Rampa S, Wilson FA, Stimpson JP. Legal mapping analysis of state telehealth reimbursement policies. Telemed J E Health. 2017;23(10):805–14. https://doi.org/10.1089/tmj.2017.0016.

26. Capistrant G. Medicare coverage and reimbursement policies. In: Rheuban K, Krupinski EA. eds. Understanding telehealth [Internet] New York City: McGraw-Hill; 2018. [cited 2020 Jul 20] Available from: https://accessmedicine.mhmedical.com/content.aspx?bookid=2217&sectionid=187795807

27. Lenert L, BY MS. Balancing health privacy, health information exchange, and research in the context of the COVID-19 pandemic. J Am Med Inform Assoc. 2020;27(6):963–6. https://doi.org/10.1093/jamia/ocaa039.

28. Senate Bill 2209 2013 (Mississippi).

29. Senate Bill 2646 2014 (Mississippi).

30. Chuchvara N, Patel R, Srivastava R, Reilly C, Rao BK. The growth of teledermatology: Expanding to reach the underserved. J Am Acad Dermatol. 2020;82(4):1025–33. https://doi.org/10.1016/j.jaad.2019.11.055.

Worldwide Rural Dermatology Health Services Research

16

James E. Roberts, Meredith E. Thomley, Manoj Sharma, and Vinayak K. Nahar

Living in a rural setting exposes you to so many marvelous things – the natural world and the particular texture of small-town life and the exhilarating experience of open space.

—Susan Orlean

Section 1: Common Dermatologic Conditions in Rural Areas

There are over 3000 dermatologic conditions that have been categorized and these conditions vary in presentation, pathophysiology, and etiology [1, 2]. Diagnosis remains the key to proper treatment and understanding the diseases endemic to rural areas in the United States

and around the world helps dermatologists properly assess skin findings. It's usually a horse and not a zebra, except in areas where zebras are prevalent! In fact, certain dermatologic conditions are found more frequently in certain rural settings. Studies conducted at the Regional Dermatology Training Centre in Tanzania demonstrated that 60–90% of all dermatologic manifestations that were identified were due to only ten conditions [3]. Rural occupations, such as livestock herders and farmers, were associated with higher frequencies of deep fungal infections, cutaneous anthrax, and zoonotic fungal infections in these resource starved rural areas of Africa [3]. A study by dermatologists over a 4 day span in rural Laos, Vietnam, revealed that 53% of participants had an active skin disease, with the most common dermatoses being eczema, dermatophyte infections, acne, scabies, melasma, and pityriasis versicolor [4]. In rural areas of the Western Cape in South Africa, chronic diseases were among the most common dermatologic conditions which included eczema, vitiligo, and psoriasis, whereas zoonotic infections and infestations were lower than they had antici-

J. E. Roberts
Department of Medicine, University of Mississippi Medical Center, Jackson, MS, USA
e-mail: jroberts5@umc.edu

M. E. Thomley
University of Alabama at Birmingham School of Medicine, Birmingham, AL, USA
e-mail: thomley@uab.edu

M. Sharma
Department of Environmental & Occupational Health, School of Public Health, University of Nevada, Las Vegas, NV, USA
e-mail: manoj.sharma@unlv.edu

V. K. Nahar (✉)
Department of Dermatology, University of Mississippi Medical Center, Jackson, MS, USA

pated [5]. On the other hand, some dermatoses, including acne vulgaris, atopic dermatitis, and contact dermatitis, were seen with a higher frequency in urban areas [6]. Recognizing that the frequency of dermatoses vary in rural areas in foreign countries, suggests that the same variation could exist in the United States of America.

Unfortunately, there has been very little research in the United States focused on the variations in skin disease that are likely to exist in different regions of the country or in rural America. The climate of the United States is as varied as anyplace in the world with ecosystems in the southwest US, the Deep South, the Rocky Mountains, and the Appalachians being vastly different. Similarly, microenvironments in cities are dissimilar to rural areas. While we know that conditions like coccidioidomycosis are common in the Southwest US and histoplasmosis in the Mississippi River Valleys, there have been no systematic studies assessing variation of common dermatologic manifestations across the varied rural regions of America.

Some insights can be garnered by assessing diagnoses made during teledermatology referrals from rural settings. Common dermatologic conditions identified included eczema, contact dermatitis, psoriasis, drug reactions, verruca vulgaris, and non-melanoma skin cancers [7]. Skin cancers are particularly common in rural settings [8]. Delays in diagnosis are expected since fewer dermatologists practice in rural areas and some skin cancers may be misdiagnosed by primary care providers who have not had sufficient dermatology training. Understanding the needs of rural populations is important since, delays in skin cancer diagnoses, especially malignant melanoma, can have profound effects on morbidity and mortality. For instance, migrant farm workers represent a special subset of rural workers whose health demographics including those related to dermatologic conditions are largely unknown [9]. In summary, the issues related to dermatologic health care in rural America can only be properly attacked after information is collected from focused epidemiologic studies in inform strategic planning.

Section 2: Rural Dermatology Workforce and Its Needs

Despite considerable resources focusing the needs of rural communities across the United States, rural communities remain to have an unfulfilled need for healthcare providers. This is especially true for the field of dermatology, as it has become clear that rural areas have workforce needs that exceed urban areas. Vaidya et al. [10] report that fewer than 10% of board-certified dermatologists ultimately practice in rural communities. One nationwide study of the distribution of dermatologists showed that while the dermatologist density improved from 3.02 to 3.65 per 100,000 people from 1995 to 2013 (21%), the urban and rural gap widened with a greater percentage of dermatologists residing in urban communities [11]. Uhlenhake et al. [12] performed a study that demonstrated that per 100,000 people, heavily populated cities have an average of 6.9 medical dermatologists, whereas rural areas may only have 4.4 medical dermatologists. Considering this disparity, the solution to providing adequate dermatologic care to rural communities must be multifaceted. Ultimately, these efforts must result in an increase in the number dermatology providers, as well as skin-savvy primary care physicians, in order to provide quality care to patients in need of dermatologic services in rural America.

Shortages in the number of practicing providers remain the greatest challenge to the workforce of dermatology within rural areas of the United States. In these counties located miles from urban hubs, 88% of people do not have a dermatologist [10]. The benefits of having specialized dermatological care and providers available to rural patients in remote areas are invaluable. Further health services research focused on the aforementioned potential solutions to the shortage of specialized care, is required if quality care is to be provided to rural patients.

Problems with the provision of dermatologic care in rural areas are not limited to the United States. Locations from the Tropics and Tanzania to Ethiopia and South Africa, demonstrate limited access to dermatologic care in rural areas resulting in delayed diagnoses and inadequate treatment [5, 13–15]. The problem is magnified in countries where greater than 70% of the population reside in rural areas. In the tropics, the burden of disease largely lies with bacterial and fungal infections, most of which are "not only preventable but also curable with simple, low-cost, and effective medication." [13] The remote nature of rural communities results in significant morbidity that could have otherwise been avoided. In Debre Markos, Ethiopia, 22 dermatologists are available to service 80 million residents [15]. Without expert care, the community turns to often ineffective home remedies and nonprescription drugs to treat skin problems. In rural Overberg, South Africa, a nurse-led, rotational dermatology service has focused their attack on prompt diagnosis and treatment of transmissible diseases [5]. With these services, the burden on tertiary centers is lifted and more hospital beds are available for patients with other conditions. Transport costs and workload of urban dermatologic practitioners are also reduced.

Overberg's efforts demonstrate that intentional dermatologic care in rural areas results in improved outcomes. Through the implementation of local education, the adoption of telemedicine, and the migration of dermatologists, similar successes have been achieved in the United States. Naafs et al. 2009 describe that while "rural" means different things to each country, (i.e. desert land vs. rainforest setting or socialist vs. capitalist societies), the prevailing factor is poverty [13]. Recognizing that the United States has its own unique needs when compared to the Tropics and rural Africa, the expansion of healthcare and availability of physicians to all populations is a universal principle required to guarantee optimal dermatologic care.

Section 3: The Use of Teledermatology to Extend Reach of Dermatologists into Rural Areas

The use of teledermatology has been an important tool in helping patients in rural areas obtain adequate and specialized care of their dermatologic conditions. Teledermatology is the use of audio and visual technology to provide dermatological services. This technology has allowed for greater access to dermatologists and improved specialized care by primary care providers in rural areas. The field of dermatology is particularly amenable to the use of telemedicine services since most dermatologic conditions can be diagnosed with visualization of conditions using either live-video or still images.

There are two main types of teledermatology: "Store and Forward" and "Synchronous" (Live action). The "store and forward" method requires providers or patients to take still images and/or video content using a device with a camera and forwarding these images along with clinical history to a dermatologist for review. The importance of de-identification and secure transmission is critical for this type of teledermatology service. Synchronous teledermatology, on the other hand, utilizes audio-visual technology which permit the dermatologist to visualize the patient's dermatologic condition in real time. The former method has proven to be the most efficient for dermatologists, though the latter was most utilized during the COVID-19 pandemic when in-person visits were severely limited.

Teledermatology has been particularly useful in specific settings: prisoners, hospitalized patients, migrant farmers, Pacific Islanders, veterans through the Veterans Health Administration, Native Americans, rural Missouri [16–19]. Not only does teledermatology allow for the efficient provision of care to underserved patients, it also has been shown to decrease patient expenses. In one study, patients being treated by their primary care physician for a six-month period spent more money compared to the eight-month period after receiving teledermatology services [7]. Cost-consciousness is especially important to rural patient with limited financial resources. Primary care providers also benefit from utilizing teledermatology services for their patients as they gain competency to more easily recognize common dermatologic conditions and implement treatments [16, 18, 20].

A focus on the dermatologic education of rural primary care providers is also central to a form of case-based distant learning called Project ECHO. Dermatology Project ECHO, started by the department of dermatology at the University of Missouri, incorporates discussion of real patient clinical cases with education through CME-approved didactic presentations [16]. Receiving feedback on difficult or complex dermatology leads to increased competency in diagnosis and management without physicians needing to leave their offices.

A similar program was developed by the Veterans Health Administration (VHA) utilizes rural physicians and imaging technicians to treat veterans in primary care clinics where dermatological services were not previously offered [18]. "Store-and-forward" teledermatology visits, in-person lectures with subsequent online educational modules, competency examinations and skill practicums comprise the bulk of training provided for these primary care providers. The program was designed to decrease travel to-and-from distant facilities and decrease the amount of referrals to overwhelmed dermatologists since basic dermatologic care would be provided by local primary care physicians. In fact, at the end of one study, the primary care physicians made fewer referrals to distant facilities and performed basic procedures including punch biopsies, shave biopsies, and even excisions in their clinics [18].

The use of teledermatology is sure to evolve over time. With continuing health services research, it is hoped that teledermatology will expand the reach of dermatological care to patients in rural areas that currently have limited access to specialized care.

Efforts to expand services and fill the vast need for rural dermatology care is not without challenges. The greatest barriers to equitable care include issues raised by health care research in the areas noted above: the availability of physicians, effective implementation of teledermatology, and the need to improve health literacy in rural populations.

Section 4: Overcoming Barriers and Challenges to Rural Dermatological Care

Solving Barriers Related to Dermatology Manpower

Solving dermatology manpower needs requires physician attraction and retention to rural communities. Unfortunately, the number of trainees produced annually via residency programs will not meet the populations needs [21]. Rural job opportunities must effectively compete with large urban centers to draw excellent physicians to their towns. There are two solutions to this challenge: first, to increase the number of available dermatologists and second, to provide incentives to practice in rural areas to provide compelling offers.

While medical schools have increased enrollment in the past 20 years, the number of residency training slots have been suppressed by the 1997 cap on Medicare support for graduate medical education (GME) [22]. An increase in government funding for residency programs could lead to more graduating dermatology residents since each year there are more applicants applying for dermatology residencies than available positions. Many go unmatched. A rise in dermatology residency spots would augment the available pool of trained dermatologists to practice in rural communities [12]. After the pool of providers is multiplied, then we must address the challenge of recruitment to specific areas: a financial battle, as well as a marketing one. Recent graduates of dermatology programs cite location as their primary factor for job consideration. With the persistence of underserved populations in rural communities, direct government funds or loan forgiveness may be needed to incentivize provider relocation [10, 12].

Of course, training more physicians and opening more dermatology residency slots do not represent the complete solution in the United States where less than 10% of dermatologists practice in rural areas [23]. In other words, training greater numbers of dermatologists who subsequently flock to cities will not solve rural access to care problems. There are three logical approaches to increasing the availability of dermatologic care in rural areas. First, we could train more medical students in the field of dermatology who have an interest in practicing in rural areas. This interest largely correlates with growing up in a rural area [24]. In fact, many residency programs have Rural Physician Programs that recruit rural students and incentivize them to return to rural areas, perhaps their home town, with tuition grants. This has also been used in dermatology residency programs including the University of Mississippi Medical Center in Jackson, Mississippi. A residency position was created to foster the training of dermatologists with an interest in practicing in rural areas of underserved Mississippi. Of course, these programs require motivated residency programs that are willing to broadly evaluate candidates and design effective programs that successfully produce rural physicians.

A second approach would be to entice urban or suburban dermatologists to relocate to rural areas. This may be easier in the post Covid-19 era since there appears to be some flight from cities. Thirdly, we could establish dermatology residency programs in rural areas. One study showed that approximately 43% of dermatologists choose to practice within 100 miles of the location of their dermatology residency [25]. Seeing that most dermatology residencies are located within urban areas or at academic centers, it is not surprising that rural areas of America do not reap the benefits of increasing numbers of dermatology residents.

Each dermatologist in rural America could also choose to add a mid-level provider (nurse practitioner or physician assistant) to their health care team. This has been shown to be associated with decreased wait times at dermatology offices [26]. The American Academy of Dermatology has encouraged this team approach to care [27]. More study is certainly required to assess the optimal approaches to using mid-level providers.

An entirely different approach may be the most feasible in the short term. Robust educational efforts should be designed to improve the dermatologic skills of rural primary care physicians to better handle annual skin cancer checkups and routine dermatologic care. In the United States, primary care physicians are the frontline workers in helping patients receive adequate and important medical treatment for all of their health problems. Approximately one-third of patients seen by their primary care doctor had one dermatologic condition with almost 60% of patients listing a skin problem as their chief compliant for that health care encounter [28]. Patients in rural settings depend on primary care physicians for guidance on their dermatologic conditions since specialized dermatologic care is often unavailable. Research has shown that primary care doctors are not as accurate in diagnosing dermatologic conditions when compared to dermatologists [29], thus highlighting the need for specialized care and educational efforts to improve the skills of primary care physicians regarding common dermatologic conditions. Since board certified dermatologists have been demonstrated to better detect melanoma and more accurately diagnose a variety of skin diseases, a collaborative effort between primary care providers and dermatologists would be beneficial [30]. Project ECHO© is one such program focused on distance learning for rural primary care providers that has been shown to positively impact participant and patient outcomes [16, 31]. While this may be a short-term solution to satiate demand, it is limited by a shortage of primary care physicians in rural areas.

Finally, after addressing workforce needs, it is important to tackle the specific challenges that impact the care of the rural patient population. Rural populations are spread over vast areas. Even when a new dermatology clinic is established, patients often must travel great distances to access their care. Patients are less likely to travel greater than 20 miles to see a provider, even if the services provided are free [12]. Furthermore, with distance as a barrier, patients may choose to postpone appointments [10]. Though wait times for both rural and urban settings may be equivalent, the barriers of distance to the dermatology office, provider-to-patient ratio, and severe limits on availability of subspecialty care contribute to a lower quality of dermatologic care [12].

Additionally, counties with predominantly African American, Hispanic American, and Native American populations often do not have a single dermatologist within their county. Thus, racial disparities impact access to a dermatologic care [10]. Knowing that minority physicians are more likely to work in underserved areas and treat minority populations [32], it can be assumed that an increase in non-Caucasian dermatologists may help alleviate this disparity. As of 2018, approximately 3% of dermatologists in the United States are African American, while 4.1% are Latin American [10]. Residency program selection committees need a greater focus on attracting qualified under-represented minority applicants and holistic evaluative processes to attack this problem.

Solving Barriers Related to Delivery of Teledermatology Services

Limitations of existing technology contribute to problems with access to rural dermatology care. Teledermatology can certainly bring dermatologic access to rural populations; however, limited access to state-of-the-art technology (computers, cameras, electricity, and broadband internet) and up-to-date training accentuates the digital divide between urban and rural communities [18]. Simply put, teledermatology consultations require clear video and stable audio. State and federal programs are required to implement and maintain telehealth programs in rural areas. Furthermore, the implementation of educational programs for local primary care providers and urban dermatologists can ensure consistency and quality of care in teledermatology visits.

Solving Barriers Related to Health Literacy

Finally, differences in health literacy and educational opportunities are a critical challenge in the provision of rural dermatology care. A devastating component of many skin conditions involves the associated stigma. Unfortunately, patients wear their skin disease on the outside of their bodies. This breeds isolation, especially in rural communities. For example, individuals in rural Tanzania suffering from leprosy may retreat into isolation in society due to the cultural views surrounding the disease [14]. Though the exact assumptions and stigmata of skin diseases in rural Mississippi may be vastly different than in Tanzania, the concern remains that patients in rural populations may fail to seek help due to societal pressures for conditions like psoriasis and eczema. Though it certainly takes intentionality and time to implement these changes, the effort is necessary to provide sound dermatologic care to rural America.

Section 5: Future Research

This chapter is designed to highlight the need for a significant investment in dermatological health services research focused on rural America. Although numerous studies have been performed regarding the practice of dermatology in rural settings, there is a clear-cut need for additional research. One area of research should focus on the needs and attitudes of rural dermatologists regarding practicing in rural areas with limited resources. Tackling the needs of dermatologists already practicing in rural areas will help us better understand where to focus resources to increase availability of rural dermatological care. Further studies regarding the efficacy of rural physician programs would be helpful in furthering the conversation regarding the funding and initiation of similar programs.

In addition, the need for research focused on the benefits of teledermatology would help raise awareness of the efficacy and ease of use of this technology. These studies may aid in funding of programs across rural America. Studies evaluating the various modes of initiation, technologic platforms, manpower needs, and financial viability of teledermatology services would be helpful to institutions who are wary of investing in telehealth.

Lastly, a broad-based lobbying effort will be required to encourage the federal government to fund more residency positions, which would allow the development of new dermatology training programs in rural areas and allow existing programs to increase the number of positions. In addition, combined efforts between existing residency programs and state legislatures could produce novel incentives to encourage physicians to practice in rural areas. With increased awareness and a steadfast approach, it is possible to overcome the barriers that prevent patients living in rural areas from getting the dermatological care they not only need but deserve.

References

1. Hay RJ, Johns NE, Williams HC, et al. The global burden of skin disease in 2010: an analysis of the prevalence and impact of skin conditions. J Invest Dermatol. 2014;134:1527–34.
2. Lim HW, Collins SAB, Resneck JS, Bolognia JL, Hodge JA, Rohrer TA, Van Beek MJ, Margolis DJ, Sober AJ, Weinstock MA, Nerenz DR, Begolka WS, Moyano JV. The burden of skin disease in the United States. J Am Acad Dermatol. 2017;76(5):958–972 E2.
3. Naafs B. The skin. In: Parry E, Godfrey R, Mabey D, Hill G, editors. Principles of medicine in Africa. 3rd ed. Cambridge, UK: Cambridge University Press; 2004. p. 1264–301.
4. Wootton CI, Bell S, Philavanh A, et al. Assessing skin disease and associated health-related quality of life in a rural Lao community. BMC Dermatol. 2018;18:11. https://doi.org/10.1186/s12895-018-0079-8.
5. Cloete D. Dermatology nursing in a rural area – the Overberg experience. In: CME : Your SA Journal of CPD, vol. 31. 7th ed. Western Cape, South Africa: South African Medical Association NPC; 2013. p. 254–8.
6. Naafs B, Matemera BO, Mudarikwa L, Noto S. Position paper to Ministry of Health; 1986. Harare, Zimbabwe.
7. Burgiss SG, Julius CE, Watson HW, Haynes BK, Buonocore E, Smith GT. Telemedicine for dermatology care in rural patients. Telemed J. 1997;3(3):227–33. https://doi.org/10.1089/tmj.1.1997.3.227.
8. Bram H, Frauendorfer M, Spencer S, Hartos J. Does the prevalence of skin cancer differ by metropolitan status for males and females in the United States?. 2017, December 14. Retrieved December 20, 2020, from https://preventive-medicine.imedpub.com/does-the-prevalence-of-skin-cancer-differ-by-metropolitan-status-for-males-and-females-in-the-united-states.php?aid=21301
9. Slesinger DP. Health status and needs of migrant farm workers in the United States: a literature review. J Rural Health. 1992;8(3):227–34. https://doi.org/10.1111/j.1748-0361.1992.tb00356.x.
10. Vaidya T, Zubritsky L, Alikhan A, Housholder A. Socioeconomic and geographic barriers to dermatology care in urban and rural US populations. J Am Acad Dermatol. 2018;78(2):406–8. https://doi.org/10.1016/j.jaad.2017.07.050.
11. Feng H, Berk-Krauss J, Feng PW, Stein JA. Comparison of dermatologist density between urban and rural counties in the United States. JAMA Dermatol. 2018;154(11):1265–71.
12. Uhlenhake E, Brodell R, Mostow E. The dermatology work force: a focus on urban versus rural wait times. J Am Acad Dermat. 2009;61(1):17–22. https://doi.org/10.1016/j.jaad.2008.09.008.
13. Naafs B, Padovese V. Rural dermatology in the tropics. Clin Dermatol. 2009;27(3):252–70. https://doi.org/10.1016/j.clindermatol.2008.10.005.
14. Roosta N, Black DS, Rea TH. A comparison of stigma among patients with leprosy in rural Tanzania and urban United States: a role for public health in dermatology. Int J Dermatol. 2013;52(4):432–40. https://doi.org/10.1111/j.1365-4632.2011.05226.x.
15. Murgia V, Bilcha KD, Shibeshi D. Community dermatology in Debre Markos: an attempt to define children's dermatological needs in a rural area of Ethiopia. Int J Dermatol. 2010;49(6):666–71. https://doi.org/10.1111/j.1365-4632.2009.04284.x.
16. Lewis H, Becevic M, Myers D, Helming D, Mutrux R, Fleming DA, Edison KE. Dermatology ECHO – an innovative solution to address limited access to dermatology expertise. Rural Remote Health. 2018;18:4415. https://doi.org/10.22605/RRH4415.
17. Norton SA, Burdick AE, Phillips CM, Berman B. Teledermatology and underserved populations. Arch Dermatol. 1997;133(2):197–200. https://doi.org/10.1001/archderm.1997.03890380069010.
18. McFarland LV, Raugi GJ, Taylor LL, Reiber GE. Implementation of an education and skills programme in a teledermatology project for rural veterans. J Telemed Telecare. 2011;18(2):66–71. https://doi.org/10.1258/jtt.2011.110518.

19. Morenz AM, Wescott S, Mostaghimi A, Sequist TD, Tobey M. Evaluation of barriers to telehealth programs and dermatological care for American Indian Individuals in Rural Communities. JAMA Dermatol. 2019;155(8):899–905. https://doi.org/10.1001/jamadermatol.2019.0872.

20. Basarab T, Munn SE, Jones RR. Diagnostic accuracy and appropriateness of general practitioner referrals to a dermatology out-patient clinic. Br J Dermatol. 1996;135(1):70–3.

21. Resneck J, Kimball AB. The dermatology workforce shortage. J Am Acad Dermatol. 2004;50(1):50–4. https://doi.org/10.1016/j.jaad.2003.07.001.

22. Salsberg E, Rockey PH, Rivers KL, Brotherton SE, Jackson GR. US residency training before and after the 1997 balanced budget act. JAMA. 2008;300(10):1174–80. https://doi.org/10.1001/jama.300.10.1174.

23. Yoo JY, Rigel DS. Trends in dermatology: geographic density of US dermatologists. Arch Dermatol. 2010;146:779.

24. Phillips R, Doodoo M, Pettersen S, et al. Specialty and geographic distribution of the physician workforce: what influences medical student & resident choices? http://www.graham-center.org/dam/rgc/documents/publications-reports/monographs-books/Specialty-geography-compressed.pdf. Accessed 17 Sept 2020.

25. Resneck JS, Kostecki J. An analysis of dermatologist migration patterns after residency training. Arch Dermatol. 2011;147(9):1065–70. https://doi.org/10.1001/archdermatol.2011.228.

26. Zurfley F, Mostow EN. Association between the use of a physician extender and dermatology appointment wait times in Ohio. JAMA Dermatol. 2017;153(12):1323–4. https://doi.org/10.1001/jamadermatol.2017.3394.

27. Jalian HR, Avram MM. Mid-level practitioners in dermatology: a need for further study and oversight. JAMA Dermatol. 2014;150(11):1149–51. https://doi.org/10.1001/jamadermatol.2014.1922.

28. Lowell B, Froelich C, Federman D, Kirsner R. Dermatology in primary care: prevalence and patient disposition. J Am Acad Dermatol. 2001;45:250–5. https://doi.org/10.1067/mjd.2001.114598.

29. Ramsay DL, Fox AB. The ability of primary care physicians to recognize the common dermatoses. Arch Dermatol. 1981;117(10):620–2. https://doi.org/10.1001/archderm.1981.01650100022020.

30. Federman DG, Concato J, Kirsner RS. Comparison of dermatologic diagnoses by primary care practitioners and dermatologists. A review of the literature. Arch Fam Med. 1999;8(2):170–2. https://doi.org/10.1001/archfami.8.2.170. PMID: 10101989.

31. Zhou C, Crawford A, Serhal E, Kurdyak P, Sockalingam S. MHPE the impact of project ECHO on participant and patient outcomes: a systematic review. Acad Med. 2016;91(10):1439–61. https://doi.org/10.1097/ACM.0000000000001328.

32. Marrast LM, Zallman L, Woolhandler S, Bor DH, McCormick D. Minority physicians' role in the care of underserved patients: diversifying the physician workforce may be key in addressing health disparities. JAMA Intern Med. 2014;174(2):289–91. https://doi.org/10.1001/jamainternmed.2013.12756.

Rural Dermatology Private Practice: A Life Worth Living

17

Caroline P. Garraway and Adam C. Byrd

We are here to add what we can to life, not get what we can from Life

—William Osler, MD

Diversity is a highly desirable touchstone of the world we live in today. For someone living in the city with diverse entertainment, dining, educational and professional options what would lead someone to choose a home on a farm? Perhaps it is the autonomy associated with wide-open spaces or the rugged individualism that drives many Americans. In fact, many rural Americans find it is easy to justify returning home, if they ever left in the first place. When asked the reasoning behind returning, the most common answer: "It's just home." For individuals who have spent most of their lives in an urban area, it may be difficult to see the benefits of living and working in a small town. However, there are distinct advantages for the practicing physician/dermatologist to choose a rural life related to: cost of living;

crime rates; mental health; financial incentives; diverse clinical practice, and factors that affect overall happiness in a less pressured rural life. The COVID-19 pandemic has also led to an exodus from urban areas. This chapter discusses all of these factors and unabashedly encourages young dermatologists to spend some time in a rural dermatology office before joining a practice in the city.

Cost of Living

On average a physicians' hard-earned dollar goes further in a rural community. According to the Bureau of Labor Statistics, in 2011, the average urban household spends $50,348/year, which was 18 percent more than the $42,540/year spent by rural households. These costs include insurance/pensions, education, entertainment, health care, transportation, apparel/services, housing, food, and other expenditures. It was determined that on average, urban households had higher expenditures on food, housing, apparel, and education, while rural households spent more on

C. P. Garraway
School of Medicine, University of Mississippi
Medical Center, Jackson, MS, USA
e-mail: cgarraway@umc.edu

A. C. Byrd (✉)
Department of Dermatology, University of
Mississippi Medical Center, Jackson, MS, USA
e-mail: acbyrd@umc.edu

© The Author(s), under exclusive license to Springer Nature Switzerland AG 2021
R. T. Brodell et al. (eds.), *Dermatology in Rural Settings*, Sustainable Development Goals Series,
https://doi.org/10.1007/978-3-030-75984-1_17

transportation, health care, and entertainment [1]. Homes, undeveloped land, and rent are all less expensive in rural areas. This advantage is significant. The market value of the average urban home was $153,147 in 2011, while the market value of a rural home was $129,111, a 16% difference [1]. Urban renters reported pay $699, for average monthly rent, while rural renters pay $354, a 50% difference [1].

While the cost of living for rural physicians is low, they earn just as much as their counterparts in urban areas. WRVU-based reimbursement models used by Medicare, Medicaid, and most insurance plans, pay the same amount regardless of the practice setting, thus as long as the physician can fill his/her schedule, the practice income is the same. From our personal experience, the population density of the actual city and county in which one starts a dermatology practice does not necessarily dictate patient volume. In other words, rural dermatologists in small towns can have very busy practices with patients driving long distances to see a dermatologist! Thus, when planning for patient volume, the rural dermatologist can extend their catchment area out by 60–90 miles and can be certain that – as long as they don't have overly restrictive insurance acceptance protocols – they will very soon have a very robust panel of patients. At our clinic in Louisville, MS we have consistently billed over 12,000 WRVUs despite the fact that our county only has a population of 19,000.

Crime

Another important element in a person's decision to move to a new area, or leave an existing one, is crime. Fear is a strong driving force behind many of life's major decisions, and the fear that crime will impact you or a loved one is no exception. In 2014, data gathered from the FBI's Uniform Crime Reporting (UCR) Program and the National Crime Victimization Survey demonstrates that crime rates are higher in urban areas than rural or suburban areas. During that year, it was determined that 55% of all rapes and sexual assaults, 50% of robberies, and 31% of aggravated assaults occur in urban areas [2]. Suburban areas accounted for 51% of all aggravated

assaults. Victims from rural areas accounted for only 10% of rapes and sexual assaults, 6% of robberies, and 18% of aggravated assaults [2]. Household victimization and personal victimization rates per 1000 people were consistently higher in urban areas when compared to rural areas. Law enforcement to citizen ratio was also higher in rural areas. Metro counties had 2.6 officers/1000 people and non-metro counties have 2.8 officers/1000 people [3]. While there may be some geographic variability with crime committed and crime reported, the available data does demonstrate a clear divide in the incidence of crime between rural and urban areas.

Mental Health

"Urbanization and Mental Health," an article published in the *Industrial Psychiatry Journal* details the effects of urbanization on mental health. Increased stressors, overcrowded and polluted environments, high levels of violence, and reduced social support, negatively impact mental health in urban communities. Urban microenvironments, such as living close to highways or airports, increases exposure to traffic noise and pollution and is associated with higher levels of stress and other downstream mental health consequences [4]. Furthermore, the concentrated population and limited resources available in urban areas increase the risk of poverty. Low socioeconomic status is also associated with development of mental health disorders [4]. A meta-analysis by Reddy and Chandrashekhar demonstrated that there was a significantly higher prevalence of depression and other mental health disorders in urban vs rural areas (80.6% and 48.9%, respectively) [5]. Of course, it is difficult to determine causation in this relationship between urbanicity and poor mental health.

There are other things that support the mental health of rural dermatologists: A short commute is an urbanite's dream! In rural areas, there are often wonderful places to live within minutes of work. One of us (ACB) has a 3-minute commute to work which compares to a 6 mile drive taking 45 minutes during residency in Minneapolis. Two of our local family physicians live on 12 acre lots, yet within *walking distance* of their hospital and clinics! It is certainly nice to be able to go home for lunch!

Happiness

Happiness is something that people strive to achieve. Evidence is mounting that happiness can be measured and that rural populations are happier when compared to their urban counterparts. In one study of 1200 Canadian Neighborhoods and Communities it was determined that people in urban areas are significantly less happy [6]. Happiness was measured using variables depicted in Table 17.1. Specifically, relative to people in the most miserable quintile of Canadian communities, the happiest quintile live in places with nearly 90 percent fewer people per square mile [7]. A "urban-rural happiness gradient" reveals that rural dwellers, suburb inhabitants, and those in small and large cities, run from happiest to least happy, in that order [8]. Physician burnout has also been shown to be lower for individuals in rural practice [9].

Incentives

It is no secret that there is an overwhelming need for physicians in rural areas. Often, patients in rural areas travel hours to receive the care that they need. Many states have attempted to mitigate this need by creating tuition assistance in

Table 17.1 Variables that contribute to happiness[6]

Relative to the "most miserable" quintile of Canadian communities, people in the "happiest" quintile of communities…
Live in areas with nearly 90 percent fewer people per square mile
Have commutes that are 4.8 minutes shorter
Are 13 percentage points less likely to spend over 30 percent of their income on housing
Are 22 percentage points less likely to be foreign-born
Are 11 percentage point more likely to have lived in the area for more than 5 years
Are 11 percentage points more likely to be religious
Are 10 percentage points more likely to have a sense of belonging in their communities

the form of grants, scholarships, and low interest loans, for those attending medical school with the intent to practice in a rural area. However, a lot can change throughout 4 years of medical school, and some individuals who intended to practice in a rural area may change their mind limiting the number of medical students willing to accept these incentives. Fortunately, there are also incentive programs for medical students who decide to practice in a rural area after graduating from medical school. These include loan repayment or loan forgiveness for individuals willing to practice in rural parts of the state. According to the American Association of Medical Colleges (AAMC), Georgia offers loan repayment through The Physicians for Rural Areas Assistance Program which pays up to 25,000 dollars per year for a maximum of 4 years to a physician that will practice a minimum of 40 clinical hours per week in a Georgia county with a population of less than 35,000 [10]. Other states have similar programs. While some of these programs are limited to specific specialties, Oklahoma, Utah and Montana do not have specialty requirements.

Hometown Hero

While it may be hard to imagine yourself becoming a superhero, you may become a "hometown hero" for serving a small community as their physician. Everyone in a small community knows everyone by name, and the respect of the entire community leads to special bonds of trust between patient and physician that are worth more than accolades. Dr. Katrine Poe experienced firsthand what it is like to be a hometown hero practicing medicine in rural Mississippi [11]. She returned to her hometown after finishing her family medicine residency at the University of Mississippi Medical Center to become the first African American to practice in the area and the second female to ever practice in the area. After a short amount of time, she became the sole practicing physician in the community, and her passion is credited with saving the local hospital from the wrecking ball. "Small-town Doctor Looms Larger Than Life for Residents,

Family," is an article detailing Dr. Poe's life and work and states: "About 4 years into her practice, in 2005, she was named Country Doctor of the Year by Staff Care Incorporated, an insurance company for physicians [11]. The honor is reserved for doctors who dedicate their careers to serving rural communities. Now 46, she was the youngest, and first African American, to receive the award [11]. Linda Turner, a former clinic nurse now living in Chicago, had put together an album of testimonials from more than 40 patients and staff; this amounted to Poe's nomination letter. 'I was shocked when I won,' Poe said. 'It was a proud moment; it was for the whole town.' The people of Kilmichael threw her a parade and staged the award presentation in the high school auditorium; it was packed." The town celebrated their hero, a physician, who has dedicated her life to providing care to all of its citizens. This is just one example of a rural physician's ability to impact an entire town, to be a hero to so many.

COVID-19 and the Return to Rural Living?

In the spring of 2020, the world quite literally stopped. The COVID-19 pandemic cancelled dinner reservations, shows, sporting events, and work at the office. As people grew tired of surfing the web for entertainment, they fled to the outdoors. City dwellers were essentially imprisoned in high rise buildings. So, it only makes sense that they find a way out once the smoke settled. In recent months, there has been a rural real estate boom. For example, Gallatin County, Montana, which includes Bozeman, saw a nearly 50% increase in sales during the month of September [12]. Great Britain has also seen the rural market boom. The rural district of Ryedale in North Yorkshire saw sales up 63% in 6 months, compared to the same period in 2019 [13]. The COVID-19 pandemic gave its push, but with technological advances and the ability to work from home, fresh air and happiness are driving continued interest in life outside of the city limits. (Figs. 17.1, 17.2, and 17.3).

Fig. 17.1 One of us (AB) with a dermatology resident after an exhilarating off-road trek following a busy day of practice

Fig. 17.2 A dermatology resident landed a Mississippi lunker bass on a weekend morning fishing trip at a lake within minutes of the University of Mississippi Medical Center rural dermatology practice where he was rotating

Diversity of Practice

One thing is true of rural dermatologic practice: the diversity of medical diseases and surgical procedures handled in one's practice is as broad as it can be. Depending upon the individual, this can be an advantage or a disadvantage. Because the rural dermatologist is "the only show in town," they will be faced with the practice of pediatric dermatology, complex medical dermatology, surgical dermatology, cosmetics, and more. This will certainly be daunting to some who prefer to subspecialize and stay within a narrow "comfortable" zone of expertise. For those willing to push their capabilities, it can be a tremendous opportunity. A rural dermatologist must have a predisposition toward a surgical practice because the number of skin cancers, cysts, irritated nevi, and other lumps and bumps that present daily is large. Since there is rarely a dermatologic surgeon nearby, the rural dermatologist is the most available and best trained for the job.

Fig. 17.3 One of us (AB) with a dermatology resident after a morning of pheasant hunting within minutes of the University of Mississippi Medical Center rural practice location in Louisville, Mississippi

Conclusion

All medical professionals have a positive impact on their community whether they are living in rural or urban areas. This chapter was designed to highlight the unique nature of life in rural areas that magnify the role of a dermatologist in their community. Although there are pros and cons to living in both rural and urban communities, the incentives to choosing a rural community are clear. Before joining a large group in a big city, take a few days to visit rural practices in your state. You may be surprised at what you find and just maybe become part of the solution of rural access to care problems in dermatology. I'll keep the lights on for you!

References

1. William Hawk, Consumer Expenditure Program. Expenditures of urban and rural households in 2011 [Internet]. Bls.gov. 2013 [cited 2020 Dec 23]. Available from: https://www.bls.gov/opub/btn/volume-2/expenditures-of-urban-and-rural-households-in-2011.htm
2. The National Center for Victims of Crime. Urban and Rural Victimization [Internet]. 2017 [cited 2020 Dec 23]. Available from: https://www.ncjrs.gov/ovc_archives/ncvrw/2017/images/en_artwork/Fact_Sheets/2017NCVRW_UrbanRural_508.pdf
3. The National Center for Victims of Crime. Urban and Rural Crime [Internet]. [cited 2020 Dec 23]. Available from: https://www.ncjrs.gov/ovc_archives/ncvrw/2016/content/section-6/PDF/2016NCVRW_6_UrbanRural-508.pdf

4. Gruebner O, Rapp MA, Adli M, Kluge U, Galea S, Heinz A. Cities and mental health. Dtsch Arztebl Int. 2017;114(8):121–7.
5. Srivastava K. Urbanization and mental health. Ind Psychiatry J. 2009;18(2):75–6.
6. Helliwell J, Shiplett H, Barrington-Leigh C. How happy are your neighbours? Variation in life satisfaction among 1200 Canadian neighbourhoods and communities. Cambridge, MA: National Bureau of Economic Research; 2018.
7. Ingraham C. People who live in small towns and rural areas are happier than everyone else, researchers say. Washington post (Washington, DC: 1974) [Internet]. 2018 May 17 [cited 2020 Dec 23]; Available from: https://www.washingtonpost.com/news/wonk/wp/2018/05/17/people-who-live-in-small-towns-and-rural-areas-are-happier-than-everyone-else-researchers-say/
8. Jstor.org. [cited 2020 Dec 23]. Available from: https://daily.jstor.org/urban-rural-happiness-gradient/
9. Hogue AL, Huntington MK. Family physician burnout rates in rural versus metropolitan areas: a pilot study. South Dakota J Med. 2019;72(7):306–9.
10. Loan repayment/forgiveness/scholarship and other programs [Internet]. Aamc.org. [cited 2020 Dec 23]. Available from: https://services.aamc. org/fed_loan_pub/index.cfm?fuseaction=public. welcome&CFID=1&CFTOKEN=84AB33BA-B9D7-0F0F-F2719E496922FCEA
11. Pettus G, editor. Mississippi Medicine [Internet]. Vol. 8. Division of Public Affairs; 2017. Available from: https://www.umc.edu/news/Miscellaneous/2017/March/images/2017-Medicine-Winter-online.pdf
12. Gopal P. Rich Buyers Seeking Open Spaces Fuel a Housing Boom in U.S. west [Internet]. Bloomberg Quint. 2020 [cited 2020 Dec 23]. Available from: https://www.bloombergquint.com/onweb/rich-buyers-seeking-open-spaces-fuel-a-housing-boom-in-u-s-west
13. Fenton A. Coronavirus: Rural markets boom as Brits flock to countryside [Internet]. Yahoo Finance. 2020 [cited 2020 Dec 23]. Available from: https://finance. yahoo.com/news/coronavirus-rural-markets-boom-as-brits-flock-to-countryside-145607073.html?guce_referrer=aHR0cHM6Ly93d3cuZ29vZ2xlLmNvbS88&guce_referrer_sig=AQAAACxFts_ugZbDR8LK-gwkEWGBFeKh26t3GALge84VjGq56_uWEy2b-saY6zUN40CKCB957nR9OYFX4hNutVucu0s_kiwTytekY6ZZuwnkFSMjt0E98yjMuJ-c-VhN1pch5SvCXQqamQpLYp4ZYgPehmP8tW5rcRcDXH56klFstwwHSQ&guccounter=2

Attracting Dermatologists to Rural America

18

Cindy Firkins Smith, Gabriel Amon, and Amelia Amon

"The Secret of happiness: Find something more important than you are and dedicate your life to it."

—Daniel C Dennet

Introduction

What can we do to attract dermatologists to live and practice in rural America? While we know that dermatologists preferentially choose to practice in suburban and urban geographies rather than rural, we are challenged to alter that paradigm. The challenge of recruiting to rural communities is not one unique to dermatology. This is a challenge that confronts many medical specialties and professionals in all fields. Dermatology can learn from available scientific and market data generated by other medical specialties and diverse industries. After all, dermatologists are likely to be impacted by the same variables and enticed by the same incentives that increase the likelihood that a physician of any specialty would choose to practice in a rural environment. This chapter extrapolates this information and focuses it on the recruitment and retention of rural dermatologists.

The Challenges of Rural Recruitment

The need for more rural dermatologists is not in question and has been well documented in previous chapters. The overall demand for dermatology services has outpaced the training of dermatologists, and with insufficient dermatologists completing residency programs to fill that need, those completing training have had little difficulty finding a job in their preferred geographic setting [1–3]. Most dermatologists, and in fact, most physicians in general, choose to live and practice in urban areas for a combination of personal and professional reasons and women, who now make up nearly half of the dermatology

C. F. Smith (✉)
Department of Dermatology, University of Minnesota, Minneapolis, MN, USA

Rural Health, CentraCare, St. Cloud, MN, USA

Carris Health, Willmar, MN, USA
e-mail: cindy.smith@centracare.com

G. Amon
Department of Dermatology, University of Minnesota, Minneapolis, MN, USA
e-mail: amonx015@umn.edu

A. Amon
University of Minnesota, Duluth, Duluth, MN, USA
e-mail: amonx018@d.umn.edu

workforce, state a desire to "live near family" as a high priority [1].

We already know from Chap. 11 that having a rural connection is the strongest predictor of practicing in a rural area. Multiple studies from around the world have demonstrated that medical students who grow up in rural areas are up to five times more likely to choose rural areas to practice [4–19]. However, the number of medical students with a rural background is decreasing, and in 2017, "rural" students represented less than 5% of incoming medical students. For the number of rural medical students entering medical school to become proportional to the share of rural residents in the US population, the rural matriculants would need to quadruple [20]. Even if efforts were initiated to dramatically increase the number of rural medical school matriculants, it would be many years before this paradigm changes. Therefore, medical students who hail from non-rural backgrounds will be required to fill this need.

A 2017 report from Becker's Hospital Review listed the following "Seven Challenges of attracting physicians to rural environments:"[21]

- Shortage of rural academic medical programs
- Lack of family and an established social network
- Fewer employment opportunities for spouse/significant other
- Fewer entertainment opportunities (e.g., concert venues and family attractions) and dining options (e.g., upscale restaurants and variety of ethnic foods) than urban locales
- School systems with fewer offerings than those situated in metropolitan areas
- Less diversity in religion, which could impact provider's ability to practice his or her faith
- Inclination of newly trained physicians finishing residency and fellowship to remain close to the site where they trained

These challenges will be addressed individually to achieve the goal of establishing more dermatology practices for the rural patients who need access to care.

The Shortage of Academic Medical Programs in Rural America

While having a rural background is the prime factor in choosing a non-urban practice, for those without this background, an exposure to rural medicine during medical school or residency is a critical factor in choosing a rural career [5]. To address this, rural dermatologists could volunteer to host medical students and residents for elective rotations to create this experience. Since most rural physicians are born and not made, rural dermatologists should mentor local students and residents creating opportunities for them to learn more about rural practice and, more specifically, rural dermatology.

These shadowing experiences might even start in high school or college. The 28% reduction in rural medical student matriculation, compared to their urban counterparts, is most likely the result of rural students applying in smaller numbers. There appears to be a growing mismatch between the qualifications of rural applicants and medical schools' admission priorities [20]. Rural applicants' Medical College Admission Test (MCAT) scores are lower, they tend to perform "more poorly" on multiple mini-interviews, and they have less research experience [22]. This is a gap that needs to be bridged by supplying opportunities that rural students need to better compete under the current admissions criteria. Rural dermatologists could also volunteer to serve on admissions committees for medical schools and as affiliate faculty in residency programs in their region. Cognitive biases, such as the in-group bias (the tendency to favor others who we perceive to be like ourselves), play a key role in decision-making. This is challenging to overcome [23–25]. Thus, admissions committees heavily populated by academic or research physicians may intentionally or unintentionally select medical school matriculants that resemble themselves. Rural committee members can speak to the needs of rural medicine in candidate selection.

Next, the current criteria being applied to student applicants to determine their admissibility to medical school and dermatology residencies should be challenged. For instance, MCAT scores may not be an important predictor of physician success! Undergraduate grades combined with MCAT scores have been shown to be most predictive of successfully progressing through the academic rigors of medical school [26–28]. Higher MCAT scores have also been shown to be predictive of higher scores on Clinical Clerkship National Board Subject Examinations but have not been shown to be associated with higher Faculty Assessments of Clinical Performance [29, 30]. In addition, MCAT scores have not been shown to correlate with stronger performance during the Post-Graduate-Year (PG-Y)-1 clinical year, nor have they been correlated with performance on the Step 2 Clinical Skills (CS) subscores [31]. Based on available data, high MCAT scores correlate with the later ability to achieve higher exam scores during medical training, but not with the ability to score well in demonstrating clinical acumen [31].

Similarly, USMLE scores are often used as a stratifier for dermatology residency programs. Candidates who have interest or clinical competency in the specialty use perceived "inadequate" board scores to self-deselect from applying to dermatology and concomitantly, dermatology programs may apply a minimum score to candidates it will consider [32]. High scores on the USLME have been demonstrated to correlate with higher scores on in-service tests, but do not predict the ability to successfully complete the American Board of Dermatology Certifying Exam [33]. With the USMLE Step 1 exam being modified from a numerical scoring system to pass/fail, there may be a de-emphasis on applying standardized scores as a major barometer for ranking applicants for residency positions. This may prove to be a good variable to follow for monitoring the number of dermatologists choosing rural practice, as well as positive step toward diversity in a specialty which currently is ranked the second least-diverse specialty behind orthopedic surgery [34]. In summary, the individuals that determine admission criteria, interview, and ultimately select medical students and residents for admission are responsible for determining the physician workforce for our nation. These committees must include individuals from diverse backgrounds, including rural physicians.

Lack of Family and an Established Social Network When You're "Not from Around Here"

It is not possible to overcome the fact that Grandma lives 500 miles away. The attraction of rural America, however, IS the community. When physicians move to rural America, they have come to understand that caring for friends and neighbors is highly gratifying because they NEED us! There is incredible "purpose" in this act. Researchers have determined that happiness is related to satisfaction and meaning in one's life. Happiness is unrelated to privilege and money. It is about the ability to connect with others, to have meaningful relationships and to have strong connections with community members [35].

As recruiters, the first contact is used to create this kind of special relationship. Arrange a "first visit" that best fits their schedule. Phone, televisit, or make an in-person visit to their community. Share literature, websites, links, social media and ensure that everything is updated, attractive, professional, and compelling. Emphasize physician collegiality, both at work and after work. Physicians want to know that they have a strong professional community, so emphasize the how, but also share the collegiality that extends beyond work; the support physicians extend to each other, staff, and their families in the community. In rural areas, physicians share the same schools with their colleagues and patients, support the same sports programs, and participate in the same recreational activities. If these exchanges are authentic and genuine, resident physicians will notice. Also, remember that the entire family must be recruited. More than one of our colleagues chose us because we were their children's first choice. Involve the entire department, clinic, hospital, and community. Plan for tours of schools, parks, playgrounds, and local amenities. Secure a community home tour with a local realtor. Arrange a second visit, or third if necessary. Finally, all questions should be answered honestly and transparently. Ultimately, the right physician must be recruited; a physician who will love their rural practice, their colleagues and the new community and whose families will be happy! This physician will remain over the long term [21, 36].

Ensuring Employment Opportunities for Spouse/ Significant Other

Professional opportunities for the life partners of physician-candidates must not only be considered but must be given the same priority assigned to the physician member of the couple. Physicians are not the only professional that is challenging to recruit to rural areas. While there may be limited options, the talent pool is also limited. The key to successfully finding positions for physician significant others is to create a network with local businesses, chambers of commerce, colleges, and other institutions so you know what opportunities are available in the community and region [37]. Just one match is needed! Don't be afraid to widen the search radius. People who are moving from urban areas where a 60-minute commute in rush hour traffic occurs daily, may be willing to commute on scenic rural roads for the same period of time. In addition, the advancement of telecommuting options, allows for far more opportunities than were available just a few years past. The Covid-19 pandemic, in fact, changed the world of telecommuting, perhaps forever. It may have also changed the world of rural physician recruitment.

Will Covid-19 Change Rural Physician Recruitment?

The effects of the SARS-Cov-2 (COVID-19) pandemic on the long-term geographical practice patterns of dermatologists cannot yet be predicted. The short-term effects on the practice viabilities of physicians of multiple specialties have been significant. National, state and specialty medical societies and associations across the United States surveyed members to measure this impact. The Medical Group Management Association found that Covid-19 had a negative

financial effect on 97% of the 724 medical practices it surveyed [38]. During the course of the pandemic, practices suffered significant reduction in patient volumes and revenues and were forced to furlough or layoff staff. Physicians and administrative staff accepted salary reductions or deferment. Practices that suffered the most were single specialty, non-primary care groups, though few were immune from revenue challenges [38]. In a snapshot of US dermatologists comparing pre-Covid volumes (February 17, 2020 versus March 16, 2020), dermatologists reported a weekly reduction of 58% in patient volumes, 26% in practice days and a 61% reduction in biopsies. Mean estimated telemedicine visits overall during this period was 37.8%. with University/academic/government dermatologists significantly more likely to use telemedicine (57.1%) than private practitioners (35.5%). Older physicians were less likely than younger physicians to use teledermatology [39]. Small, independent dermatology practices may never return to a pre-Covid "normal." In fact, it is certain that not all practices will survive [39, 40].

Even before COVID the practice patterns and demographics of dermatologists were changing. While dermatologists had previously practiced in a solo setting, that percentage had dropped from 44% in 2005 to 35% in 2014. On the other hand, 46% of dermatologists in rural areas are still in solo practice, when compared to 31% in urban areas [41]. Over the past two decades US physicians have been increasingly combining their practices, affiliating with hospitals, insurance companies, and specialty management firms or going to work for such organizations directly. The reasons for this are complex. Younger physicians report looking for a set schedule and salary, especially in the face of declining insurance/payer reimbursement and are intimidated by the complexity and expense of managing a practice and attempting to negotiate independently with payers [42]. Changing reimbursement models, such as accountable care organizations or pay-for value contracts incentivize practice consolidation. Population demographics in rural America are another variable that tends to incentivize health care system consolidation. Approximately 20% of

Americans age 65 and older live in rural America [43]. Both poverty and disability rates are higher in rural than urban areas [44]. Reimbursement rates are often lower. Consolidation and integration of practice settings allows for consolidation of cost centers such as administration, finance, human relations, marketing, supply chain and others which creates leverage in negotiating contracts with non-government payers. Only time will tell whether changing practice demographics transform rural dermatology and whether it will impact the attractiveness of rural practices.

Are there Fewer Entertainment and Dining Opportunities in Rural Areas?

It is important to address the (real or perceived) lack of great theatre, large athletic venues or 5-star dining in rural areas. Focus on the things that make rural America VERY special. Rural communities may not have a Ritz-Carlton, but they also don't have congestion, rush hour traffic, stop lights at every corner and competition for parking. Rural areas have natural beauty, safe spaces for families, clean air, friendly people, and an incredible quality of life. The costs-of-living in rural areas tends to be significantly lower. When a rural physician and his or her family choose to pursue urban-focused activities they can better afford to do it [45].

Limited School Systems, Educational or Other Opportunities

Physicians value education and want to live near excellent school systems. Emphasize the variety of schools and programs offered within the community. Outline additional on-line or local college opportunities for grade- or high school students and the successes achieved by motivated rural students in both educational and extracurricular endeavors. Particularly impactful are the educational success stories of physician colleague's children. Rural education does not equate to lesser talent or skills. In some cases, students who come

from a rural background can share life experiences that help their college applications stand out from those who hail from the suburbs.

This is the time to share experiences for opportunities for partners and families that might never have been considered: the doctor's spouse who took a career break from his college professor's job to be a full-time parent and run for the school board, or the spouse who switched careers, attended law school, and became a local judge, or the other one who gave up her legal career to do pro bono work for underserved kids (all true stories!) Working in a rural area may mean that one income is sufficient, creating the freedom to discover new passions.

Are Rural Communities less Diverse?

Traditionally much of rural America has been white, and identified as cis-gender, heterosexual and Christian. Rural America, however, is undergoing a diversity transformation. Though rural America remains less diverse than urban areas, recent minority population gains suggest a rising diversity. The Hispanic population grew fastest in rural areas, while non-Hispanic white growth rates are minimal and slowing. Rural African American growth rates are modestly increasing. Immigration to non-metropolitan areas is on the upswing and immigrants are dispersing more widely [46]. With increasing numbers of immigrants (and urban transplants) comes a concomitant increase in religious practices and attitudes, though biases persist. Cultural acceptance of sexual orientation and gender identify diversity likewise tend to face pockets of bias but are improving, especially within hospitals and medical centers.

The Tendency of Newly Trained Physicians to Remain Close to Where they've Trained

As the world continues to shrink, the definition of "close" will expand. With the advent of social media, virtual consults, and virtual conferences, one can be geographically distant but remain socially and professionally close. We can leverage this "new world" as we recruit to rural locales.

What About the Money?

Offering a competitive salary and benefits package is essential to recruitment and retention. According to Credible.com, a student loan refinancing website, eight of ten 2019 medical school graduates borrowed to earn their degree, most taking on six-figure debt with 18% borrowing $300,000 or more, and an average education debt after medical school of $251,600 [47]. In addition to addressing their debt, young physicians are looking to purchase a home, start or expand a family and pursue some of the activities that they have long delayed. Offering an attractive compensation package is not always easy, given that many rural practices, hospitals, and health systems serve the poor and elderly, and receive concomitantly lower second party reimbursements. It is, however, essential.

When determining the appropriate level of compensation, it's important to know the market for the geographical region. An October 2020 article in HealthCare Finance News reported that rural physicians are often compensated at rates higher than their urban counterparts and this was especially true in the upper Midwest where physicians received higher compensation than the national average. Compensation for rural physicians averaged 10–15% higher than urban comparisons despite producing work relative-value-units (wRVU) that were typically 20–25% lower. The urban-rural pay disparity was attributed to the greater difficulty in recruiting new talent to rural communities [48].

Money alone will not recruit or keep a rural physician, however, and if a physician is joining a rural practice simply for financial reasons, this physician may not be the optimal physician for your community. Money is one of many "extrinsic motivators" which are defined as anything promised in recompense for work. Other examples include bonuses, recognition, awards, and anything other than the joy of doing the work itself [49]. While we can leverage extrinsic motivators to incentivize any behavior, it is less important than incentives based on the simple joy or purpose derived from the activity itself. In fact, an overemphasis on external motivators can negatively impact the powerful effect of internal motivation [50]. Leveraging internal motivation to achieve end outcomes can best be achieved by focusing on five characteristics of a fulfilling job. These include: skill variety (the variety a job brings), task identify (the degree to which we can see what we do produces an outcome), task significance (the degree to which what we do affects the lives of others), autonomy (the freedom we have to choose our work and how we do it), and feedback (seeing/hearing the results of our work) [49]. Rural dermatology offers physicians a varied practice with significant self-direction or autonomy, and incredible purpose. Our patients need us, because if we are not here, there is no one else to take care of them. Rural patients are the physicians' friends and neighbors; they see and hear about the results of their work everywhere they go.

So, What Can Rural Dermatologists Do to Attract Rural Dermatologists?

To attract dermatologists to rural America, all available data suggests the effort must begin well before the end of their residency. An early start may lead to growing a dermatologist from rural stock or slowly convincing an urbanite to understand the advantages of rural practice.

Mentor. Create opportunities for local students Grow our own dermatologist. Don't wait until college or even high school but look at ways to expose the youngest students to the world of Science, Technology, Engineering and Math, and of course, health care and the life of purpose it offers.

Create rural exposure opportunities for medical students and residents While working to grow rural physicians, expose the non-rural born to the attractions of rural medicine.

Serve on admissions committees for your state medical school and residency programs While medical schools have embarked on a much-needed endeavor to increase the matriculation of students from underrepresented backgrounds, this commitment has primarily concentrated on students from underrepresented minority and lower socioeconomic strata. A similar commitment has not been made to attracting students from rural backgrounds. The individuals that determine admission criteria, interview, and determine admission for medical students and residents must be from diverse backgrounds and must include those working in rural America. Cognitive biases influence every decision an admission committee makes. The rural affiliate faculty member can help the committee recognize in-group bias, the tendency to favor others who we perceive to be like ourselves. Just a few rural students returning home after graduation can make a difference [23–25].

Invest in our communities Physicians and their families want to come to a community that is vibrant, welcoming and has opportunities for business, education, indoor and outdoor entertainment, environmental beauty, exercise, and growth. Rural physicians can ensure a better future by joining the Chamber of Commerce, school board, volunteer organizations, or organizing a sports league to build and maintain a better rural community.

Offer purpose A growing body of literature suggests that having a strong sense of purpose in life leads to both physical and mental health and enhances overall quality of life [51]. Living life with purpose makes you *happy*. Rural America is in desperate need of dermatologists to care for them. There is incredible purpose in this service.

Be a leader There are geographical and physical barriers that make it difficult for rural physicians to engage with the academic and political centers that reside in urban areas. Unfortunately, this is where decisions are made. Rural dermatologist must expend the additional time and energy to overcome these barriers. Overcome them we must! The advancement of virtual communica-tion, meetings and conferences may speed the collapse of these barriers. Medical students and residents must see that choosing to practice in rural America does not equate to isolation from power and influence. They need to know that, in fact, rural America needs its physicians to lead…. to be a voice for the health care challenges and needs of rural patients and communities. If not us, then who will speak for them?

Final Thoughts

There is a strong misconception by those who have never experienced rural dermatology that the scope of practice offered is limited or mundane….that serving in a rural area equates to acceptance of a constricted professional career. It does not. Rural dermatology is varied, fun, challenging, and some days, exhausting. This is NOT different than urban practice. Imagine being a dermatologist in a rural community offering the opportunity to live, work and raise a family where many people choose to vacation! Rural dermatologists are academically inclined, politically active, and innovative. They are thought leaders in their individual practices, health systems, communities, specialty, states, and country, and they serve as a voice for their patients. They live lives of purpose. What could be better than that?

References

1. Porter ML, Kimball AB. Predictions, surprises, and the future of the dermatology workforce. JAMA Dermatol. 2018;154(11):1253–5. https://doi.org/10.1001/jamadermatol.2018.2925.
2. Resneck J Jr, Kimball AB. The dermatology workforce shortage. J Am Acad Dermatol. 2004;50(1):50–4.
3. Cheng CE, Kimball AB. The canary seems fine. J Am Acad Dermatol. 2010;63(2):e23–8.
4. Jaret P. Attracting the next generation of physicians to rural medicine. AAMC website. February 3, 2020. https://www.aamc.org/news-insights/attracting-next-generation-physicians-rural-medicine. Accessed 20 Sept 2020.
5. Chan BT, Degani N, Crichton T, et al. Factors influencing family physicians to enter rural practice: does rural or urban background make a difference? Can Fam Physician. 2005;51(9):1246–7.

6. Rourke J. Strategies to increase the enrollment of student of rural origin in medical school: recommendations from the Society of Rural Physicians of Canada. CMAJ 2005; 172(1): 62–65.

7. Rabinowitz HK, Diamond JJ, Hojat M, Hazelwood CE. Demographic, educational, and economic factors related to recruitment and retention of physicians in rural Pennsylvania. J Rural Health. 1999;15:212–28.

8. JTB R, Incitti F, Rourke LL, Kennard M. The relationship between practice location of family physicians in Ontario and rural background and rural medical education. Can J Rural Med. 2005;10(4):231–40.

9. Brooks RG, Mardon R, Clawson A. The rural physician workforce in Florida: a survey of US- and foreign-born primary care physicians. J Rural Health. 2003;19(4):484–91.

10. Laven GA, Beilby JJ, Wilkinson D, McElroy HJ. Factors associated with rural practice among Australian-trained general practitioners. Med J Aust. 2003;179(2):75–9.

11. Wilkinson D, Laven G, Pratt N, Beilby J. Impact of undergraduate and postgraduate rural training, and medical school entry criteria on rural practice among Australian general practitioners: a national study of 2414 doctors. Med Educ. 2003;37:809–14.

12. Rabinowitz HK, Diamond JJ, Markham FW, Paynter NP. Critical factors for designing programs to increase the supply and retention of rural primary care physicians. JAMA. 2001;286:1041–8.

13. Easterbrook M, Godwin M, Wilson R, Hodgetts G, Brown G, Pong R, et al. Rural background, and clinical rotations during medical training: effect on practice location. CMAJ. 1999;160(8):1159–63.

14. Fryer GE Jr, Stine C, Vojir C, Miller M. Predictors, and profiles of rural versus urban family practice. Fam Med. 1997;29(2):115–8.

15. Canadian Medical Association. Report of the advisory panel on the provision of medical services in underserviced regions. Ottawa: The Association; 1992.

16. Strasser RP. Attitudes of Victorian rural GPs to country practice and training. Aust Fam Physician. 1992;21:808–12.

17. Stratton TD, Geller JM, Ludtke RL, Fichenscher KM. Effects of an expanded medical curriculum on the number of graduates practicing in a rural state. Acad Med. 1991;66:101–5.

18. Rabinowitz HK. Relationship between US medical school admission policy and graduates entering family practice. Fam Practice. 1988;5(2):1442–4.

19. Carter RG. The relation between personal characteristics of physicians and practice location in Manitoba. CMAJ. 1987;136(4):366–8.

20. Shipman S, et al. The decline in rural medical students: a growing gap in geographic diversity threatens the rural physician workforce. Health Aff. 2019;38(12):2011–8.

21. Hall K. Attracting and retaining physicians in rural America. Becker's Hospital Review. September 25, 2017. https://www.beckershospitalreview.com/hospital-physician-relationships/attracting-and-retaining-physicians-in-rural-america.html. Accessed 20 Sept 2020.

22. Raghavan M, Martin BD, Burnett M, Aoki F, Christensen H, MacKalski B, Young DG, Ripstein I. Multiple mini-interview scores of medical school applicants with and without rural attributes. Rural Remote Health. 2013;13:2362. www.rrh.org.au/journal/article/2362

23. Ingroup Bias (Definition + Examples) Practical Psychology. https://practicalpie.com/ingroup-bias-definition-examples/. Accessed September 20.2020.

24. Ariely D. Predictably irrational. New York, NY: Harper Collins Publishers; 2009.

25. Kahneman D. Thinking fast and slow. New York, NY: Farrar Straus and Giroux; 2011.

26. Gauer JL, Wolff JM, Jackson JB. Do MCAT scores predict USMLE scores: an analysis on 5 years of medical student data. Med Educ Online. 2016;21(1):31795.

27. Ellen J. Validity of the medical college admission test for predicting medical school performance. Acad Med. 2005;80(10):910–7.

28. Busche K, et al. The validity of scores from the new MCAT exam in predicting student performance: results from a multisite study. Acad Med. 2020;95(3):387–95.

29. Casey PM, Palmer BA, Thompson GB, et al. Predictors of medical school clerkship performance: a multispecialty longitudinal analysis of standardized examination scores and clinical assessments. BMC Med Educ. 2016:128. https://doi.org/10.1186/s12909-016-0652-.

30. Silver B, Hodgson. Evaluating GPAs and MCAT scores as predictors of NBME I and clerkship performances based on students' data from one undergraduate institution. Acad Med. 1997;1997:394–6.

31. Saguil A, Dong T, Gingerich R, et al. Does the MCAT predict medical school and PGY-1 performance? Mil Med. 2015;180(40):4–11.

32. Chen D, Priest K, Batten J, et al. Student perspectives on the "Step 1 climate" in preclinical medical education. Acad Med. 2019;94(3):302–4.

33. Fening K, Horst AV, Zirwas M. Correlation of USMLE Step 1 scores with performance on dermatology in-training examinations. J Am Acad Dermatol. 2011;64(1):102–6.

34. Isaq N, et al. Taking a "step" toward diversity in dermatology: de-emphasizing USMLE step 1 scores in residency applications. Int J Woman Dermatol. 2020;6(3):209–10.

35. Dolan P. Happiness by design: change what you do, not how you think. New York, NY: Plume Publishing. (Penguin Publishing Group); 2014.

36. Charbonneau G. Recruiting physicians to rural communities. Can Fam Physician. 2018;64:622.

37. Godson A. Let's get rural: 5 tips for sourcing talent in isolated areas. https://www.cielotalent.com/insights/5-tips-for-sourcing-and-recruiting-talent-in-rural-areas/. Accessed 27 Sept. 2020.

38. Covid-19 Financial Impact on Medical Practices. Medical Group Management Association . https://

www.mgma.com/getattachment/9b8be0c2-0744-41bf-864f-04007d6adbd2/2004-G09621D-COVID-Financial-Impact-One-Pager-8-5x11-MW-2.pdf. aspx?lang=en-US&ext=.pdf. Accessed 19 Sept 2020.

39. Litchman GH, Rigel DS. The immediate impact of COVID-19 on US dermatology practices. J Am Acad Dermatol. 2020;83(2):685–6.

40. Rubin R. COVID-19's crushing effects on medical practices, some of which might not survive. JAMA, 2020;323(4):321–3.

41. Ehrlich A, Kostecki J, Olkaba H. Trends in dermatology practices and the implications for the workforce. J Am Acad Dermatol. 2017;77(4):746–52.

42. Kirchhoff S. Physician practices: background, organization, and market consolidation. Congressional Research Service Report for Congress. January 2, 2013.   https://digitalcommons.ilr.cornell.edu/cgi/viewcontent.cgi?article=2007&context=key_workplace. Accessed 27 Sept 2020.

43. Smith S, Trevelyan E. In some states, more than half of older residents live in rural areas. United States Census Bureau. October 22, 2019. https://www.census.gov/library/stories/2019/10/older-population-in-rural-america.html. Accessed 27 Sept 2020.

44. Weiler S. 6 charts that illustrate the divide between rural and urban America. PBS News Hour. March 17, 2017.   https://www.pbs.org/newshour/nation/six-charts-illustrate-divide-rural-urban-america. Accessed 27 Sept 2020.

45. New Census Data Shows Differences Between Urban and Rural Populations United States Census Bureau. December 08, 2016. https://www.census.gov/newsroom/press-releases/2016/cb16-210.html. Accessed 27 Sept 2020.

46. Rural America Undergoing a Diversity of Demographic Change. Population Reference Bureau. May 1, 2006. https://www.prb.org/ruralamericaundergoingadiversityofdemographicchange/. Accessed 27 Sept 2020.

47. Carter M. Average Student Loan Debt for Medical School. Credible. https://credible.com/blog/statistics/average-medical-school-debt/. Accessed 03 Oct 2020.

48. Lagasse J. Physician compensation at Midwest rural hospitals is higher than the national average. HealthCare Finance News. October 09, 2020. https://www.healthcarefinancenews.com/news/physician-compensation-midwest-rural-hospitals-higher-national-average. Accessed 11 Oct 2020.

49. Burkus D. Extrinsic vs Intrinsic Motivation at Work: What's the Real Difference? Psychology Today. April 11, 2020. https://www.psychologytoday.com/us/blog/creative-leadership/202004/extrinsic-vs-intrinsic-motivation-work. Accessed 03 Oct 2020.

50. Wrzesniewski A, et al. Multiple motives don't multiply motivation. Proc Natl Acad Sci. 2014;111(30):10990–5. https://www.pnas.org/content/111/30/10990/. Accessed 03 Oct 2020.

51. Alimujiang A, Wiensch A, Boss J, et al. Association between life purpose and mortality among US adults older than 50 years. JAMA Netw Open. 2019;2(5):4270.

Free Rural Clinics: City Folk Making a Difference in Rural America

19

Hannah Hoang, Hannah McCowan,
Morgan Pfleger, and Nancye McCowan

There is still one place where medicine is not a business, where healthcare providers can be grateful that they can cut out (most of) the administrative work. This is the free clinic.

—Richard Gibbs, MD, *co-founder of the San Francisco Free Clinic*

Introduction

A rural area is classified by the United States Department of Agriculture (USDA) as an area with both geographic isolation and small population size. It is important to focus on rural healthcare because poor health outcomes and high rates of socioeconomic disparity are common in rural areas. Rural areas, home to 20% of the U.S. population, have higher mortality rates, higher hospitalization rates, and higher occurrences of chronic diseases [7, 9]. These findings are associated with barriers to accessing care faced by rural populations. A significant hurdle is the shortage of accessible rural health-care providers. In fact, the lack of access to healthcare providers is thought to be the primary reason that rural citizens have a 40% higher prevalence of preventable hospitalization rates and a 23% higher rate of mortality than their urban counterparts. In addition, there are geographic challenges for patients to reach healthcare providers [9]. Finally, rural clinics often do not have the resources of those in urban areas. They may not have "in house" laboratory services, long-term care facilities, or access to the digital technology. The lack of resources is ameliorated by collaborating with regional urban facilities that provide these necessary services; however, this leads to longer wait time for results [8]. In addition, there is strong evidence that poor health disproportionately affects regions characterized by lower economic status or high rates of poverty. Rural citizens may experience poor sanitation services, deficient housing options, and less access to fresh produce, all of which accentuate the poor health gap [4]. Unfortunately, these problems are getting worse as rural hospitals are closing at an alarming rate. This seemingly contradictory problem is due to financial problems at these facilities leading to the loss of rural physicians moving to urban areas [7].

H. Hoang · H. McCowan
School of Medicine, University of Mississippi
Medical Center, Jackson, MS, USA
e-mail: hhoang2@umc.edu; hmccowan@umc.edu

M. Pfleger
School of Graduate Studies in the Health Sciences,
University of Mississippi Medical Center,
Jackson, MS, USA
e-mail: mpfleger@umc.edu

N. McCowan (✉)
Department of Dermatology, University of
Mississippi Medical Center, Jackson, MS, USA
e-mail: nmccowan@umc.edu

Currently, 50 million people are uninsured in the United States, and the number of underinsured people has been steadily rising since the year 2000 [17]. Many of these individuals live in rural areas. Of the uninsured, approximately 10% are children and 30.6% receive coverage from governmental programs such as Medicare and Medicaid [17]. The uninsured population has a significant impact on the United States' economy; poor health outcomes lead to shorter life spans of the uninsured which costs the US economy an estimated $207 billion per year [17]. Additionally, chronic poor health in a population is costly to treat, and has great risk for long-term complications [5]. Therefore, addressing healthcare concerns and disparities for both the uninsured and underinsured is critically important.

Legislation resulting in government-funded programs have attempted to address rural access to care issues. These are primarily directed towards increasing access to emergency and primary care facilities. However, increasing numbers of uninsured and underinsured patients could shift patients to seek access to care in emergency rooms and urgent care centers, while insured patients receive care from primary providers and specialists of their choice [17]. For some specialties, the volume of patients seeking care can outstrip supply. In 2009, a study from the American Academy of Dermatology (AAD) found that 38% of dermatologists report provider shortages that limit access to specialty care in our field, and the shortages were greatest in rural areas [17]. These challenges will continue to grow in the upcoming years due to America's aging population which results in a greater demand for healthcare services. In fact, it is projected in the year 2060, the population of Americans over the age of 65 will double. An aging population also leads to more physicians transitioning to retirement exacerbating physician shortages. The field of dermatology is not immune from these effects: more than 25% of Americans required treatment for skin diseases and 23,000 individuals die from skin diseases annually. Since the success rate of treating skin diseases depends on preventative screenings, and early detection, limited access to dermatological care produces increased morbidity and mortality [3].

The origin of the pockets of uninsured individuals stems from the privatization of healthcare care [4]. Dating back to the 1940s, Americans have favored rugged individualism and opposed socialism and the welfare state. Unfortunately, privatization of health insurance enforces an employer-based insurance market which makes it more difficult for self-employed and under employed individuals to obtain coverage. In fact, about one fifth of adults are rejected by insurance carriers for medical reasons and approximately 58% of individual insurance seekers unable to find affordable coverage [4]. The cost of health care limits access to proper health care services for many people without health insurance who don't qualify for Medicare or Medicaid.

Throughout the country, free clinics have been established in rural areas. They provide healthcare to uninsured and underinsured members of a community at little or no cost to the patient. This chapter reviews the history of free clinics, the clinic financial structures that have proved to be sustainable and provides an analysis of the impact of free clinics in rural America. long long-term impact of free clinics on patient care and the impact of dermatology mission trips around the world will also be considered.

History of Free Clinics

Free clinics, or nonprofit walk-in clinics, were formed with the goal of becoming a reliable safety net of the healthcare system to aid any patient regardless of insurance or economic status. Until there is a comprehensive universal healthcare system in the US, uninsured or underinsured individuals will seek free care where they can [4]. While services provided by free clinics were originally intended for low-income citizens without access to any form of healthcare, they also provide services to all community members including unauthorized immigrants and individuals who have high deductible health insurance. Common services provided by free clinics include primary care, preventative examinations, prenatal care, and pharmaceutical assistance with some free clinics providing additional services such as dental care, vision care, mental health services, urgent care, and immunizations. These additional services are either provided on-site, or through an established referral system [2, 14]. Some clinics provide on-site educational services such as mental health counseling and disease management education [11]. Few clinics are able to provide onsite laboratory or radiology services, however, most have developed arrangements for patients to receive these services free of cost.

The first free clinic in America was opened in 1967 in San Francisco by Dr. David Smith [4]. This clinic was created to help individuals without access to healthcare services due to their inability to buy health insurance. Currently, approximately 1,400 free clinics are operating in the US providing health care services to the 15 million uninsured or underinsured individuals each year [13, 21]. Of the 15 million uninsured, free clinics directly provide care to about 10% of these individuals (1.8 million) [14]. Free clinics operate either through a full volunteer basis or a partial volunteer basis.

Physicians make up a majority of the volunteer staff with 85% of volunteers serving as doctors [14]. Other common volunteer health professionals include nurses, physician assistants, nurse practitioners, social workers, and psychologists. Approximately 77% of free clinics have part-time or full-time paid staff positions for administrative or executive duties. Apart from a few clinics that receive partial funding from government grants or agencies, most free clinics are privately funded and heavily dependent on donations from community members or philanthropic groups [1, 2, 14]. The median operating budget for free clinics is $125,000 a year, and a majority of clinics have multiple revenue sources. In a 2010 survey sent to over 1,000 U.S. free clinics, a majority 59% of clinics received no funding from the government and only 3% of clinics depended on the government for more than 75% of their revenue from the government.

Free clinics can also be connected to organizations such as medical centers, churches, universities, homeless shelters, or charity organizations. While a majority of 57% of clinics are independent of organizational affiliation, 30% are directly a part of another organization, and 13% are affiliated with a larger organization [14]. Services from free clinics are often completely free, however, some clinics earn some income by operating on a sliding scale payment system based on a patient's income. This system considers the patient's annual income and the size of the family. It reflects the patient's income in comparison to the poverty level, which then dictates clinic service prices.

Clinics can be student-based volunteer clinics, healthcare professional-based, or a mixture of both. Since most free clinics run on a small budget, they operate out of low-cost buildings either privately owned, rented, or connected to a larger organization such as a hospital [2]. Free clinics provide care an average of 18 hours/week, however, their schedules are highly variable with some clinics open on weekends, weekdays, or nighttime hours. On opposite sides of the spectrum, a total of 29% of clinics are open part-time at fewer than five hours a week while 25% of clinics are open full time at 40 hours per week or more. Thus, while free clinics are diverse in their methods of operation, they all strive to accomplish a similar goal: providing health care services to vulnerable patients [14].

A more comprehensive understanding of free clinic models can be achieved by examining two common structures of free clinics: part-time, student-run, and full-time volunteer-based models. The student-run free clinic is a common model across America. These free clinics are usually associated with medical universities, are only open for a limited number of hours a week, and can be incorporated into a medical school's curriculum. In student-run free clinics, medical students work in teams to provide primary care to patients under the supervision of health professionals. There are over one hundred student-run free clinics in America that serve approximately 36,000 patients per year. This popular model of free clinics has spread to Mexico and the Caribbean. Student-run free clinics provide healthcare to uninsured patients and also benefit students by teaching valuable professional skills unable to be taught in classrooms [6]. Another unique free clinic structure is the full-time clinics such as the Barrier Islands Free Medical Clinic (BIFMC) in Charleston, SC. This clinic is open full-time on weekdays for a minimum of eight hours a day and is staffed with a five-person executive committee, a twenty-two-person board of directors, and seven full and part-time paid positions. This clinic has a large volunteer force composed of 138 established health professionals which contrasts with the limited number of supervisory physicians at student-run free clinics. Most of the volunteers for the BIFMC are retired doctors, nurses, and administrative helpers. Both styles of free clinics work to serve individuals who seek access to healthcare services, however, clinics vary in the structure to which this common goal is accomplished [13].

Patient satisfaction at free clinics is generally high. Most patients heard about the clinic through family and friends and described services received with positive words such as great, friendly, nice, and clean [12]. Patients also expressed feeling "comfortable" visiting free clinics for their healthcare needs [11]. There are, however, four primary areas that are frequently reported by patients as requiring improvement. First, long "wait times" are commonly reported both regarding the date when appointments are scheduled and time spent in the waiting room on the day of the appointment. Funding issues and/or the limited number of volunteers make these problems difficult to avoid. Secondly, patients required many services that are not offered on-site due to limited resources. The most common "off-site" services were dermatology, orthopedics, gynecology, vision care, and mammograms services. The referral system may be limited by the number of physicians willing to see non-paying patients at their offices, and some patients do not have the time or money to travel to these offices. The third recommendation from free clinic patients is to provide educational opportunities for participants to improve their health through knowledge and lifestyle changes. Finally, patients expressed the need for more time at the end of their visit to better understand h medication instructions and many did not have a phone number to contact the clinic if side effects arise [12].

Free clinics are been demonstrated to be an effective approach to providing healthcare for uninsured or underinsured individuals. Implementing early and preventive patient care increases the quality of life, and reduces overall treatment costs. In areas with emergency rooms (ERs) neighboring free clinics, there is a 33% decrease in low-complexity ER visits since many patients use free clinics to access primary care providers. Decreasing visits to the hospital ER is cost-effective and serves to improve patient health by identifying disease at early stages. Patients are also more comfortable visiting free clinics which reduces barriers to receiving care [11]. While most of the time, services at free clinics lead to improved health outcomes, this is not always the case. Follow-up appointments are difficult to schedule at free clinics since long appointment wait times are the norm. Receiving treatment after one appointment without the benefit of a follow-up visit can lower the quality of healthcare since monitoring and adjusting medications are critical to optimal results. Additionally, patients who miss follow up appointments have higher instances of high blood pressure, depression, and improper medication use. Also, since many free clinic providers are volunteers, patients often meet with different providers during follow-up visits, adversely affecting continuity of care [10].

Free Clinics and Dermatologic Care Across the United States

Free clinics across the United States generally function as primary care or family care safety-net facilities for the under- or uninsured populations. Clinics serve their communities by managing some chronic conditions, refilling prescription medications, and providing other general health-related services [15]. One of the drawbacks of seeking care at a free clinic in the United States is the relative inability to access specialty care by board-certified physicians. People with low income, who lack insurance, or who are underinsured can therefore have difficulties accessing consistent specialty care, including dermatological care [15, 16].

While treating patients regardless of ability to pay is the primary goal of a free clinic, understanding the makeup of the patient population may impact the quality of care provided to a rural population. Even though care quality has generally been increasing, access to care and healthcare disparity gaps have not improved [17]. One clinic which treated patients regardless of background, immigration status, or other healthcare services found most patients were female (55%), white, under the age of 65, and uninsured (81%) [15]. In another study, the majority of patients seeking dermatologic care were white [19]. Since many underinsured patients are Black, there may be other barriers to access despite simply being uninsured even for those who attend free clinics [19]. Free dermatology clinics in the San Antonio area determined that 69% of their patients identified as Hispanic, 17.4% as Caucasian, and 5.8% as African American/Black [22]. In a Massachusetts study, 93% of patients noted insurance or financial reasons as the purpose for attending a free clinic, and 74% of patients in a study reported not having a consistent source of care outside of emergency rooms and free clinics [15]. Understanding demographic information related to patients attending free clinics can allow clinics to better tailor culturally sensitive healthcare to patients by understanding the needs of a local population.

Accessibility to proper dermatological care is essential as many people experience skin-related health problems. One study found that "the incidence of childhood eczema in the United States is approximately 10.7%" [17]. Some skin conditions, including atopic dermatitis, are more common in rural and minority populations [24]. A study in free clinics found 8% of patients had a dermatologic chief complaint [19]. Eczema has a higher incidence in the insured population compared to the uninsured population, but this could be attributed to not being able to diagnose cases in the uninsured population [17]. One study found that patients seen by dermatologists are predominantly white (84%) compared to physicians in other healthcare areas (84%) [17]. These numbers likely reflect a variety of factors includ-

ing access to dermatologists and healthcare coverage.

People in rural areas face challenges when trying to access primary care, but especially when attempting to find specialty care [18]. In the U.S., the number of dermatologists per 100,000 people has increased from 1.9 to 3.5 between 1970 and 2002; however, despite this nearly twofold increase, dermatologists are still in short supply [25]. The decreased access to specialists is aggravated by both the physician and patient issues. Specialists are rarely offered financial incentives to move to rural areas and a smaller regional population may be unable to support a specialist [18]. Additionally, rural citizens may find it difficult to find transportation to specialists or to afford to pay for their care [18]. Since "up to 42% of the United States' populations reside in areas underserved by dermatologists," identifying solutions to providing dermatological care to rural areas is critical [18]. One solution to providing dermatological care to rural areas is through the development of free clinics.

Dermatology complaints are common in free clinics. One study in Tampa demonstrated that 8% of patients were seen for dermatologic chief complaints in a free clinic [19]. Dermatology patients were also found to have a visit rate of 3.0 compared to the 2.1 visit rate for other patients in the studied year [19]. The most common chief complaints for dermatology patients were localized rashes, genital rashes, generalized rashes, questionable lesions, and genital lesions in that order [19]. A combination of free dermatology clinics in San Antonio, Texas, identified the following chief complaints: atopic dermatitis (18.3%), acne (9.2%), seborrheic keratosis (4.9%), epidermal inclusion cyst (4.8%), and alopecia areata (4.8%) [22]. The median age for these patients was 39 years, and patients were most likely to be white (21.6%) or African American (6.1%), female (57.9%), and employed (32.9%) [19].

There are several approaches to providing dermatologic care in free clinics. Where there is a high incidence of dermatologic complaints, volunteer dermatologists could attend the free clinic on specified days scheduling patients in advance and/or allowing "walk in" appointments. For example, a dermatology clinic serving a rural area was established by the University of Mississippi Medical Center Department of Dermatology in the Mississippi Delta at a nurse practitioner primary care clinic. Dermatology appointments are scheduled once a month with both dermatology providers and residents in attendance at Rolling Fork High School. Each of 12 dermatology faculty and nurse practitioners in the Department of Dermatology volunteer to supervise this clinic once per year. In a three-year period starting in May 2017, the clinic completed 187 patient visits including 250 diagnoses [20]. Of these visits, the most common chief complaints have been acne vulgaris (34.4%), intrinsic (allergic) eczema (8%), and neoplasm of uncertain behavior of skin (6.8%) [20]. Since the dermatology providers travel to a rural area, the clinic is more accessible than the university dermatology clinics 90 miles away. The regularly scheduled dermatology clinic days permit follow-up appointments to be scheduled to track a patient's progress through a treatment. This clinic also serves as a rapid access clinic supporting a wider store-and-forward teledermatology program. About 80% of teledermatology visits submitted by rural primary care physicians in the Mississippi Delta can be effectively handled through this consultative service. The other 20% can be referred for additional work-up or treatment to this monthly clinic.

Timely healthcare is crucial for the best outcomes. Delays in care in underinsured dermatology patients are associated with worse outcomes, increased hospitalizations, and longer hospital stays [23]. However, nearly 60% of free clinic users in one study reported delays in seeking care [15]. When considering dermatology-related care, these delays can be concerning since both increased age and lack of insurance are associated with negative skin cancer related outcomes [17]. For example, while melanoma is more common for those with a high socioeconomic status (SES), the underinsured with lower SES are more likely to present with advanced stage melanomas and have higher mortality rates [17]. Utilizing free clinics could help detect skin cancers earlier in these vulnerable populations and prevent unnecessary deaths.

Including undergraduates, medical students, and residents in the dermatological treatment of underserved communities, especially by treatment at free clinics, fosters a sense of volunteerism, interest in dermatology, and a desire to help underserved communities [16]. Encouraging volunteerism remains especially important for rural areas which may have more limited access to specialty care. In fact, one of the primary factors discouraging physicians from practicing in rural areas is the cultural perception of low job satisfaction in rural areas [26]. One study, however, found medical professionals have similar job satisfaction in both rural and nonrural areas [27]. Exposing residents and future physicians, especially those in specialties, to this reality could help to address the United States' healthcare rural coverage gap. Another approach would be to identify medical students with an interest in service to patients in free clinics. When dermatology residency directors were questioned, 52% thought practicing in underserved areas was important but only 23% used this as a criterion for recruiting residents [26].

Although free clinics have shown great promise by providing healthcare to over 15 million Americans, they still require some polishing [21]. Free clinics are difficult to establish and patients report they are difficult to find. In one study, clinics were contacted three times within a span of 2 weeks and still 38% of clinics did not respond [21]. Another 17% of contacted clinics were either closed or no longer provided healthcare services [21]. Finding dermatologic care at free clinics can be an even greater challenge. One study found 28% of free clinics offered no dermatologic services or referrals, and 16% of clinics had dermatologists available only occasionally on site [21]. When evaluating the scope and use of free clinics across the United States, the lack of uniform or systemic reporting of healthcare services poses a considerable challenge for analysis. There is no board or organization which tracks or examines the number of existing free clinics or their operations [15]. Free clinics around the U.S. utilize different methods for tracking patient care, and these methods vary not only from state to state but also from clinic to clinic. Despite these issues, free clinics represent a viable safety net for the un- and underinsured populations which can help decrease the negative impact of this population on the U.S. healthcare system and economy especially by addressing specialty needs [23].

International Dermatology and Medical Mission Trips

In developed countries where Universal Health Coverage (UHC) is available, a system is in place where "all people and communities can use the promotive, preventive, curative, rehabilitative and palliative health services they need, of sufficient quality to be effective, while also ensuring that the use of these services does not expose the user to financial hardship [28]." It would seem that free clinics would not be necessary. In fact, there is very little research focused on international free clinics and even less on dermatology in such clinics. Of course, these systems do not ensure that everyone is getting the health care they need. Health disparities, including those in dermatology, are certainly present in these countries and need. Needs in underdeveloped countries are much more critical leading to the emergence of medical mission trips to bring health professionals to people who do not have access to quality healthcare.

While skin diseases are often not considered urgent when compared to medical conditions such as HIV/AIDS, tuberculosis, and pneumonia [29], they should not be disregarded when it comes to international medical missions. Skin problems are among the most common conditions seen by primary care providers in tropical areas [29]. In the World Health Organization's 2001 report on the "global burden of disease", skin diseases produced mortality rates of 20,000 in Sub-Saharan Africa which are "comparable to mortality rates attributed to meningitis, hepatitis B, obstructed labor, and rheumatic heart disease in the same region [29]." A series of studies assessing the prevalence of skin diseases in rural areas in Ethiopia, Nepal, Mali, Tanzania, Indonesia, and Brazil have demonstrated: "a high

prevalence of skin disease, usually exceeding 50% and sometimes as much as 80% of the population, depending on whether there are local endemic conditions such as scabies, onchocerciasis or tinea capitis, all of which lead to significant increases in case numbers [30]." A study in western Ethiopia revealed that 47–53% of people in two rural communities reported a skin disease; however, physical examination identified a treatable skin condition in 67% of patients [29].

Medical missions provide clear benefits for participating health care providers. They represent unique learning experiences for medical students and other medical trainees. Half of the medical schools in the United States offer global health experiences in their curriculum [31]." Medical students and their faculty learn to provide patient care and resource allocation in new environments and are exposed to cultures that may be foreign to them [32]. In addition, they may treat health conditions that would not typically be seen in a developed country [32].

There is considerable controversy, however, in determining whether the overall health care status of a country is improved since medical missions do not provide permanent service availability. There is often no consideration given to the provision of follow-up care. Furthermore, there is a temptation for volunteers to work beyond their capabilities which puts patients at risk especially when there are too few critical-care physicians and surgeons in low-income countries [31, 32]. Perhaps most importantly, medical missions can result in dependency on visiting international medicine. For example, Liberia had about 200 physicians responsible for the care of its four million citizens before an outbreak of Ebola in 2014. Many of these physicians worked for missionary and humanitarian groups. The threat of this disease to the health of providers caused 3/4 of these providers to leave the country. Hospitals and clinics closed when they were most needed [33]. Thus, chronic dependency on foreign doctors can leave a country with an unreliable and unstable healthcare system.

While there is a need for dermatology diagnosis and treatment in certain areas, persons on medical mission trips should approach them with caution and a plan recognizing that it is difficult to change lives in a week [33]. It almost certainly should include a component of education for local health care providers. This has been successfully done by the Regional Dermatology Training Center in Tanzania. This center was established in 1992 [34] and aims to serve as a "supra-regional training, research and clinical center providing facilities for the care of patients with skin disorders, leprosy and sexually transmitted infections and for training health care professionals from different African countries" [35]. This facility simultaneously improves the health condition of the citizens of Tanzania and educates medical personnel to create a sustainable system.

References

1. Amenta MM. Free clinics change the scene. Am J Nurs. 1974;74(2):284–8.
2. Geller S, Taylor BM, Scott HD. Free clinics helping to patch the safety net. J Health Care Poor Underserved. 2004;15(1):42–51.
3. Brodell R. Bridging the gap in rural health access. Health Affairs Blog. 2019. Available from: https://doi.org/10.1377/hblog20191106.337502.
4. Kamimura A, Christensen N, Tabler J, Ashby J, Olson LM. Patients utilizing a free clinic: physical and mental health, health literacy, and social support. J Community Health. 2013;38(4):716–23.
5. Hall J, Schmidt G, and Kress J. Principles of critical care. Fourth ed. New York: McGraw-Hill Education. 2015 [cited 2021 July 27]. Available at: https://ummc.on.worldcat.org/oclc/906700899.
6. Ramos MA, Rotenstein LS, Mata DA. Student-run free clinics: the USA's psychiatry recruitment solution? Lancet Psychiatry. 2016;3(4):321–2. Available from:. https://doi.org/10.1016/S2215-0366(16)00099-7.
7. Weil AR. Rural Health. Health Affairs (Project Hope). 2019;38 (12): 1963–63. Available from: https://doi.org/10.1377/hlthaff.2019.01536.
8. Mosca L. Rural health. Am J Nurs. 2011;111(12):12.
9. Marcin JP, Shaikh U, Steinhorn RH. Addressing health disparities in rural communities using telehealth. Pediatr Res. 2015;79:169–76. Available from:. https://doi.org/10.1038/pr.2015.192.
10. Mallow JA, Theeke LA, Barnes ER, Whetsel T, Mallow BK. Free care is not enough: barriers to attending free clinic visits in a sample of uninsured individuals with diabetes. Open J Nurs. 2014;4(13):912–9. Available from:. https://doi.org/10.4236/ojn.2014.413097.
11. Hwang W, Liao K, Griffin L, Foley KL. Do free clinics reduce unnecessary emergency department visits?: the Virginian experience. J Health Care Poor Underserved. 2012;23(3):1189–204. Available From:. https://doi.org/10.1353/hpu.2012.0121.
12. Kamimura A, Ashby J, Trinh H, Prudencio L, Mills A. Uninsured free clinic patients' experiences and perceptions of healthcare services, community resources, and the patient protection and affordable care act. Patient Exp J. 2016;3(2):12–21. Available from: https://pxjournal.org/journal/vol3/iss2/4
13. Moores, Carrie. Email sent to: Hannah Hoang. 3rd August 2020.
14. Darnell JS. Free clinics in the United States: a nationwide survey. Arch Intern Med. 2010;170(11):946–53. Available from:. https://doi.org/10.1001/archinternmed.2010.107.
15. Mott Keis R, Gifford DeGeus L, Cashman SB, Savageau J. Characteristics of patients at three free clinics. J Health Care Poor Underserved. 2004;15(4):603–17. Available from:. https://doi.org/10.1353/hpu.2004.0062.
16. Pyles MN, Nkansah N, Sun BK. Patient satisfaction in dermatologic care delivered by a medical-student-run free clinic. J Am Acad Dermatol. 2016;74(6):1265–7. Available from:. https://doi.org/10.1016/j.jaad.2015.12.036.
17. Buster KJ, Stevens EI, Elmets CA. Dermatologic health disparities. Dermatol Clin. 2012;30(1):53–9. Available from:. https://doi.org/10.1016/j.det.2011.08.002.
18. Barton M. Access to dermatology care in rural populations. J Nurse Pract. 2012;8(2):160–1. Available from:. https://doi.org/10.1016/j.nurpra.2011.12.012.
19. Ayoubi N, Mirza A, Swanson J, Hamoui R, Mhaskar R. Dermatologic care of uninsured patients managed at free clinics. J Am Acad Dermatol. 2019;81(2):433–7. Available from:. https://doi.org/10.1016/j.jaad.2019.03.010.
20. McCowan, N. K. Email sent to: Hannah Kyle McCowan. 5th August 2020.
21. Madray V, Ginjupalli S, Hashmi O, Sams RW, Johnson JA, Stoff B. Access to dermatology services at free medical clinics: a nationwide cross-sectional survey. J Am Acad Dermatol. 2019;81(1):245–6. Available from:. https://doi.org/10.1016/j.jaad.2018.12.011.
22. A retrospective chart review of the most common dermatologic diagnoses at a Hispanic-serving dermatology free clinic and efforts to increase supply order efficiency. J Am Acad Dermatol. 2019;81(4):AB11. Available from: https://doi.org/10.1016/j.jaad.2019.06.079.
23. How to navigate dermatology care for the uninsured. J Am Acad Dermatol. 2019;81(4): AB220. Available from: https://doi.org/10.1016/j.jaad.2019.06.809.
24. Spears CR, Nolan BV, O'Neill JL, Arcury TA, Grzywacz JG, Feldman SR. Recruiting underserved populations to dermatologic research: a systematic review.

Int J Dermatol. 2011;50(4):385–95. Available from:. https://doi.org/10.1111/j.1365-4632.2010.04813.x.

25. Resneck J, Kimball AB. The dermatology workforce shortage. J Am Acad Dermatol. 2004;50(1):50–4. Available from:. https://doi.org/10.1016/j.jaad.2003.07.001.

26. Banco G, Vasquez R, Nezafati K, Allensworth A, Bernstein IH, Cruz PD Jr. How residency programs can foster practice for the underserved. J Am Acad Dermatol. 2012;67(1):158–9. Available from:. https://doi.org/10.1016/j.jaad.2011.11.954.

27. Bae S. Nurse practitioners' job satisfaction in rural versus nonrural areas. J Am Assoc Nurse Pract. 2016;28(9):471–8. Available from:. https://doi.org/10.1002/2327-6924.12362.

28. World Health Organization. Universal health coverage and health financing. Available from https://www.who.int/health_financing/universal_coverage_definition/en/.

29. Hay R, Bendeck S, Chen S, Estrada R, Haddix A, McLeod T, et al. Skin diseases. In: Jamison DT, Breman JG, Measham AR, et al., editors. Disease control priorities in developing countries. 2nd ed. New York: Oxford University Press; 2006. Accessible from https://www-ncbi-nlm-nih-gov.ezproxy2.umc.edu/books/NBK11733/.

30. Hay R, Fuller L. The assessment of dermatological needs in resource-poor regions. Int J Dermatol. 2011;50:552–7. Available from http://skincareforall.org/wp-content/uploads/2012/10/2.-The-assessment-of-dermatological-needs-in-resource-poor-regions.pdf.

31. Stoff B, McMichael J. Short-term international volunteerism in dermatology: ethical considerations. J Am Acad Dermatol. 2014;71(4):822–5. Available from https://www-sciencedirect-com.ezproxy2.umc.edu/science/article/pii/S0190962214012602.

32. Thaller BS GK, Thaller S. Surgical mission trips as an educational opportunity for medical students. J Craniofac Surg. 2015;26(4):1095–6. Available from https://ummc.on.worldcat.org/search?databaseList=638&queryString=medical+mission+trips#/oclc/5842705503.

33. Bartelme T. Medical missions: do no harm? Physician Leadersh J. 2015:8–13. Available from http://web.a.ebscohost.com.ezproxy2.umc.edu/ehost/pdfviewer/pdfviewer?vid=1&sid=cbfacb51-63a9-4624-b4c9-bccea7b97c05%40sdc-v-sessmgr02.

34. Mponda, Kelvin, and John Masenga. Skin diseases among elderly patients attending skin clinic at the Regional Dermatology Training Centre, Northern Tanzania: A Cross-Sectional Study. Bmc Research Notes. 2016;9(1):1–5. Available from https://doi.org/10.1186/s13104-016-1933-6.

35. Regional Dermatology Training Center. History. Available from https://rdtc.go.tz/about-us/history/.

Index